T0305828

TRILOGY OF NUMBERS AND ARITHMETIC

Book 1

History of Numbers and Arithmetic: An Information Perspective

World Scientific Series in Information Studies
(ISSN: 1793-7876)

Series Editor: Mark Burgin *(University of California, Los Angeles, USA)*

International Advisory Board:

Søren Brier *(Copenhagen Business School, Copenhagen, Denmark)*
Tony Bryant *(Leeds Metropolitan University, Leeds, United Kingdom)*
Gordana Dodig-Crnkovic *(Mälardalen University, Eskilstuna, Sweden)*
Wolfgang Hofkirchner *(The Institute for a Global Sustainable Information Society, Vienna)*
William R King *(University of Pittsburgh, Pittsburgh, USA)*

More information on this series can also be found at https://www.worldscientific.com/series/wssis

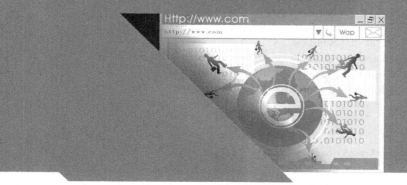

World Scientific Series in Information Studies — **Vol. 12**

TRILOGY OF NUMBERS AND ARITHMETIC

Book 1

History of Numbers and Arithmetic: An Information Perspective

Mark Burgin

University of California, Los Angeles, USA

World Scientific

NEW JERSEY · LONDON · SINGAPORE · BEIJING · SHANGHAI · HONG KONG · TAIPEI · CHENNAI · TOKYO

Published by

World Scientific Publishing Co. Pte. Ltd.
5 Toh Tuck Link, Singapore 596224
USA office: 27 Warren Street, Suite 401-402, Hackensack, NJ 07601
UK office: 57 Shelton Street, Covent Garden, London WC2H 9HE

Library of Congress Control Number: 2022011713

British Library Cataloguing-in-Publication Data
A catalogue record for this book is available from the British Library.

World Scientific Series in Information Studies — Vol. 12
TRILOGY OF NUMBERS AND ARITHMETIC
Book 1: History of Numbers and Arithmetic: An Information Perspective

ISBN 978-981-123-683-9 (hardcover)
ISBN 978-981-123-684-6 (ebook for institutions)
ISBN 978-981-123-685-3 (ebook for individuals)

For any available supplementary material, please visit
https://www.worldscientific.com/worldscibooks/10.1142/12273#t=suppl

Desk Editors: Vishnu Mohan/Rok Ting Tan

Typeset by Stallion Press
Email: enquiries@stallionpress.com

Printed in Singapore

Preface

The more extensive a man's knowledge of what has been done,
the greater will be his power of knowing what to do.

Benjamin Disraeli

In contemporary society, mathematical or quantitative literacy is as important as language literacy, and mathematical literacy begins from operation with numbers. Together with operations, numbers constitute arithmetic forming, in such a way, basic intellectual instruments of theoretical and practical activity of people and offering powerful tools for representation, acquisition, transmission, processing, storage, and management of information. Numbers persist in all walks of people's life providing information about the diversity of things and people — from information about dates and time to information about weights and heights to information about weather and finance to scientific and engineering data to mention but a few fields.

Calculators and computers were invented as information processing devices to help people working with numbers. Modern science and technology are impossible without operation with numbers. Political, social and economic applications of mathematics are now multiplying with a great speed and the majority of these applications involve numbers and arithmetic operations. People need numerical information and tools to operate with it because they cannot always trust their intuition to make effective decisions. At the same time, it is important to understand that numeracy may not be an assurance of rationality in the face of big volumes of quantitative information.

To afford, process and store all numerical information in society, a variety of number classes, such as whole or real numbers, and their arithmetics are utilized by people. Although people discovered or created the first number systems and elaborated rules for operation in these systems very long ago, demands of practice inspired mathematicians to introduce more and more new kinds of numbers and their arithmetics. This process continues now with the immense intensity. As a result, numbers and their arithmetics have a fascinating history, exposition of which is the main goal of this book.

In addition, various information issues related to numbers and arithmetic are reflected here, and it is explained how different types and kinds of numbers came into the being. The fascinating history of numbers and their representations is described and analyzed from ancient times when people were only becoming intelligent beings and to our time in the 21st century. As numbers are inseparable from the arithmetic, in which they live and function, the described history of numbers is, at the same time, the history of arithmetic.

This history is full of intellectual adventures and creativity feats. Mathematicians from different cultures, countries and continents have created this history. The reader will be able to get acquainted not only with the existing diversity of numbers and arithmetics but also with the great mathematicians and scientists who developed mathematics discovering new kinds of numbers and creating novel arithmetics.

About the Author

Mark Burgin received his MA and PhD in mathematics from Moscow State University and Doctor of Science (DSc) in logic and philosophy from the National Academy of Sciences of Ukraine. He was a Visiting Professor at UCLA; Professor at the Institute of Education, Kiev, Ukraine; at International Solomon University, Kiev, Ukraine; at Kiev State University, Ukraine; the Head of the Assessment Laboratory of the Research Center of Science at the National Academy of Sciences of Ukraine, and the Chief Scientist at the Institute of Psychology, Kiev. Currently he is affiliated with UCLA, Los Angeles, California, USA. Mark Burgin is a member of the New York Academy of Sciences, a Senior Member of IEEE, of the Society for Computer Modeling and Simulation International, and of the International Society for Computers and their Applications, as well as an Honorary Professor of the Aerospace Academy of Ukraine and Vice-President of the International Society for The Studies of Information. Mark Burgin was the Chair of the IEEE San Fernando Valley Computer and Communication Chapter and the Secretary of the International Society for Computers and their Applications. He was the Editor-in-Chief of the international journals *Integration, Information,* and the *International Journal of Swarm Intelligence & Evolutionary Computation* and is an Editor and Member of Editorial Boards of more than 21 journals. He was a member of Program, Organization and Scientific Committees in more than 100 international conferences, congresses, and symposia as well as the chief organizer of some of them such as Symposia "Theoretical Information Studies", "Evolutionary Computation and Processes of Life", and "Creativity in Education." Mark Burgin is involved in research and publications, and has taught courses in various areas of mathematics, information sciences, artificial intelligence (AI), computer science, system theory, philosophy, logic, cognitive sciences, pedagogical sciences, and methodology of science. He originated such theories as the general theory of information, theory of named sets, mathematical theory of schemas, theory of super-recursive algorithms, hyperprobability theory, mathematical theory of oracles, theory of non-Diophantine arithmetics, system theory of time, and neoclassical analysis (in mathematics) making essential contributions to such fields as human creativity, theory of knowledge, theory of information, theory of algorithms and computation, theory of intellectual activity, probability theory, and complexity studies. He was the first to discover Non-Diophantine arithmetics; the first to axiomatize and build mathematical foundations for negative probability used in physics, finance and economics; and the first to explicitly overcome the barrier posed by the Church–Turing Thesis in computer science and technology by constructing new powerful classes of automata. Mark Burgin has authored and co-authored more than 500 papers and 21 books, including *"Non-Diophantine arithmetics in mathematics, physics and psychology"* (2020), *"Semitopological Vector Spaces: Hypernorms, Hyperseminorms and Operators"* (2017), *"Functional Algebra and Hypercalculus in InfiniteDimensions: Hyperintegrals, Hyperfunctionals and Hyperderivatives"* (2017), *"Theory of Knowledge: Structures and Processes"* (2016), *"Structural Reality"* (2012), *"Hypernumbers and Extrafunctions: Extending the Classical Calculus"* (2012), *"Theory of Named Sets"* (2011), *"Measuring Power of Algorithms, Computer Programs, and Information Automata"* (2010), *"Theory of Information: Fundamentality, Diversity and Unification"* (2010), *"Neoclassical Analysis: Calculus Closer to the Real World"* (2008), *"Super-recursive Algorithms"* (2005), *"On the Nature and Essence of Mathematics"* (1998), *"Intellectual Components of Creativity"* (1998), *"Introduction to the Modern Exact Methodology of Science"* (1994), and *"The World of Theories and Power of Mind"* (1992). Mark Burgin is also an Editor of 10 books.

Contents

Acknowledgments

I am thankful to the staff at World Scientific and especially, to Ms. Tan Rok Ting, for their help in bringing about this publication. My appreciation also goes to all my teachers, above all my teachers of mathematics and physics, and especially, to my PhD thesis advisor, Alexander Gennadievich Kurosh, who helped shaping my scientific viewpoint and research style. The advice and help of Andrei Nikolayevich Kolmogorov from Moscow State University in the development of the holistic view on mathematics and its connections with natural and social sciences is also greatly appreciated. It is also necessary to assert that many discussions and conversations with friends and colleagues were useful for better understanding of the history of mathematics. Credit for the desire to write this book must go to the academic colleagues. I would also like to thank the Departments of Mathematics and Computer Science in the School of Engineering at UCLA for providing space, equipment, and helpful discussions.

Introduction

1. The General Preamble

What would life be without arithmetic, but a scene of horrors?

Sydney Smith

The world is immersed in numbers. In essence, all people use numbers. The role of numbers and arithmetic in human civilization has been permanently growing through ages and millennia. Now operation with numbers pervades the behavior of people in the contemporary society. Every woman and every man carry out additions, subtractions and counting many times every day. Organizations operate with huge amounts of numbers in different forms. Numbers are present in all fields of science and technology, in economy and politics, in medicine and agriculture.

If we ask the question why numbers are so important, we can see that their significance comes from their ability to provide information. Numbers bestow information about myriads of things. Numbers give information about separate people, for example, their height, weight and age are essentially reflected by numbers. Numbers supply information about nature, for example, temperature of different things, heights of mountains, of mountains, distances to the Sun and to the Moon. Measurements present information in numbers. The speed of your car, the price of your computer, the distance from your home to your work — all this information is made available by numbers.

People learned how numbers can provide valuable information about almost every aspect of life and about practically any system people encounter or create. At the same time, arithmetics essentially optimized operation with information presented by numbers. Calculators and

computers were invented as information processing devices to help people
working with numbers. Modern science and technology are impossible
without operation with numbers. Political, social and economic applications
of mathematics are now multiplying with a great speed and the majority of
these applications involve numbers and arithmetic operations. People need
numerical information and tools to operate with it because they cannot
always trust their intuition to make effective decisions. At the same time,
it is important to understand that numeracy may not be an assurance of
rationality in the face of big volumes of quantitative information.

Numbers are also used to store information about a variety of things as
well as to control information processes. For instance, numbers are used
in decision-making, planning, organization, maintenance, construction,
control, and monitoring.

Archeologists demonstrate that objects with sequential markings have
been used to store and retrieve numerical information since the beginning
of the European Upper Palaeolithic around 40–45 millennia ago and show
that numerical notations were in use among archaic hominins (d'Errico
et al., 2018).

At the same time, to process information made available through
numbers, people need arithmetic. People started working with numbers,
e.g., performing addition and then multiplication, millennia ago, while
mathematicians created arithmetic to grant rigorous and efficient rules and
algorithms for effectively handling numbers. All people who operate with
numbers use arithmetic acquiring and processing information. Computer
functioning is aimed at crunching numbers providing important information, which often helps people but also can be confusing and misleading
without correct understanding what numbers represent.

Mathematicians and logicians know about different calculi. Newton and
Leibniz created differential calculus and integral calculus, which together
form the calculus. Logicians created and utilized a variety of calculi, such
as the classical propositional calculus or the classical predicate calculus.
However, the first calculus created by people was arithmetic, which provided
efficient means for operating with numbers.

Moreover, it is possible to find arithmetic in a variety of natural
systems. For instance, Vladimir shCherbak writes:

*"The first information system emerged on the earth as primordial
version of the genetic code and genetic texts. The natural appearance of
arithmetic power in such a linguistic milieu is theoretically possible and*

practical for producing information systems of extremely high efficiency. In this case, the arithmetic symbols should be incorporated into an alphabet, i.e. the genetic code. A number is the fundamental arithmetic symbol produced by the system of numeration. If the systems of numeration were detected inside the genetic code, it would be natural to expect that its purpose is arithmetic calculation e.g., for the sake of control, safety, and precise alteration of the genetic texts. The nucleons of amino acids and the bases of nucleic acids seem most suitable for embodiments of digits." (shCherbak, 2003)

Then he explains that these assumptions were used for analyzing the genetic code and brought researchers to the following conclusion:

"*The genetic code turns out to be a syntactic structure of arithmetic, the result of unique summations that have been carried out by some primordial abacus at least three and half billion years ago. The decimal place-value numerical system with a zero conception was used for that arithmetic. It turned out that the zero sign governed the genetic code not only as an integral part of the decimal system, but also directly as an acting arithmetical symbol.*" (shCherbak, 2008)

Numbers are also used as metaphors. For instance, in the book "1984" by English novelist George Orwell (1903–1950), the arithmetical expression $2 + 2 = 5$ was treated as an obviously false dogma connoting official deception, secret surveillance and brazenly misleading terminology. However, time passed, and starting from the beginning of the 21st century, this expression along with the expression $1 + 1 = 3$ has become the symbol and metaphor of synergy — an important natural and social phenomenon when the union (sum) of the parts is larger than those parts simply taken together (Cambridge Business Dictionary; Wapnick, 2013; Grimsley, 2018). For instance, Shawn Grimsley explains that in mathematical terms, a synergy is when $2 + 2 = 5$. The Cambridge Business English Dictionary defines: "when two companies or organizations join together, they achieve more and are more successful than if they work separately [making] $2+2 = 5$ in this merger."

The simpler expression $1 + 1 = 3$, which also symbolizes synergy, is even more often used in a variety of areas such as business and industry (cf., for example, Beechler, 2013; Gottlieb, 2013; Grant and Johnston, 2013; Hattangadi, 2017; Jude, 2014; Kress, 2015; Marks and Mirvis, 2010; Phillips, 2008; Murphy and Miller, 2010; Renner, 2013; Ritchie, 2014), economics and finance (cf., for example, Burgin and Meissner, 2017), psychology and sociology (cf., for example, Boksic, 2019; Brodsky, *et al.*, 2004; Bussmann, 2013; Enge, 2017; Frame and Meredith, 2008; Jaffe, 2017; Klees, 2006; Mane, 1952; Trott, 2015), library studies (cf., for example, (Marie, 2007)), biochemistry and bioinformatics (cf., for example, Kroiss *et al.*, 2009), computer science (cf., for example, Derboven, 2011; Glyn, 2017; Lea, 2016; Meiert, 2015), physics (cf., for example, Lang, 2014), chemistry (cf., for example, Murphy, 1999), medicine (cf., for example, Archibald, 2014; Jaffe, 2017; Lawrence, 2011; Phillips, 2016; Trabacca *et al.*, 2012; Caesar and Cech, 2019), agriculture (Riedell *et al.*, 2002), pedagogy (Nieuwmeijer, 2013) and politics (Van de Voorde, 2017). Discussing similar situations, Vidya Hattangadi writes that mathematically, synergy is also represented as $1 + 1 = 3$ (Hattangadi, 2017).

Another seemingly "improbable" arithmetical equality is related to negative synergy when the union (sum) of the parts is less than those parts simply taken together. As Shawn Grimsley spells out, in mathematical terms, a negative synergy is when $2 + 2 = 3$ (Grimsley, 2018). This equality is also absurd in the conventional arithmetic.

Actually, all those who used these expressions as metaphors and many others believed that such equalities violate laws of mathematics and mathematicians, for example, will never claim that $1 + 1 = 3$. Nevertheless, mathematics in its development brought these expressions into the scope of mathematics when non-Diophantine arithmetics of natural numbers were discovered because in some of them, it is possible that $1 + 1 = 3$, $2 + 2 = 3$ or $2 + 2 = 5$ (Burgin, 1977, 1997; Burgin and Czachor, 2020).

As numbers and arithmetic are extremely important in almost all areas of life, it is natural that from the ancient times, the study of arithmetic has formed the foundation of the mathematical curriculum. As the Canadian mathematician and philosopher William Paul Byers writes, *"number* is basic to our civilization" and for a long time, mathematics was called the science of numbers (Byers, 2007).

Any society that strives towards technological and scientific proficiency needs to be able to educate its citizens in the basic mathematical machinery necessary for survival, and arithmetic is the base and root of all mathematical knowledge. Contemporary information technology embeds this knowledge into intricate calculators and powerful computers.

However, in spite of all theoretical and technological achievements related to numbers and arithmetics, it is useful to know the history of these pivotal elements of the human culture. The past can teach you about the future if you know this past and can correctly interpret it.

This book will bring you to the fascinating world of numbers and operations with them. Together with operations, numbers constitute arithmetic forming in such a way basic intellectual instruments of theoretical and practical activity of people and offering powerful tools for representation, acquisition, transmission, processing, storage, and management of information. As people say, our world is deluged with quantitative information. This is why, the history of numbers and arithmetic is the topic of a variety of books and at the same time, it is extensively presented in many books on the history of mathematics. However, all of them, at best, bring the reader to the end of the 19th century not actually including developments in these areas in the 20th century and later. Besides, such books consider and describe only the most popular classes of numbers, such as whole numbers or real numbers. In this book, which is the first of the trilogy on numbers and arithmetic, the reader will be able to find and get acquainted with the much larger diversity of kinds and types of numbers and arithmetics used in contemporary mathematics and its applications. Here the history of numbers and arithmetics covers the vast period of time from the past millennia when the numbers appeared to the beginning of the 21st century. It is informally described for the general intelligent audience, while philosophical and methodological analysis of numbers and arithmetics as well as their basic contemporary mathematics are made available in the next books of this trilogy.

One more advantage of this book is its information orientation. Mathematicians, as a rule, did not pay enough attention to this important aspect of numbers and arithmetic. However, the history of human civilization shows that numbers and arithmetic were developed to provide, keep and utilize information about the whole world, all its parts and elements

We describe the history of numbers and growth of arithmetic from ancient times to modernity, reflecting innovative directions and new results in this development. Reading this book, the reader will find how different types of numbers came into the being. The fascinating history of numbers and their representations is described and analyzed from ancient times when people were only becoming intelligent beings and to our time in the 21st century. As numbers are inseparable from the arithmetic, in which they live and function, the described history of numbers is, at the same time, the history of arithmetic including the history of the transition from arithmetic to algebra.

This history is full of intellectual adventures and creativity feats. Mathematicians from different cultures, countries and continents have created this history developing arithmetic and its applications. The reader will be able to get acquainted not only with the existing diversity of numbers and arithmetics but also with the great mathematicians who created mathematics discovering new kinds of numbers and creating novel arithmetics.

For a long time, the creation of new arithmetics was caused by discovering brand new kinds of numbers. In the 20th century, some researchers started operating with the same numbers but in a different from conventional arithmetics way. In this process, the real breakthrough happened with the discovery of non-Diophantine arithmetics. For millennia, the conventional (Diophantine) arithmetic of natural numbers existed unchallengeable while its statements, such as $2 + 2 = 4$, were treated as absolute eternal truths. However, it was found that this arithmetic is (may be the most important but still) only one element of a rich family of arithmetics of natural numbers, the majority of which are non-Diophantine and in some of which it is possible that $2 + 2 = 5$ or $2 + 2 = 3$.

The necessity in other arithmetics of natural numbers was maybe the longest-standing mathematical problem, which existed from the time when the ancient Greek philosopher Zeno of Elea (ca. 490–430 B.C.E.) formulated his *paradox of the heap* or the *Sorites paradox* demonstrating that the conventional arithmetic of natural numbers cannot correctly represent some real-life situations. The solution for this problem was obtained only at the end of the 20th century by discovering non-Diophantine arithmetics of natural numbers.

Due to the importance of this discovery, the history of non-Diophantine arithmetics is described in the second part of this book, which consists of Chapters 4 and 5. An interesting peculiarity of this history is that some elements, it is possible to say roots, of non-Diophantine arithmetics appeared in mathematics in ancient times when the Pythagorean Theorem was uncovered. However, non-Diophantine arithmetics, as rigorous mathematical systems, were discovered only at the end of the 20th century. Moreover, fractions of non-Diophantine arithmetics were implicitly exploited for quite a while in the 20th century but their explicit utilization came only in the 21st century.

In his book "Infinite in all Directions", Freeman Dyson writes about two opposite directions in science — unification and diversification. He contrasts Einstein as the utmost proponent of unification and Rutherford as a bright adherent of diversification. It is interesting that the discovery (creation) of non-Diophantine arithmetics, as well as the discovery (creation) of non-Euclidean geometry, propagates mathematics and science in both directions. On the one hand, it was diversification because instead of one absolute arithmetic of natural numbers (one absolute geometry), an infinite diversity of non-Diophantine arithmetics of natural numbers (of non-Euclidean geometries) was found. On the other hand, all those new arithmetics (geometries) were unified by definite mathematical laws governing their system.

An important peculiarity of this book is that the reader will actually see no formulas in the first part of the book containing the main exposition of the history of numbers and arithmetic given in the first three chapters and few simple formulas, in which only arithmetical operations are used, in the second part (Chapters 4 and 5) dedicated to the roots and discovery of non-Diophantine arithmetics. Thus, to understand the presented historical material, it is not necessary to know more mathematics than it is taught in secondary schools.

To conclude, we once more inform the reader that this book is the first one in the trilogy on numbers and arithmetics. Thus, the history exposed in this book will be complemented by the philosophy of numbers and arithmetics in the second book and by the basic mathematics of arithmetic in the third book of this trilogy.

2. An Overview

> *All of mathematics can be deduced from the sole notion of an integer;*
> *here we have a fact universally acknowledged today.*
>
> E. Borel

The development of mathematics in general and arithmetic in particular has been accelerating in time. It means that introduction of the new kinds of numbers and construction of novel arithmetics has been happening more often with the passing time. That is why the history of numbers and arithmetic is naturally divided into three periods, which are described in the first three chapters:

- the interval from ancient times to the second millennium;
- time from the beginning of the second millennium to the 20th century;
- the 20th and 21st centuries.

Each of the first three chapters covers one of these three periods describing the process of the development of arithmetic, which included the elaboration and utilization of numbers and numerical operations. At the same time, the rationale for introducing each class of numbers is justified elsewhere.

Because the conventional arithmetic, which is called the Diophantine arithmetic of natural numbers, was considered unique and absolute for millennia, the discovery of non-Diophantine arithmetics became rather unexpected for the vast majority of people. Even now their essence is not correctly understood. That is why the history of non-Diophantine arithmetics is described separately in Chapters 4 and 5.

To better understand the history of mathematics in general and arithmetic in particular, it is necessary to think about numbers and their arithmetics as sources of information, tools for information acquisition and designs for information storage. As Amy Goldstein and Shelli Avenevoli write, *"any fact and knowledge item that can be expressed in numbers and words is called information"* (Goldstein and Avenevoli, 2018). Expressing this idea more exactly based on the contemporary information theory, we can say that by assigning numbers to arbitrary objects, people numerically codify information about these objects. This information can be preserved, used for obtaining new information or for operating with those objects. For instance, it is possible to utilize numerical parameters of a system for maintaining this system. When engineers are planning to construct a system, they accumulate various numerical characteristics for this construction. One more example is statistical information, which as a rule, has the numerical form and is used for predicting behavior of people and social systems.

Describing the history of arithmetic, we distinguish three meanings of the term *arithmetic*:

(1) Arithmetic is the ordered system of natural numbers, that is numbers 1, 2, 3, ..., with two basic operations — addition and multiplication.
(2) Arithmetic is a mathematical structure for computations with numbers or other structures.
(3) Arithmetic is a mathematical field that studies numbers and operations with them.

Note that textbooks on arithmetic are often called simply *Arithmetic*.

In this context, natural numbers are called natural because they naturally come from counting and comparison of different sets of objects. However, it is necessary to note that the concept of natural numbers has been changing with time. At the beginning, there were no numbers that were not natural. That is why all of them were called simply numbers. The adjective natural was applied much later when other classes of numbers were introduced.

With the development of mathematics, people understood that there were infinitely many numbers, which were defined by ancient Greek mathematicians and philosophers as collections of units while 1 was the unit. That is why for a long time, 1 was not treated as a number. Moreover, for some time, 2 also was not regarded as a number, and numbers started from 3.

Much later, confusion of terminology emerged and proliferated to the 21st century. Namely, some mathematicians and scientists take for granted that natural numbers start with 1while others assume that the first natural number is 0 (cf., for example, Bourbaki, 1960; Halmos, 1974; Dijkstra, 1982; Martin, 1991; Cassani and Conway, 2018).

Here we adhere to the following terminology:

Natural numbers consist of 1, 2, 3, ...

Whole numbers consist of 0, 1, 2, 3, ..., i.e., of all natural numbers and zero.

Integer numbers (or simply, *integers*) consist of 0, 1, 2, 3, ...and −1, −2, −3, ..., i.e., of all whole numbers and natural numbers taken with the negative sign −.

As a part of integer numbers, natural numbers are called *positive integer numbers*, zero is the *neutral integer number*, and natural numbers taken with the negative sign — are called *negative integer numbers*.

As a mathematical field that studies numbers and operations with them, arithmetics includes *higher arithmetic*, which is now mostly known under the name *number theory*. Here we make the main emphasis on the arithmetic in the long-established understanding and give only an outline of the history of higher arithmetic (number theory) because detailed exposition of this theory would demand writing another book twice the size of this book.

From the information theory perspective, arithmetics are used for operation with information embedded in numbers. For instance, if we know that the population of the city A is 2 million and the population of the city B is 3 million, addition tells us that both cities together have the population of 5 million. We can also deduce which of these cities is larger. Knowing that the price of one pound of apples is \$2, we can easily find that if we want to buy five pounds, the price will be \$10. These are simple examples of information that arithmetic can afford but this and similar information is necessary in many practical situations. At the same time, numbers and arithmetics together with more advanced mathematical tools can bring knowledge about the most sophisticated mysteries of the Universe explaining the behavior of the tiniest particles and predicting the movements of stars and galaxies.

However, the operations with numbers prescribed by the conventional Diophantine arithmetic do not always correctly represent real situations and thus, are not always giving correct information. For instance, in our example of two cities, imagine that we know that the population of the city A is 2 million and the population of the city B is 3 million, while these cities have a common part, in which 500,000 people live. Then addition in the conventional Diophantine arithmetic tells us that both cities together have the population of 5 million. However, in reality, the combined population will be only 4.5 million. This brings us to the non-Diophantine arithmetic, in which $2 + 3 = 4.5$ and which gives us more correct information about the real situation than the Diophantine arithmetic. Note that there are other non-Diophantine arithmetics with different rules of addition and multiplication.

Because non-Diophantine arithmetics are less known to people, we pay special attention to the history of this relatively new mathematical field. In Chapter 4, at first, it is explained how mathematicians and other researchers found that the Diophantine arithmetic of natural numbers, was insufficient for treating many real-life situations. After this, it is described how non-Diophantine arithmetics were discovered and what properties they have.

Chapter 5 provides an exposition of the unexpectedly long history of non-Diophantine operations with numbers telling about diverse applications of these operations. It is explained that implicit (concealed) emergence of non-Diophantine operations happened when mathematician discovered the so-called Pythagorean Theorem and this event happened millennia ago. Considering different arithmetical operations, it is necessary to understand that they do not exist independently but form some kind of arithmetics while their efficiency grows when the whole arithmetic is utilized. Thus, we see that in the hidden form, elements of non-Diophantine arithmetics came to people millennia ago but were discovered only at the end of the 20th century.

This situation reminds the situation with non-Euclidean geometry. Indeed, people lived on the Earth, the global geometry of which is non-Euclidean, but mathematicians were not able to discover non-Euclidean geometries till the beginning of the 19th century. Besides, it took decades even for very good mathematicians to understand non-Euclidean geometries.

In the appendix, we give the list of arithmetics used by people at different times to show the reader how far arithmetic goes from that arithmetic the reader studied at school.

It is necessary to remark that the research related to numbers and arithmetic is extremely active, while numbers are related almost to everything. That is why we tried to reflect the most important results, ideas, issues, directions, and references to materials that exist in this area. In addition, an extended bibliography is provided to allow the reader to find more information on topics and issues presented in this book.

Chapter 1

The Story of Numbers and Arithmetic from Ancient Times to the Beginning of the Second Millennium

All that passes for knowledge can be arranged
in a hierarchy of degrees of certainty, with
arithmetic and the facts of perception at the top.

Bertrand Russell

Describing the history of numbers and arithmetic, it is natural to distinguish inner and outer directions of their evolution. In the *inner direction* of the number development, subclasses (subtypes) of the known classes (types) of numbers were determined and their properties were studied. For instance, even, odd, prime and composite numbers were defined and explored in the class of natural numbers. In the *outer direction* of the number development, mathematicians constructed new larger than previously known classes (types) of numbers and studied their properties. For instance, the class of positive rational numbers extended the class of natural numbers or the class of real numbers extended the class of rational numbers.

In the *inner direction* of the arithmetic development, existing arithmetics were studied and methods of working with numbers were improved. Many systems for a mathematical number representation (naming), which are called either *numerical systems* or *systems of numeration* or *numeric systems* or *numeral systems*, were created in different parts of the world. The development of numerical systems brought forth corresponding rules for performing arithmetical operations. The Hindu-Arabic numeral system allowed construction of much better algorithms for multiplication and division.

In the *outer direction* of the arithmetic development, new arithmetics were constructed and their properties were studied. Often creation of new arithmetics was based on emergence of new classes of numbers. For instance, the introduction of real numbers brought forth their arithmetic, and the introduction of complex numbers brought forth their arithmetic. In addition, new arithmetics were created even for existing classes of numbers. For instance, a diversity of non-Diophantine arithmetics was constructed for natural numbers at the end of the 20th century (Burgin, 1977; 1980; 1997) although almost all mathematicians and other people were completely certain that only one such arithmetic was possible.

It is also important that the history of numbers and arithmetic consists not only of the discovery of new kinds of numbers and building new arithmetics but also includes elaboration of more rigorous ways of number modeling and construction as well as improvement of the methods of operation with numbers. For instance, for a long time, natural numbers were described as the sequence 1, 2, 3, 4, 5, and so on, while their mathematically rigorous definition by axioms was elaborated only in the 19th century in works of Grassmann, Dedekind, Peirce, Peano and other mathematicians and logicians.

History of mankind shows it was a long way from the primitive forms of numbers to the creation and studies of contemporary highly abstract number systems. Forms of simple numbers and arithmetic emerged in prehistoric times becoming fundamental concepts of mathematics in the course of a long historical process. The origin and formation of numbers was going simultaneously with the birth and development of mathematics, which came out as an applied discipline growing up into a colossal theoretical field. The practical activities of humankind on the one hand, and the internal requests of mathematics on the other, have shaped the growth and expansion of the concept of a number.

However, as Eves writes,

"The concept of number and the process of counting developed so long before the time of recorded history that the manner of this development is largely conjectural." (Eves, 1990)

Now the earliest known archaeological evidence of any form of writing or counting is scratch marks on a bone from 150,000 years ago. Archaeologists suggest these marks represent counting of ancient people. However, these opinions were only conjectures because the first unyielding confirmation of counting was the Ishango bone. It was found in the 1950s by Belgian archaeologist Jean de Heinzelin de Braucourt (1920–1998) near a Palaeolithic residence in Ishango, Congo, Africa (Were, 2019). The age of the Ishango bone is around 30,000 years although there other estimates of its age in different sources. It has two equal markings of sixty scratches each on the one side and similarly numbered groups on the other. These markings are a dependable indication of counting and they mark a defining step in the development of the human civilization.

However, later in the 1970s the Lebombo bone was found in the Lebombo Mountains located between South Africa and Swaziland. It was also marked by ancient people with 29 clearly defined notches (d'Errico *et al.*, 2012). The age of the Lebombo bone is between 43,000 and 44,000 years, according to radiocarbon dating. That is why now it is considered the oldest mathematical object known. Interestingly, the Lebombo bone resembles the calendar sticks still in use by Bushmen clans in Namibia.

However, the notable French psychologist and neuroscientist Stanislas Dehaene conjectures that ability to operate with numbers emerged even before people started to speak (Dehaene, 1997). This hypothesis is supported by the evidence that animals, birds, fish, and even insects are able to process numerical information (cf., for example, Agrillo *et al.*, 2015; Bar-Shai *et al.*, 2011; Gross, *et al.*, 2009)). Deahaene calls this ability the *number sense*.

Note that in mathematics education, number sense refers to a group of key mathematical abilities an intuitive understanding of numbers, their magnitudes, relationships, and how they are affected by operations. Even more, definitions of number sense include a well-organized conceptual framework of number information that enables an individual to understand numbers and number relationships and to solve mathematical problems that are not bound by traditional algorithms (Bobis, 1991).

According to Dehaene, when people first began to speak, they were able to name only the numbers 1, 2, and perhaps 3. Oneness, "twoness", and "threeness" are perceptual qualities that human brains compute effortlessly, without counting (Dehaene, 1997). Consequently, giving them a name was probably no more difficult than naming any other sensory attribute, such as red, big, or warm.

The first three number words have a special status in many languages. For instance, in English, most ordinals end with "th" (fourth, fifth, etc.), but the words "first," "second", and "third" do not. There are also special names to denote one, two, or three objects. For one object, we have a *monad.* For two objects, we have a *dyad*. For three objects, we have a *triad* and *triple*.

Interestingly, "the Indo-European root of the word "three" suggests that it might have once been the largest numeral, synonymous with "a lot" and "beyond all others" — as in the French *tres* (very) or the Italian *troppo* (too much), the English *through,* or the Latin prefix *trans-* (Dehaene, 1997). This well correlates with the situation that the prevailing mathematical tradition in ancient Greece was that numbers started with 3. Besides, number 3 is naturally related to the fundamental triad — the most basic structure in the world (Burgin, 2011).

Although the need to count objects led to the origin of the notion of a natural or counting number, this notion appeared much later and for a long time, people gave special names to diverse groups of objects or events that they often encountered. For instance, in the languages of certain tribes, there were words for such concepts as *three men* or *three boats,* which are named numbers (cf., for example, Depman, 1959; Burgin, 2011), but there was no abstract concept *three.* In this way, probably, there arose comparatively short series of named numbers used for the identification of similar groups of people, boats, coconuts, and so on. At these early stages, only named numbers were used and there were no such objects as abstract numbers as can be judged from indirect data provided by linguistics and ethnography.

Indeed, the process of abstraction is rather sophisticated and demands definite intellectual efforts. The renowned Austrian-Ireland-Italian-American philosopher Ernst von Glasersfeld (1917–2010), who is naturally associated with Austria, Ireland, Italy, and USA because he worked in all these countries, writes:

"*What constitutes the abstract concept of number is the attentional pattern abstracted from the counting procedure. In this pattern it is irrelevant what the focused moments of attention are actually focused on. The salient features are:* (1) *the iteration of moments that are focused on some unitary items and attentional moments that are not;* (2) *that the iterated sequence itself is bounded by unfocused moments; and* (3) *that the focused moments are coordinated with number words.*

The conceptual transformation of a plurality into the concept of number has intermediary steps ...

The symbol 'number' acquires its meaning from the fact that it points to the conceptual structure of attentional iteration in an actual or potential counting situation. In my discussion of reflection and abstraction, I explained the pointing function of symbols. In the case of the concept of number, this function is crucial. Number words and all kinds of numerals point to specific instantiations of the number concept's attentional structure, but this does not entail that the indicated attentional iteration and the count have to be carried out. To understand the symbols, one merely has to know the required procedure and that it could be carried out." (von Glasersfeld, 1981; 1995)

As a result, people used named numbers, e.g., five apples, ten pounds or seven books, for millennia before they came to the concept of (abstract) numbers, i.e., numbers as they exist in mathematics now. Mathematicians separated numbers from their names, which were considered irrelevant to abstract mathematics. In contrast to this, practice demanded mathematical consideration of named numbers, and in the 21st century, a rigorous mathematical theory of named numbers was created by Mark Burgin based on the theory of named sets (Burgin, 2011).

However, even in our time more primitive societies, such as the Wiligree of Central Australia, have not even used named or abstract numbers, nor felt the need for them (Osborn, 2005). The cause for using numbers, which were named numbers, was that the Sumerians and people from other developed civilizations lived in cities and required much better organization of their life. For instance, different products, such as grain, needed to be stored and exchanged. These operations required arithmetic. In ancient China, Egypt, Babylon, and some other early civilizations, arithmetic was used for commercial purposes, records of taxation, and astronomy. As Behrooz Parhami writes, *"four thousand years ago, Babylonians knew about natural numbers and were proficient in arithmetic"* (Parhami, 2002).

Thus, historical, ethnographical and anthropological studies show that natural numbers came from counting and their arithmetic emerged in operation with counted objects. A simple observation shows that counting is an operation applied to various objects and by its nature, it is naming of objects by natural numbers providing information about the counted set of objects (Burgin, 2011). In turn, counting is based on the arithmetical operation, which taking a natural number, converts it into the next number.

When people in ancient civilizations started constructing big buildings and pyramids, selling and buying land, they applied numbers to measuring, which was of great significance from the perspective of mathematics and its applications. For instance, using standardized measures millennia ago, the Egyptians completed gigantic construction projects, such as their great pyramids, with astonishing precision. The Harappan civilization in India, which existed in the third millennia B.C.E., delved into commerce and cultural activities inventing complex systems of weights and measurements and utilizing highly developed arithmetic to work with them.

Thus, the practical need for counting, elementary measurements and calculations became the major reason for the emergence and development of arithmetic. Besides, simple arithmetic emerged and became ubiquitous as a mean of transacting business.

The first authentic data on arithmetic knowledge are found in the historical monuments of Babylon and Ancient Egypt from the third and second millennia B.C.E.

Archeologists found that even before this time, in about 4000 B.C.E., Sumerians, who lived in Mesopotamia, employed tokens as an improvement over scratches on a stick or bone to represent numbers. The next step was the discovery of how to add such tokens representing addition of numbers. In addition, people found that taking away tokens, it is possible to represent subtraction of numbers. Introduction of these basic arithmetical operations gave birth to arithmetic, which was an event of crucial significance. The Sumerian's tokens were used to assess assets, estimate profit and loss and even more significantly, to collect taxes and keep enduring records. In this manner, accounting was born. The popular belief is that in the form of scratches, notches, or tokens, numbers became the world's first writings used for preservation of and operation with information.

Mathematics became highly developed in this area within the Babylonian empire, which existed between 18th and 6th centuries B.C.E. in Mesopotamia. The Babylonians wrote their texts mostly on clay tablets. Written in Cuneiform script, such tablets were inscribed while the clay was moist, and then baked hard in an oven or by the heat of the sun. Archeologists unearthed more than half million clay tablets. That is why in contrast to the scarcity of sources in Egyptian mathematics, knowledge of Babylonian mathematics is much better. Around 400 clay tablets have been found with mathematical texts. Most of the recovered mathematical clay tablets date from 1800 to 1600 B.C.E. They cover topics that include operations with natural numbers; fractions; what may be called elementary algebra as it provided solutions of problems that can be represented by quadratic and cubic equations; and the Pythagorean Theorem. In particular, Babylonians used pre-calculated tables to assist operation with numbers. Two tablets found at Senkerah on the Euphrates in 1854, dating from 2000 B.C.E., give lists of the squares of numbers up to 59 and the cubes of numbers up to 32.

Babylonians elaborated an advanced system for representing natural numbers. In contrast to our numerical system, which has the base 10, Babylonians used the base 60 for their enumeration. It is possible to suggest that the base 10 is used because people, as a rule, have 10 fingers and fingers have been used for counting and calculation for a long time. At the same time, the base 60 used by Babylonians looks mystifying — why 60 and not 50 or 30. To answer this question, researchers suggested the following hypothesis. Babylonians had advanced astronomy, which essentially used trigonometry while the sexagesimal numerical system came from trigonometry because numbers 60 reflected an important trigonometric property: namely, the ancient mathematicians found that in any circle, the chord of 60° was equal to the radius of the circle.

It is necessary to understand that solving problems that can be represented by quadratic and cubic equations did not mean emergence of algebra because these problems were solved by arithmetical methods without involving operation with abstract symbols.

After the Sumerians, some of the most advanced ancient works on mathematics come from Egypt. Egyptians developed impressive numerical systems for representation of natural numbers. In the first period of the Egyptian history, their numerical system had the base 10 as our numerical system but it represented numbers in a different way utilizing distinct hieroglyphs as digits for powers of 10, that is, for the numbers 1, 10, 100, and up to 1,000,000.

Later Egyptians introduced hierarchical numerals using 10 hieroglyphs as digits for the first 10 numbers, different 10 hieroglyphs as digits for the numbers from 10 to 90, different 10 hieroglyphs as digits for the numbers from 100 to 900, and so on. This system is called pseudo-positional.

Egyptians wrote their texts on papyrus, which is much more vulnerable than clay tablets. That is why now we have much less mathematical text from ancient Egypt in comparison with ancient Babylonia. The most preserved artifacts of Egyptian mathematics are the Rhind Mathematical Papyrus and the Moscow Mathematical Papyrus. The more famous Rhind Papyrus has been dated to approximately 1650 or 1550 B.C.E. but it is thought to be a copy of an even older scroll (Robins and Shute, 1987). This papyrus essentially played the role an early textbook for Egyptian students starting with the claim that it provides "a thorough study of all things, insight into all that exists, knowledge of all obscure systems" (cf. Burton, 1997). It show the high esteem of mathematics or more exactly, or arithmetic at those time because actually it was a mathematical handbook teaching how to solve various practical problems including business management with the help of arithmetic.

The first part of the Rhind papyrus consists of reference tables and a collection of 21 purely arithmetic and 20 related to problems in algebra, which can be described by linear equations. The second part of the Rhind papyrus contains geometry problems, which are usually entail operation with numbers. The third part of the Rhind papyrus consists of general problems also involving numbers. In addition to giving area formulas and rules for arithmetical operations with natural numbers and working with unit fractions, it also contains evidence of other mathematical knowledge, for example, about composite, perfect and prime numbers; arithmetic, geometric and harmonic means; and methods for solving problems that can be reduced to linear equations. As David Burton mentions, "the Egyptian arithmetic was essentially 'additive' meaning that its tendency was to reduce multiplication and division to repeated additions" (Burton, 1997).

There are also problems involving simple number series in the Rhind papyrus. A *number series* is an expression obtained from numbers by application of addition many times. Namely, the smallest number series consists of two numbers connected by the sign plus. The number series with three elements consists of three numbers connected by two signs plus and so on. Thus, any finite series is an arithmetic expression obtained from numbers by addition. The result of performing all additions in a series is called the *sum* of this series. Finding the sum of a series is called *summation*. Thus, people started operating with finite series in the 2nd millennia B.C.E. Infinite number series came to mathematics much later.

The Moscow Mathematical Papyrus was most likely written down in the 13th dynasty of Egyptian pharaohs, which ruled Egypt approximately from 1800 B.C.E. till 1650 B.C.E. The text was based on older material probably dating to the 12th dynasty of Egyptian pharaohs, that is, around 1850 B.C.E. The papyrus contains 25 mathematical problems with solutions

Historians of mathematics think that mathematics of ancient Egypt never reached the level attained by the Babylonian mathematics (Eves, 1990). For instance, the problems treated by Egyptians involve only linear equations while the problems solved by Babylonians demand representation by quadratic and cubic equations.

Babylonian mathematics was essentially based on what previously had been developed first by the Sumerians and later by the Akkadians. Besides, there are much more Babylonian mathematical texts, which were preserved till our time, than Egyptian because Babilonians wrote on clay tablets which were much better preserved than papyruses on which Egyptians wrote.

To operate with numbers, it is necessary to have names for numbers. Such names exist in natural languages. For instance, the number 3 is called *three* in English, *tres* in Spanish, *trios* in French, *tre* in Italian, and *drei* in German. However, these names are not good for operating with numbers. That is why mathematicians elaborated *numerical systems*, which provide mathematical names for numbers. These names are called *numerals* and use *digits* (elementary symbols) to construct numerals. We already discussed the Egyptian numerical system.

Note that the contemporary technology uses physical objects, e.g., electrical signals, and their states as names of numbers. For instance, numbers are represented by the states of the cells in the computer or calculator memory.

Babylonians also developed and advanced numerical system inheriting ideas of previous civilizations — Sumerians and Akkadians. The greatest achievement of Babylonian mathematicians was building their numerical system on positional principles. It means that the meaning of digits used for building numerals depends on their position. For instance, in our decimal positional system, if we take the number 333, we see that the first 3 from the right means 3, the second 3 means 30, and the third 3 means 300. This meaning depends on the base of the numerical system, which is 10 in our case and thus, uses 10 digits 0, 1, 2, 3, 4, 5, 6, 7, 8, and 9.

In contrast to this, the base of the Babylonian positional numerical system is 60. However, Babylonians did not use 60 elementary digits. They invented a technique allowing them using only 2 elementary digits to represent all numbers. The clue was the hierarchical construction of numerals. On the first level, sixty composite digits are constructed from the elementary digits, while on the next level, numerals are formed from the composite digits.

Historical sources also show that ancient mathematicians believed that numbers are sums of units, that is, they utilized only natural numbers. Fractions came later due to the necessity for measuring parts of objects. This situation brought forth positive fractions, which were expressed through natural numbers but did not have the status of numbers.

Now in many textbooks, it is possible to read that fractions are numbers. However, ancient mathematicians were right — fractions are not numbers but only names or representations of rational numbers. For instance, 1/2, 2/4, and 3/6 are names of one and the same rational number.

According to the historical evidence, simple fractions were used by the Egyptians around 1500 B.C.E. They employed unit fractions, i.e., fractions with the numerator 1, directly while utilizing interpolation tables to approximate the values of the other fractions.

As according to historical, ethnographical and anthropological studies, positive rational numbers came from measuring, while their arithmetic emerged in the course of operation with measured objects.

Other ancient civilizations also developed mathematics and especially arithmetic as the intellectual tool for acquiring, processing, and preserving information in the practical activity of people.

According to the known sources, mathematics in China emerged later, namely, around the 11th century B.C.E. (Yong, 1996). Chinese mathematicians independently developed the arithmetic of integer numbers using numerical system with base 2 and base 10, arithmetic algebra, geometry, number theory and trigonometry (Chemla, 2015). Besides, negative number became ubiquitous in China very early. In particular, already in the second century B.C.E., Chinese used a number rod system, in which positive numbers were represented by red rods and negative numbers by black rods.

The most ancient documented numerical system was used in the second millennium B.C.E. It had ten different elementary digits from which composite digits were constructed to represent numbers.

The most important work in the history of Chinese mathematics is Jiǔzhāng Suànshù (which means *The Nine Chapters on the Mathematical Art*), which was composed by several generations of scholars from the 10th century B.C.E. to the 2nd century C.E. It contains solutions of arithmetic and geometric problems related to the duties of the civil administration such as surveying fields (areas), levying taxes according to various types of grains (ratios), determining wages for civil servants according to their position in the hierarchy (unequal sharing), measuring planned earthworks to determine labor needs and granaries to determine storage capacity (volumes), levying fair taxes (problems combining various proportions), and so on.

Another ancient Chinese mathematical book is Suànshùshū (which means *Book on Numbers and Computation*), which was written during the early Western Han Dynasty, sometime between 202 B.C.E. and 186 B.C.E. It also contains solutions of various practical problems in arithmetic.

One more ancient Chinese mathematical book *Zhoubi Suanjing* (which means *Arithmetical Classic of the Gnomon and the Circular Paths of*

Heaven) appeared during the Zhou Dynasty (1046–256 B.C.E.) but its compilation and addition of materials continued during the Han Dynasty (202 B.C.E.–220 C.E.). It contains arithmetic and geometric algorithms and proofs including a proof of the Pythagorean Theorem.

One more place where mathematics in general and arithmetics in particular achieved high level was India. For instance, while in Europe, the existence of infinite numbers was essentially rejected until the 19th century (cf., for example, Arthur, 2001; Brown, 2000; Nachtomy, 2011), ancient Indian mathematicians, namely, those who belonged to the Jainas, not only described actual infinity in a numerical way but even differentiated different types of numbers in general and infinite numbers, in particular. Jainas fascination with large numbers brought them to the conception of infinity represented by a number, which provided important information about infinity. As a result, the Jaina mathematicians were the first to abandon the idea that all infinities were the same or equal, contrary to the idea widespread in Europe till the 19th century when mathematicians started studying the growth of functions and Georg Cantor created set theory. The Jaina mathematical text Surya Prajnapti (dated 3rd–4th century B.C.E.) classifies all numbers into three sets: *enumerable*, *innumerable*, and *infinite*. Each of these groups was further subdivided into three groups:

- Enumerable numbers consist of *lowest enumerable*, *intermediate enumerable*, and *highest enumerable* numbers.
- Innumerable numbers consist of *nearly innumerable*, *truly innumerable*, and *innumerably innumerable* numbers.
- Infinite numbers consist of *nearly infinite*, *truly infinite*, *infinitely infinite* numbers.

In addition, two basic types of infinite numbers are distinguished in this work. On both physical and ontological grounds, a distinction was made between *asamkhyāta* ("countless, innumerable") and *ananta* ("endless, unlimited"). For instance, the set of real numbers in the interval $[0, 1]$ is bounded but countless, while the set of all natural numbers is unlimited but countable.

This once more validates the following statement of the great German mathematician David Hilbert (1862–1943).

"*The infinite! No other question has ever moved so profoundly the spirit of man.*" (Hilbert, 1967)

David Hilbert was born in Wehlau, near Königsberg, Prussia (now Kaliningrad, Russia) and studied in the University of Königsberg and

Heidelberg University. He worked at the University of Königsberg and as the chair of mathematics at the University of Göttingen. Hilbert made important contribution to many fields of mathematics — invariant theory, the calculus of variations, functional analysis, commutative algebra, the foundations of geometry, and algebraic number theory. In particular, in functional analysis, he introduced metric vector spaces of infinite dimensions, which later acquired the name *Hilbert space* becoming an important tool for quantum mechanics and field theory. In number theory (higher arithmetic), he solved the Waring problem and unified algebraic number theory. Many mathematical systems have his name. Examples are: the Hilbert cube, Hilbert cyrve, Hilbert matrix, Hilbert ring, Hilbert spectrum and so on. Hilbert also contributed to physics developing general relativity almost at the same time as Albert Einstein.

Based on the contemporary mathematical knowledge, it is possible to give the following interpretation for the notions *enumerable*, *innumerable*, and *infinite* concepts from Jaina mathematics.

- *Enumerable numbers* are natural numbers that an individual can build by adding ones, e.g., numbers 5, 100 and 2000 are enumerable.
- *Innumerable numbers* are natural numbers that an individual cannot build by adding ones, e.g., numbers 10^{100} and $10^{10^{10}}$ are innumerable.
- *Infinite numbers* are, for example, transfinite numbers from set theory or infinite whole hypernumbers (Burgin, 2012).

Interestingly, the Jaina classification of enumerable numbers into lowest enumerable, intermediate enumerable, and highest enumerable numbers is intrinsically parallel to the classification of natural numbers suggested by the great mathematician of the 20th century Andrei Nikolayevich Kolmogorov (1903–1987), who was born in Tambov, Russia, studied at Moscow State University and obtained his first important results when he was only 18 years old. During his life he made essential contributions to probability theory, topology, information theory, theory of algorithms, computational complexity, intuitionistic logic, integration theory, turbulence, and classical mechanics demonstrating originality, a breadth of approach, and depth of thought in many of his works.

In his work on relations between automata and living beings, Kolmogorov distinguished *small, medium, large,* and *super-large* numbers (Kolmogorov, 1961, 1979). He described his classification in the following way.

A natural number k is *small* if it is possible in practice to list and work with all combinations and systems such that are built from k elements each of which has two inlets and two outlets.

A natural number m is *medium* if it is possible to count to and work directly with this number. However, it is impossible to list and work with all combinations and systems that are built from m elements each of which has two or more inlets and two or more outlets.

A natural number n is *large* if it is impossible to count a set with this number of elements. However, it is possible to elaborate a system of denotations for these elements.

If even this is impossible, then a natural number is called *super-large*.

The most ancient Indian mathematical text available is the Baud-hayana's Shulbasutra developed by the Indian mathematician Baudhayana (800–740 B.C.E.). It contains geometric solutions of arithmetical problems that can be reduced to linear or quadratic equations with one unknown variable as well as quite accurate, approximate value for $\sqrt{2}$ (Kumari, 1980).

It is nothing known about another Indian mathematician Apastamba (ca. 600 B.C.E.) except that he was the author of a Sulbasutra which is certainly later than the Sulbasutra of Baudhayana.

However, the main contribution of Indian mathematics to the human culture is the decimal positional numerical system. As at first the out-standing French mathematician and physicist Pierre-Simon Laplace (1749–1827) and then the American mathematician and historian of mathematics Howard Eves (1911–2004) write:

"It is India that gave us the ingenious method of expressing all numbers by means of ten symbols, each symbol receiving a value of position as well as an absolute value; a profound and important idea which appears so simple to us now that we ignore its true merit. But its very simplicity and the great ease which it has lent to computations put our arithmetic in the first rank of useful inventions; and we shall appreciate the grandeur of the achievement the more when we remember that it escaped the genius of Archimedes and Apollonius, two of the greatest men produced by antiquity." (Eves, 1988)

There were different numerical systems in India, which according to the historians of mathematics have roots in the Brahmi numerals, which were used from around the middle of the 3rd century B.C.E. up to the 4th century C.E. There are different hypotheses how these numerals originated.

The next Indian numerical system was formed of Gupta numerals, which was in use from the 4th century C.E. up to the 7th century C.E. when Nagari numerals emerged. There is also indirect evidence that Indian mathematicians already had the decimal positional numerical system in the 1st century C.E.

Working with numbers and developing mathematics, mathematicians discovered some special quantities, which later were described as numbers. The first was the famous number π (pi), which came from geometry reflecting an invariant property of circles, namely, that the ratio of the length of a circle to its diameter is always equal to π. Its notation by the Greek letter goes back to such Greek words $\pi\varepsilon\rho\iota$-, which means around, about; $\pi\varepsilon\rho\iota\mu\varepsilon\tau\rho o\nu$, which means perimeter, circumference; and $\pi\varepsilon\rho\iota$-$\varphi\acute{\varepsilon}\rho\varepsilon\iota\alpha$, which means circumference, arc.

Now we know that π is an irrational number and thus, cannot be represented by fractions. However, ancient mathematicians operated only with natural numbers and fractions. As the result, they found only approximations of the number π.

Here is the chronology of getting approximations of π (Schepler, 1950; Eves, 1990):

The Babilonians (ca. 1900–1700 B.C.E.) took π equal to the product 25/8 giving two correct digits in the decimal representation of π.

In the Rhind papyrus (ca. 1650 or 1550 B.C.E.), π is taken equal to the product $(4/3)(4/3)(4/3)$ giving two correct digits of the decimal representation of π.

The Greek mathematician Archimedes (ca. 287–212 B.C.E.) found that π is between 223/71 and 22/7 giving three correct digits of the decimal representation of π.

The Chinese mathematician Zhang Heng (78–139) found that π is equal to 92/29 or $\sqrt{10}$ giving two correct digits of the decimal representation of π.

The Greek astronomer and mathematician Claudius Ptolemy of Alexandria (ca. 100–170 C.E.) determined that π is equal to 377/120 giving almost five correct digits of the decimal representation of π.

The Chinese mathematician Wang Fan (219–257) calculated that π is equal to 142/45 giving two correct digits of the decimal representation of π.

The Chinese mathematician Liu Hui (219–257) calculated that π is equal to 157/50 giving five correct digits of the decimal representation of π and providing an algorithm to calculate π with arbitrary accuracy.

The Chinese mathematician Zu Chogzhi (429–500) calculated that π is equal to 355/113 giving almost eight correct digits of the decimal representation of π.

The Indian astronomer and mathematician Aryabhata (ca. 476–550) calculated that π is equal to 62832/20000 giving four correct digits of the decimal representation of π.

The Indian astronomer and mathematician Madhava of Sangamagrama (ca. 1340–1425) developed a method of representing π by infinite series, which allowed getting 12 correct digits of the decimal representation of π using 21 terms of such a series.

The Muslim mathematician Jamshid Mas'ud al-Kashi (1380–1429) calculated the number π to sixteen correct decimal places.

In comparison, now computers can calculate the decimal representation of π up to many trillions of correct digits.

Thus, we can see that several ancient civilizations in Babylonia, China, Egypt and India utilized numbers and developed arithmetic as the most ancient component of mathematics. However, the real breakthrough in mathematics in general and in arithmetic in particular was made in ancient Greece due to the following reasons.

First, historians of mathematics came to the conclusion that the first proofs of mathematical statements appeared in ancient Greece and this was extremely important. Indeed, proofs form the backbone of the contemporary mathematics.

Second, before mathematics was an entirely applied field while in ancient Greece theoretical mathematics originated. For instance, Plato in his dialogue *Philebus* maintains: *"Arithmetic is of two kinds, one is popular, and the other philosophical"* (Plato, 1961) where philosophical means theoretical.

Third, the whole area of mathematical knowledge was reshaped into two mathematical disciplines — arithmetic and geometry providing an efficient structure for mathematics as a science.

That is why it is believed that contemporary mathematics as a theoretical science originated in ancient Greece. The same is true for arithmetic as it is understood now and for number theory as the higher arithmetic. That is, arithmetic as a theoretical science originated in ancient Greece.

As a result, the first systematic study of numbers as abstractions as well as the theoretical development of arithmetic is often attributed to the ancient Greek mathematicians and philosophers, the most famous of which were Pythagoras and Archimedes.

Greeks also has their own numerical systems. At first, they used the acrophonic system used in the 1st millennium B.C.E. In it, symbols for the numerals (digits) were the first letters of the number name. Besides, Greeks used specific numerals to represent named numbers.

Later Ionic, also called Ionian, Milesian, or Alexandrian, numerals, which use the letters of the Greek alphabet as the digits for building numerals, came into being. Numbers were represented using the additive principle. For instance, the number 12 was written as $\iota\beta$ because ι denoted 10 and β denoted 2. In modern Greece, these numerals are still used for ordering notation in the same way as Roman numerals are still used elsewhere in the West.

The birth of the Greek mathematics is usually attributed to one of the most famous mathematicians of all times Pythagoras (ca. 569–500 B.C.E.), who was born on the island of Samos, off the Anatolian coast. Actually, he is a legendary person because nothing is known for sure about his life. However, different ancient philosophers wrote about him and this is what they told. According to the influential Neoplatonic philosopher Iamblichus (ca. 245–325), Pythagoras was introduced to mathematics by the famous philosopher Thales of Miletus and his pupil Anaximander. Besides, in his youth, Pythagoras traveled extensively throughout the ancient world visiting Babylon, Egypt and may be, even India. There he learnt achievements of their mathematicians. Upon his return, Pythagoras moved to the Greek colony Crotona situated in southern Italy originating there a school of philosophy and mathematics with the main emphasis on the study of numbers. This school existed in the form of a secret society, which was also active in politics and for some time seized power in Crotona. According to the legend, Pythagoras also coined the words *philosophy*, which means *love of wisdom* in Greek, and *mathematics*, which means *knowledge* or *that which is learned* in Greek.

Now Pythagoras is treated as a famous mathematician and philosopher. However, in ancient Greece, Pythagoras was recognized as (Kahn, 2001):

- an authority on the soul who described that the soul was immortal and went through a series of reincarnations;
- a connoisseur of religious rituals;
- a wonder-worker who had a thigh of gold and who could be two places at the same time;
- the adherent of a strict way of life with dietary restrictions, religious ritual, and rigorous self discipline.

Pythagoras and his followers studied different types of natural numbers: odd and even numbers, triangular numbers, and perfect numbers. For instance, triangular numbers are described as numbers that can be represented by pebbles arranged in the shape of an equilateral triangle.

Here are representations of the triangular numbers 3 and 6:

·

· ·

and

·

· ·

· · ·

The triangular and other figurative numbers demonstrated the complementary character of arithmetic and geometry, two basic mathematical disciplines in antiquity.

Besides, Iamblichus ascribed the discovery of *amicable*, or *friendly*, numbers to Pythagoras (cf., Eves, 1990). Two numbers are called amicable if each of them is the sum of the proper divisors of the other. For instance, numbers 220 and 284 are amicable. It is the smallest pair of amicable numbers.

Pythagoreans believed that numbers have important mystical meaning and spiritual significance. For instance, they believed that 3 is male and 4 is female (Stewart, 2008). Some philosophers assumed that the Pythagoreans argued that numbers were corporeal while sense-perceptible things are, or are made of, numbers. Pythagoras was quoted to say: "Number is the ruler of forms and ideas and the cause of gods and daemons" (Sol, 2021)

Pythagoras attached special meaning to the first four numbers, which were represented by the Pythagoreans as four rows of ten pebbles arranged in the form of the following equilateral triangle:

·

· ·

· · ·

· · · ·

The triangle symbolized the principle "*many*, i.e., a set of objects, *in one*, i.e., forming unity" highlighting that the sum of the first four numbers was equal to the triangular number 10, which was venerated as the foundation of all numbers. Interestingly, Confucius, like Pythagoras, regarded the numbers 1, 2, 3, and 4 as the source of all perfection (Jeans, 1968).

The principal doctrine of Pythagoras and his school was that mathematics ruled the universe, which ran on numerical harmony. The pivotal motto of Pythagoreans was *"All is number"*. It was the basis of their metaphysical "science" and ideology (Graves-Gregory, 2014). Their numerology was an endeavor to find the relationships of the universe, which they believed could be described entirely in terms of natural numbers and ratios of natural numbers (fractions). In other words, it is possible to consistently interpret the statement *"All is number"* as the assumption that everything can be characterized by numbers or that numbers give information about everything.

The system of ascribing different (often mystical) meanings to numbers is called *numerology*. It came into being in different cultures before Pythagoras lived and survived till our time. For instance, according to Mateo Sol, 0 is the representative of the primordial void and the realm of potential symbolizing, at the same time, totality, the eternal force and the Divine Spirit (Sol, 2021). In a similar way, 3 is typically seen as a lucky and powerful number representing the divine principle that underlies life (Sol, 2021).

In many religions, there are three principal gods. For instance, Zeus, Poseidon, and Hades are topmost gods in the Greek mythology; Anu, Bel (Baal), and Ea are highest gods in the Babylonian mythology; or Brahma, Vishu, and Shiva are uppermost gods in Indian mythology. Plato related number 3 to the triangle conceiving that the world was built from triangles (Plato, 1961). Naturally, 3 is the number associated with the triad and the fundamental triad is the most basic structure in the world (Burgin, 2017b).

Numerology is related to the superstition of many people who believe in lucky (good) and unlucky (bad) numbers. However, meaning of numbers varies in different cultures. For instance, in the Western culture, 13 is unlucky (bad) number. At the same time, in Tibetan culture and Judaism, number 13 is a lucky (and even holy for Tibetan people) number.

Even more, there are cultures that consider the same number being lucky and unlucky. For instance, in Chinese tradition, number 3 is considered lucky as it sounds as the word *life* in Chinese. On the other hand, number 3 is considered unlucky as it sounds as the word *separation* in Chinese.

Some people think that being lucky or unlucky number depends on an individual — what is good for one person can be bad for another one and vice versa.

It is necessary to remark that the majority of mathematicians and scientists deem numerology as a pseudoscience (cf., for example, Stewart, 2008).

The Pythagoreans represented numbers by pebbles or *calculi*. That is why the terms *calculus* and *calculation* come from the word *calculi* used by the Pythagoreans.

In the contemporary mathematics, there are two basic meanings of the term *calculus*. The first of them comprises the calculus, which was originated by Newton and Leibniz being later developed by many mathematicians. The calculus consists of two parts:

- the *differential calculus*, which deals with derivatives of functions defining them, studying their properties and applying them to the study of functions;
- the *integral calculus*, which deals with integrals of functions defining them, studying their properties and applying them to the study of functions.

It is possible to say that derivatives describe local behavior of functions while integrals measure areas or volumes determined by functions.

The second meaning of the term *calculus* comes from mathematical logic. In logic, a calculus is a system, which consists of a priory given axioms, rules of inference (deduction) and statements (theorems, propositions and lemmas) deduced by these rules from axioms (cf., for example, Hilbert and Bernays, 1968, 1957; Ershov and Palyutin, 1986; Burgin, 2016). Examples of logical calculi are propositional calculus and predicate calculus.

Returning to ancient Greece, it is necessary to remark that by *numbers* the Pythagoreans meant only natural numbers larger than one. Although they knew ratios, which later were corresponded to rational numbers, Pythagoreans as other Greek mathematicians of that period did not treat ratios as numbers. That is why the discovery of simple geometrical objects that could not be expressed by numbers in the sense of Pythagoreans was devastating for Pythagoras and his school.

Here is the legendary history of that discovery. Some sources tell that the Pythagorean philosopher Hippasus of Metapontum did this committing what was considered the supreme crime in the school of Pythagoras. Yet the evidence relating this discovery to Hippasus is very perplexed. Different myths also tell us dissimilar versions of the aftermath of this discovery. Some sources claim that Hippasus was simply expelled from the society. According to other legends his punishment was much more severe. One narrative implies that Pythagoras himself strangled or drowned the traitor, although this version contradicts the conclusion of some historians that Hippasus lived in the late 5th century B.C.E., i.e., about a century after the time of Pythagoras. Another tale describes how the Pythagoreans dug a grave for Hippasus when he was still alive and then mysteriously caused him to die. Yet another legend holds it that Hippasus was set afloat on a boat that was then sunk by the members of the society.

These stories vividly demonstrate how majority usually reacts to the revolutionary innovations.

The famous Pythagorean Theorem has been credited to Pythagoras although it had been known to mathematicians in ancient Babylonia, Egypt and India long before Pythagoras lived. For instance, it is possible to find this theorem in ancient Indian texts, the Baudhayana's Sulbasutra (800 B.C.E.) and the Shatapatha Brahmana (8th to 6th centuries B.C.E.).

In addition, the ancient Chinese mathematical book *Zhoubi Suanjing* also included a proof of the Pythagorean Theorem. However, the book appeared between 1046 and 256 B.C.E. while its compilation and addition of materials continued till 220 C.E. It means that there is a possibility that the proof of the Pythagorean Theorem in China also appeared much earlier than Pythagoras lived and proved his theorem.

Note that although the Pythagorean Theorem describes triangles, the description involves numbers. This means that it is possible to treat the Pythagorean Theorem as the statement about some properties of and relations between numbers.

Returning to Europe, it is worth mentioning that Theaetetus of Athens (ca. 417–368 B.C.E.), a friend of Socrates and Plato, studied irrational lengths, what was later included in Book X of Euclid's *Elements*, and proved that there were precisely five regular convex polyhedrons — tetrahedron, cube, octahedron, dodecahedron and icosahedrons. Note that in ancient Greece lengths were considered magnitudes while now they are represented by numbers. This means that Theaetetus investigated geometrical representation of irrational numbers and their arithmetic.

It is known that Theaetetus was a student of another Greek mathematician Theodorus of Cyrene, which was a prosperous Greek colony on the coast of North Africa in what is now Libya. Theodorus lived during the 5th century B.C.E. and had explored the theory of incommensurable quantities, classifying various types of magnitudes and implicitly operated with irrational numbers according to the way they are expressed as square roots. Later these and similar results were presented in great detail in Book X of Euclid's *Elements* (Euclid, 1956).

One of the greatest philosophers of all times and all nations Plato (ca. 427–347 B.C.E.), or more exactly, Platon (Πλάτων) in Greek, did not develop mathematics but he taught arithmetic and geometry in his Academy and discussed the essence of numbers and the role of arithmetic in some of his dialogues (Plato, 1961). Although the name of Plato is familiar to any intelligent person, we present here a brief synopsis of his life.

The exact time and place of Plato's birth are unknown but most scholars assume that he was born into an aristocratic and influential family. As Diogenes Laertius writes, Plato's father Ariston traced his descent from the king of Athens, Codrus, and the king of Messenia, Melanthus (Laertius, 2018).

In his maturity, he called himself Platon (Πλάτων). This name originated from the adjective *platýs* (πλατύς) meaning *broad* and later was transformed to Plato, but according to Diogenes Laertius, his original name was Aristocles, which meant *best reputation* (Laertius, 2018). As his family was wealthy, Plato got a very good education, and in addition he was a wrestler participating in the Isthmian games as some historians suggest. Ancient sources portray Plato as a bright though modest youth who excelled in his studies. In philosophy, Plato was a student of the great Greek philosopher Socrates (ca. 469–399 B.C.E.) who also lived in Athens. After Socrates was tried and executed by Athenian democratic authorities, Plato, who was in his thirties at that time, and several other student of Socrates left Athens. In general during his life, Plato mostly lived in Athens traveling to other places from time to time. According to ancient sources, Plato visited different cities in Greece, Italy, Sicily and perhaps Egypt. Returning from his trips at the forty years of age, he founded the *Academy*, which was one of the earliest known organized schools in Western Civilization and continued to function for several centuries after Plato. Some thought that the name of the Academy came from the ancient hero Academus, while others assumed that it came from a supposed former owner of the plot of land, an Athenian citizen whose name was also Academus. While not a mathematician, Plato held mathematics in high esteem and taught mathematics in the Academy.

Plato contributed to many fields of philosophy but his main idea was about the World of Ideas/Forms. Namely, he taught that the world as a whole consisted of two realms: the *physical world*, which people could comprehend with their five senses, and the *World of Ideas* or *Forms*, which people could comprehend only with their intellect. This teaching on Ideas or Forms is at the heart of Plato's philosophy shaping his views on knowledge, ethics, esthetics, psychology and political teachings. For millennia, philosophers either rejected the existence of the World of Ideas/Forms or tried to understand what these Ideas/Forms are and where this World exists. Only in the 21st century, a rigorous scientific answer to these questions was obtained by interpreting Ideas/Forms as ideal structures (Burgin, 2017b).

Throughout his later life, Plato became entangled with the politics although not in the Athens but in the city of Syracuse in Sicily, which he visited several times invited by its rulers. He tried to improve the political system but this only endangered his life — he was sentenced to death but then was spared and sold into slavery. His friends saved him and he returned to Athens.

Plato is one of the most insightful, wide-ranging, creative, and influential authors in the history of philosophy as well as one of the most dazzling and skilful writers in the Western literary tradition. As the noted English mathematician and philosopher Alfred North Whitehead (1861–1947) wrote,

"the safest general characterization of the European philosophical tradition is that it consists of a series of footnotes to Plato." (Whitehead, 1978)

Another great Greek philosopher Aristotle (384–322 B.C.E.) paid special attention to mathematics making *number* one of the basic categories in his ontology (Aristotle, 1984). Although the name of Aristotle is familiar to any intelligent person, we present here a brief synopsis of his life.

Aristotle, or more exactly, Aristotélēs ('Αριστοτέλης) in Greek, was born in the city of Stagira in Northern Greece. According to Diogenes Laertius, Aristotle's father, Nicomachus, served as private physician to the Macedonian king Amyntas (Laertius, 2018). Unfortunately, his father died when Aristotle was a child, and he was brought up by a guardian. At seventeen or eighteen years of age Aristotle came to Plato's Academy in Athens and remained there until the age of thirty-seven. After Plato died, Aristotle was not chosen as the head of the Academy and left Athens. Later he was invited by the king Philip II of Macedon to tutor his son Alexander the Great.

After Alexander the Great started his military campaign in Asia, Aristotle moved to Athens and founded his own school the Lyceum, the Peripatetic school of philosophy, receiving substantial financial support from Alexander the Great. However, Alexander's death revived anti-Macedonian sentiment in Athens leaders and they reportedly denounced Aristotle for impiety. As he did not want to end his life as Socrates, who was killed by the Athenian democracy, Aristotle preferred to flee to Chalcis, on Euboea.

Aristotle is often considered the father of Western science while his works continue to be a subject of contemporary philosophical discussion. From his teachings, the Western culture inherited its intellectual lexicon, as well as problems and many methods of inquiry influencing many generations of philosophers. Being a polymath, he wrote on many subjects including physics, biology, zoology, metaphysics, logic, ethics, aesthetics, poetry, theatre, music, rhetoric, psychology, linguistics, economics, politics, and government.

The prominent ancient Greek mathematician and astronomer Eudoxus of Cnidus (ca. 390–337 B.C.E.) was a student of the Pythagorean astronomer Archytas (428–347 B.C.E.) and of the great philosopher Plato (ca. 427–347 B.C.E.). Although all works of Eudoxus are lost, we know about some of his results from other ancient authors. For instance, according to Archimedes, Eudoxus introduced the method of exhaustion for area determination and used the assumption that is now called the Archimedes axiom or property. Besides, Eudoxus improved the Pythagorean theory of proportions, so that it could also treat incommensurable magnitudes. At that time, mathematicians did not know real numbers. So, when it was necessary to use them, mathematicians employed the concept of a magnitude, which was expressed in geometrical terms.

Eudemus of Rhodes (ca. 370–300 B.C.E.) was born on the isle of Rhodes, but spent a large part of his life in Athens, where he studied philosophy at Aristotle's Peripatetic Lyceums. Eudemus's collaboration with Aristotle was long-lasting and rather close to the extent that being generally considered as one of Aristotle's best pupils, he and Theophrastus of Lesbos were regularly called not Aristotle's "disciples", but his "companions" (ἑταῖροι). Eudemus wrote on histories of arithmetic, geometry and astronomy but all his works including History of Arithmetics ('Αριθμητική ἱστορία) are now lost.

Aristarchus of Samos (ca. 310–230 B.C.E.) was an ancient Greek astronomer and mathematician who presented the first known model that placed the Sun at the center of the known universe with the Earth revolving around it. He was influenced by Philolaus of Croton and identified the "central fire" with the Sun situating the other planets in their correct order of distance around the Sun. That is why Nicolaus Copernicus (1473–1543) attributed the heliocentric theory to Aristarchus. Like Anaxagoras before him, Aristarchus suspected that the stars were just other bodies like the Sun, albeit further away from Earth. He was also the first one to deduce the rotation of earth on its axis. Aristarchus used numbers and arithmetic to find distances from the Earth to the Moon and Sun. However, his astronomical ideas were often rejected in favor of the incorrect geocentric theories of Aristotle and Ptolemy.

The great ancient Greek mathematician Archimedes of Syracuse (ca. 287–212 B.C.E.) is one of the top mathematicians of all times and all nations (Bell, 1937). In addition, he was a physicist, engineer, inventor, and astronomer being regarded as one of the foremost scientists in classical antiquity.

Archimedes was born in the seaport city of Syracuse, Sicily, at that time a self-governing colony in Magna Graecia, located along the coast of Southern Italy. In his work *The Sand Reckoner*, Archimedes wrote that his father was astronomer Phidias (Archimedes, 1980) but nothing else was known about him. The prominent Greek historian Plutarch wrote that Archimedes was related to King Hiero II, the ruler of Syracuse, and helped him to defend Syracuse from Romans inventing different military mechanisms.

In his mathematical works, Archimedes anticipated modern calculus by implicitly applying the concept of an infinitesimal and the method of exhaustion to derive and prove a range of geometrical theorems, finding formulas for the area of a circle, the surface area and volume of a sphere, and the area under a parabola (Archimedes, 1980).

To better understand achievements of Archimedes, it is necessary to note that infinitesimals explicitly but not rigorously came to mathematics in the calculus created by Newton and Leibniz at the end of the 17th century, that is, almost two millennia after Archimedes lived and implicitly used infinitesimals. However, for almost two centuries after Newton and Leibniz, infinitesimals were useful in spite of being casually defined mathematical concepts. In the 19th century, some accurate approaches to infinitesimals were elaborated but only in the 20th century, infinitesimals from the calculus became precise numbers in the setting of nonstandard analysis, which harbors a huge variety of different infinitesimals and will be described later.

So, what is an infinitesimal? Informally, an infinitesimal, or an infinitely small number, is a number that is less that any real number but larger than zero. The contemporary approach treats infinitesimals as nonstandard numbers (Robinson, 1966).

Archimedes also calculated an accurate approximation of the number π proving that the value of π lies between $3 + 1/7$ (which is approximately, 3.1429) and $3 + 10/71$ (which is approximately, 3.1408), gave a very accurate estimate of the value of the square root of 3, defined and studied the spiral bearing his name, and using exponentiation created a system for expressing very large numbers. Archimedes also found a formula for turning a sphere into a cylinder, which was later used to take a globe and turn it into a flat map.

Historians of mathematics think that Archimedes was the first to work with infinite number series and performing their summation (Bidwell, 1993). Summation of series extends addition in arithmetic assigning one number to an infinite series, which is called the sum of the series. As a result, mathematicians come to the arithmetic with infinite operations. Because series constitute one of the central concepts of the calculus, the latter becomes an extension of arithmetic. As the great French mathematician Jules Henri Poincaré (1854–1912) wrote,

"Weierstrass leads everything back to the consideration of series and their analytic transformations; to express it better, he reduces analysis to a sort of prolongation of arithmetic." (Poincaré, 1905)

Infinite series came to many areas of mathematics and through it to physics, finance, computer science, and economics. Many mathematicians elaborated different methods of series summation.

In spite of its early beginning, the classical summation of series was made rigorous only in the 19th century. However, mathematicians found that the classical summation had too many limitations and started using divergent series, i.e., series that did not have the classical sum. Different summability methods for divergent series were suggested by such mathematicians as the great Norwegian mathematician Niels Hendrik Abel (1802–1829), the Italian mathematician Ernesto Cesàro (1859–1906), and the great Swiss/Russian/Prussian mathematician Leonhard Euler (1707–1783). However, all these methods did not give the complete solution to the problem and many series still remained divergent. Only introduction of hypernumbers in the 20th century allowed finding sums of arbitrary number series making series summation a total operation with numbers (Burgin, 2008b, 2017). In essence, it is possible to treat series summation as an infinite addition, i.e., as a derived arithmetical operation.

Returning to ancient Greece, we also find that Archimedes was one of the first to apply mathematics to the exploration of physical phenomena, in particular, discovering the principle of the lever and founding such fields of physics as statics and hydrostatics. As an engineer, he is credited with designing innovative mechanisms, such as his screw pump, compound pulleys, and defensive war machines to defend his native Syracuse from Roman invasion. Nevertheless, it did not help. Romans defeated Syracuse and killed Archimedes in 212 B.C.E. thereby impeding the development of mathematics.

There were other fine mathematicians in ancient Greece. History tells us about many of them but here we mention only those whose work was related to arithmetic.

One of most renowned Greek mathematicians was Euclid or Euclides in Greek, who lived around 300 B.C.E. and is often referred to as the "founder of geometry" or the "father of geometry". Sometimes he was called Euclides of Alexandria to distinguish him from Euclides of Megara. Euclid was active in Alexandria during the reign of Ptolemy I (323–283 B.C.E.) when he wrote one of the most (if not the most) influential works in the history of mathematics *Elements* (Euclid, 1956). This book served as the model of a rigorous scientific publication, and was used as the main textbook for teaching mathematics, especially, geometry, from the time of its publication until the early 20th century. In the *Elements*, Euclid studied geometry as an axiomatic system deducing the theorems of what is now called Euclidean geometry from a small set of axioms and postulates. In addition, he wrote works on perspective, conic sections, spherical geometry, number theory, and rigor in mathematics.

Books 7, 8 and 9 of Euclid's *Elements* mostly contain materials on arithmetic in the sense in which the word was employed in ancient times but what now is called the theory of numbers. These books contained 102 propositions. In particular, Euclid demonstrated infinity of prime numbers, presented or developed what is now called the *Euclidean algorithm*, and proved the so-called Fundamental Theorem of Arithmetic, commutativity of multiplication and it is distributivity with respect to addition. He also gave an exposition of the theory of proportions, which are now represented by fractions. In other books of his great treatise, we can find the general theory of relations between magnitudes, which is in some sense the beginning of the theory of real numbers.

Euclid represented natural numbers by line segments named by one letter or by two letters, which indicated the ends of a line segment. Products of two numbers, he called *plane numbers* while products of three numbers, he called *solid numbers*. In spite of this geometrical representation, Euclid did not use geometry for his proofs.

In his work, both in geometry and arithmetic, Euclid was building on the results of his predecessors although it was a great work to organize all of those materials in his magnificent tract *Elements*. Besides, it is possible to suggest that some of the material from *Elements* belong to Euclid.

Eratosthenes of Cyrene (ca. 276–195 B.C.E.) was a Greek mathematician, geographer, poet, historian, astronomer, and music theorist. He was the foremost scholar of his days, who was invited at the age of 30 by the king Ptolomey III to serve as tutor to his son and later as the chief librarian at the Library of Alexandria. As a mathematician, Eratosthenes explored numbers and used his sieve algorithm to swiftly find prime numbers. He also solved the famous Delian problem of doubling the cube although the tools of his solution were not restricted to straightedge and compass, which were considered the only upright tools for geometrical construction.

In addition, Eratosthenes originated the discipline of geography introducing terminology used today and creating one of the first maps of the world based on the available geographic knowledge of his era.

However, Eratosthenes is best known for being the mathematician who calculated with remarkable accuracy the circumference of the Earth by comparing altitudes of the mid-day sun at two places and being the first to determine the tilt of the Earth's axis. His result was much more accurate than all previous estimates. It is also believed that Eratosthenes accurately calculated the distance from the Earth to the Sun and invented the leap day.

Nicomachus of Gerasa (ca. 60–120 C.E.) was one more important ancient mathematician best known for his works *Introduction to Arithmetic*, which is his only extant work on mathematics, and *Manual of Harmonics* (Nicomachus, 1926). He was born in Gerasa, in the Roman province of Syria (now Jerash, Jordan), and was strongly influenced by Aristotle and Pythagoras becoming a neo-Pythagorean, who concentrated on the mystical properties of numbers explaining that natural numbers and basic mathematical ideas are eternal and unchanging in an abstract realm.

Stressing importance of mathematics, Nicomachus wrote that it is possible to be a philosopher only if one had enough mathematical knowledge.

In his works, Nicomachus distinguished between the wholly conceptual immaterial numbers, which he called the *"divine numbers"*, and the numbers which measured material things, calling them the *"scientific numbers"*. He wrote considerably on numbers, especially on the significance of prime numbers and perfect numbers arguing that arithmetic is ontologically prior to the other mathematical sciences, which consisted at that time of music, geometry, and astronomy, being their cause. The book *Introduction to Arithmetic* (*Arithmetike eisagoge*) is the only preserved work on mathematics by Nicomachus. Boethius' book *De institutione arithmetica* is mostly a Latin translation of this work (Boethius, 1983).

Nicomachus also wrote other books on arithmetic, which were lost. In particular, *Art of Arithmetic* is his larger work on arithmetic, mentioned by Photinus. In his *Theology of Arithmetic*, which was also mentioned by Photinus, Nicomachus described the Pythagorean mystical properties of numbers. Nicomachus also constructed one of the earliest Greco-Roman multiplication tables.

Similar ideas were championed by St. Augustine of Hippo (354–430), who was a theologian, philosopher influencing the subsequent Western philosophy and the bishop of Hippo Regius in Numidia, Roman North Africa. St. Augustine wrote:

"Numbers are the Universal language offered by the deity to humans as confirmation of the truth." (Augustine, 2018)

However, as Nicola Graves-Gregory writes about mathematics of ancient Greeks:

"The term 'number' ('arithmos', αριθμος) was used only for the integers greater than one. 'One' (the 'monas', μονας) was the unit and unity, and was considered the principle or beginning of number and as such had a different status from the numbers which it generated. 'Two' was sometimes similarly excluded from the realm of number; the reason for this stems from the Pythagorean identification of 'one' and the odd numbers with 'limited', opposing 'two' (or the 'dyad') and the even numbers as 'unlimited'; but this perception of two was not observed so strictly. Fractions were not considered to be numbers, since the 'one', the source of numbers, was essentially indivisible; only ratios of integers were allowed in arithmetike." (Graves-Gregory, 2014)

Thus, the prevailing mathematical tradition in ancient Greece was that numbers started with 3. Now natural numbers start with 1 but we know that the least polygon, triangle, has three sides. In Greek, a triangle is called *triagon.* There are no polygons with one or two sides. Interestingly, the approach of ancient Greek mathematicians to numbers has definite similarity to ideas of the ancient Chinese philosophy. Indeed, the great Chinese philosopher Laozi (ca. 6–5 century B.C.E.) in his classical work *Tao Te Ching* wrote (cf. (Mair, 1990)):

> The Way (Tao) gives birth to one;
> one produces two;
> two generates three;
> and three brings about the whole world.

In turn, this vision of Laozi finds its counterpart in contemporary science and mathematics where the fundamental triad is the most basic structure (Burgin, 2011).

Laozi is a semi-legendary figure because the earliest reference is found only in the 1st century B.C.E. and exact dates of his life are not known. It is supposed that Laozi was an older contemporary of another great Chinese philosopher Confucius. It is possible to find contradictory information about his life in different sources. According to some of them, Laozi was the Keeper of the Imperial Archive. In other sources, he was a court astrologer. In any case, it is definitely assumed that Laozi founded the philosophy of Taoism becoming something like a deity in Taoism as one of two main religions in China.

Returning to Greece, we can see the further development of arithmetic. Indeed, the famous Greek astronomer, mathematician, geographer and astrologer Claudius Ptolemy of Alexandria (ca. 100–170 C.E.), who was influenced by Hipparchus and the Babylonians, was using a special symbol for zero, which had the form of a small circle with a long over bar. It appeared in his work on mathematical astronomy called the *Syntaxis Mathematica,* also known as the *Almagest.* Ptolemy is famous for his geocentric model of the universe, which was later substituted by the more correct heliocentric model Nicolaus Copernicus (Mikołaj Kopernik in Polish) (1473–1543).

One of the most outstanding mathematicians of ancient Greece is Diophantus, who is habitually called the "father of algebra" and is best known for his *Arithmetica,* which is often considered as a treatise on the theory of numbers. As Schappacher writes:

"Diophantus's Arithmetica is one of the most influential works in the history of mathematics. For instance, it was in the margin of his edition of Diophantus that Pierre de Fermat, some day between 1621 and 1665, wrote the statement of his so-called Last Theorem (which was proved only a few years ago). But Fermat was not the first to derive inspiration from Diophantus's collection of algebraic/arithmetic problems: the Arabs had profited from reading the Arithmetica when developing Algebra as a mathematical discipline." (Schappacher, 1998)

There is even an opinion that Diophantus implicitly introduced a new concept of number. For instance, the famous German historian and philosopher Oswald Manuel Gottfried Spengler (1880–1936) wrote:

"A new Zahlengefühl, a new notion of number ... is at work in him. What an undetermined number a, an unnamed number 3 is — both neither quantity, nor measure, nor line segment — a Greek would have not been able to say." (Spengler, 1923)

Diophantus worked in the great city of Alexandria in Egypt. At that time, Alexandria was the center of mathematical learning. However, very little is known of Diophantus' life and there has been much debate regarding the time period at which he lived.

Nonetheless, there are limits that can be put on the dates of Diophantus's life by analyzing his references. On the one hand, Diophantus quotes the definition of a polygonal number from the work of Hypsicles, which means that Diophantus wrote his work later than 150 B.C.E. On the other hand, Theon of Alexandria, the father of the famous scholar Hypatia, quotes one of Diophantus's definitions, which means that Diophantus wrote his works not later than 350 C.E. Nonetheless, this leaves a very big span of 500 years for life of Diophantus. The reason for this is that while many references to the work of Diophantus have been made, Diophantus himself, made very few references towards works of other mathematicians making the process of pinpointing the dates of his life rather hard. However, some researchers found some evidence that Diophantus lived in the third century, e.g., around 250 C.E. as it is argued in the Introduction to the book (Diophantus, 1974). According to some sources, he lived 84 years and had one son.

Diophantus was the first mathematician who made an essential contribution specifically to arithmetic as a mathematical field collecting and extending previously existed knowledge while making the first steps in transformation of arithmetic into algebra. Some historians assumed that in his work, Diophantus used negative quantities introducing them formally by some rules of operation and calling them *shortage* in contrast to positive quantities called *excess*. He employed the sign to denote negative quantities in the same way as we use the sign "−" (minus) for negative numbers. However, the idea of negative quantities was not accepted by other European mathematicians for a long time after Diophantus. That is why some historians of mathematics renounce this interpretation of the work of Diophantus.

In their studies of natural numbers, Greek mathematicians discerned odd, even and prime numbers proving infinity of prime numbers. They also introduced other classes of natural numbers. For instance, based on the previous mathematical works, Euclid defined the following classes of natural numbers (Euclid, ca. 300 B.C.E./1956):

- odd numbers
- even numbers
- prime numbers
- composite numbers
- relatively prime numbers
- relatively composite numbers
- square numbers
- plane numbers
- solid numbers

An interesting peculiarity of Greek mathematics was the utilization of letters to denote numbers. The first Greek letter α denoted 1, the second Greek letter β denoted 2, the third Greek letter γ denoted 3 and so on. Relation between letters and numbers brought forth investigation of relation between numbers and words as well as relations between words induced by relations between words and numbers. This activity and its rules were called *isopsephy* in Greek.

Hebrew alphabet was also used for naming numbers and a similar activity of corresponding numbers and words was developed in Jewish culture under the name *gematria*.

Gematria (isopsephy) as a process has three forms: (1) assigning numbers to words and texts based on numerical values of letters, (2) assigning words and texts to numbers based on numerical values of letters, and (3) the combination of the first two forms. It is possible to call the first form by the name *textual numeration* and the second form by the name *numeric textualization*.

The latest documented use of gematria is known from an Assyrian inscription dating to the 8th century B.C.E., commissioned by Sargon II, and stating: "*the king built the wall of Khorsabad 16,283 cubits long to correspond with the numerical value of his name.*" (Luckenbill, 1927)

Although people have used gematria for a long time, there is no enough evidence of its reliability not speaking about its scientific relevance. It is also known that gamatria is not mathematics.

Returning to Greece, we can see that in spite of many great mathematical achievements, ancient Greek mathematicians also had definite shortcomings in understanding numbers.

First, for a long time, Greek mathematicians did not consider 1 to be a number calling 1 by the name *unit* (μονάς). As a result, they assumed that there were only natural numbers, which started at first with the number 3 and then with the number 2. For instance, Diophantus begins his exposition of arithmetic with definitions of *number* (ἀριθμός) and *unit* (μονάς) or one, "all numbers are made up of some multitude of units" (Diophantus, 1974; Tannery, 1893/1895).

Second, although they used fractions representing positive rational numbers, Greek mathematicians did not call them numbers or representatives (names) of numbers but only *ratios of numbers* and did not establish relations between fractions and the corresponding rational numbers. According to some historians of mathematics, Diophantus was the first to treat ratios as valid numbers (Diophantus, 1974). Interestingly, this approach correlates with the situation in the contemporary mathematics. Namely, in spite of many arguments that fractions are not numbers but only names of rational numbers, even now many people including some mathematicians believe that fractions are numbers. This opinion is expressed in numerous textbooks and other educational sources. Here are some examples:

- In Parker and Baldridge (1999), it is written (p. 133): "*The key points are that a fraction is a single number…*"
- In Spector (2021), it is specified that "*A fraction is a number we need for measuring.*"
- In Gaskin (2007), it is explained "*Every fraction is a single number.*"
- In Gugoiu (2006), it is clarified "*The concept of fractions and the relations between fractions and other types of numbers, like many abstract mathematical concepts, is not always easy to understand.*"

Third, although they encountered and worked with some irrational numbers, Greek mathematicians also did not call them numbers but treated them as magnitudes strictly discerning them from numbers. For instance, Aristotle explained that, although *number* and *magnitude* are quantities, they belong to distinct species: *number* is discrete and *magnitude* is continuous (Aristotle, 1984; Fernandes, 2017).

In addition, the main body of Greek mathematicians knew only natural numbers and representations of positive rational numbers in the form of fractions. They did not explicitly discover zero and negative numbers although Diophantus had a special symbol for zero and used negative quantities introducing them formally by the rules of operation.

Diophantus was maybe the last Greek mathematician of antiquity because after Greece and other regions with the Greek dominance had become parts of the Roman Empire, mathematics as a scientific discipline began deteriorating because Romans were utilitarian people, who despised mathematical theory and were concerned only in applications of arithmetic and geometry to their engineering projects of building aqueducts, roads, bridges, public buildings, and to land surveys. Naturally, there were no Roman mathematicians of note.

Nevertheless, theoretical mathematics and its practical applications originated and flourished not only in ancient Greece. Other regions where theoretical mathematics in general and arithmetic in particular were also developed in ancient times were India and China.

The ancient Indian mathematician Pingala (ca. 3rd/2nd century B.C.E.) wrote the treatise *Chandaḥśāstra* (also called *Pingala-sutras*). It was the earliest known book on Sanskrit prosody (Plofker, 2009). In it, Pingala presented the first known description of a binary numerical system in connection with the systematic enumeration of musical meters with fixed patterns of short and long syllables. The combinatorics of musical meters involves the binomial theorem with a description of the Pascal's triangle called *meruprastāra* in Sanskrit. Besides, Pingala's work contains a presentation of the Fibonacci numbers, called *mātrāmeru* in Sanskrit. Sometimes original utilization of the number *zero* is sometimes ascribed to Pingala due to his discussion of binary numbers, which are now usually represented using 0 and 1. However, Pingala used terms *light* (*laghu*) and *heavy* (*guru*) rather than 0 and 1 to describe syllables.

For a long time, the binary representation of numbers played no role in mathematics and its applications. At the beginning of the 18th century, the great German mathematician and philosopher Gottfried Wilhelm Leibniz (1646–1716) tried to attract attention to *binary arithmetic* (Leibniz, 1703) but only with the advent of computers, binary arithmetic became vitally important to society because all basic digital devices, such as computers, calculators and cell phones, use it. Indeed, all operations inside these devices are performed employing binary representations. As Ronald Kneuser writes,

"In the end, computers work with nothing more than binary numbers which is really quite interesting to think about: all the activities of the computer rely on numbers represented as ones and zeros." (Kneusel, 2015)

According to John O'Connor and Edmund Robertson, Leibniz also introduced the term *function* in a 1673 letter (O'Connor and Robertson, 2005). Now the concept *function* has become one of the most fundamental in mathematics in general and in arithmetic, especially, in higher arithmetic, in particular.

Returning to the history of arithmetic in ancient times, we can see that ancient Greeks eventually encountered irrational numbers but were not able to discover/invent the whole class of these numbers. As a result, the number 0, negative numbers, integers and the class of all rational numbers were discovered/invented at other places of our planet.

From the very beginning, arithmetic of natural numbers was used in many practical areas. That is why it was important to have an economical notation for natural numbers as well as efficient algorithms for performing arithmetical operations. Now we use the decimal positional numerical system, which originated in India. With digits, which were different from those that are used now, this system and rules of operations were described in the main work *Aryabhatiya* of the ancient Indian mathematician Aryabhata (476–550 C.E.), who was the first of the main mathematician-astronomers from the classical age of Indian mathematics and astronomy. In this book, methods of solving problems described by linear equations as well as calculation of square and cubic roots are given (Volodarsky, 1977).

An interesting peculiarity of the Indian arithmetic as a mathematical field was existence of a variety of numerical systems, which included non-positional word system, positional word system, alphabetic non-positional system, alphabetic positional system, and digital positional system. Numerical arithmetics are different from number arithmetics because numerical arithmetics describe operations with numerals while number arithmetics express operations with numbers. It means that many numerical arithmetics existed in ancient India.

An important step in the development of arithmetic was the discovery/invention of zero, which is now denoted by the digit (numeral) 0. The history of zero (0) is in essence a tale of two zeroes. One zero is a symbol to represent absence of a positional value, the so-called, *placeholder*, while another zero is a full-scale number used in calculations with very specific mathematical properties. In particular, the number 0 plays a central role in mathematics as the additive identity of the whole numbers, integers, real numbers, complex numbers, and many other algebraic structures.

Names for the number 0 in English include: *zero, nil, nought* (in UK), *naught* (in USA), or — in contexts where at least one adjacent digit distinguishes it from the letter "O" — *oh* or *o*. In programming, zero is denoted by $\emptyset$. Informal or slang terms for zero are *zilch* and *zip*. Such names as *cipher, ought* and *aught* have also been historically used. The English word *zero* came from the French word *zéro*, while the word *cipher* came from the Arabic words *sifr* meaning nothing and *safira*, which meant "it was empty." In turn, the contemporary Arabic word *sifr*, meaning *zero* or *nothing*, was the translation for the Sanskrit word *sunya*, which means void or empty.

Writing about discoveries and inventions in arithmetic and number systems made millennia ago, such as the discovery/invention of zero, it is not a simple problem to find exact time when these events happened. This is the consequence of the following situation. First, the majority of mathematical works were not preserved through millennia. Second, it is impossible to accurately locate the place in time of many ancient mathematical sources. For instance, the book *Zhoubi Suanjing*, which means *Arithmetical Classic of the Gnomon and the Circular Paths of Heaven* and is an important source of arithmetic knowledge in ancient China, dates from the period of the Zhou Dynasty (1046–256 B.C.E.) with compilation and addition of materials during the Han Dynasty (202 B.C.E.–220 C.E.). This gives then span of 1200 years. Another example is an important book of Jaina mathematics *Sthananga Sutra*, which is dated to the 2nd century B.C.E. in the MacTutor History of Mathematics and to the period from the 3rd to the 4th century C.E. in the Wikipedia.

Returning to the history of numbers, we find that, at first, zero was introduced as a placeholder and not as a number due to the following reason. As we know, numbers (or more exactly, natural numbers) originated from counting, which always starts with the first counted object labeled by the number 1. As a result, it was so difficult to understand the idea that 0 can and must be considered as a number having its own denotation by a numeral. Thus, quite for a while, zero was utilized in place-value notation by the Babylonians. By the middle of the 2nd millennium B.C.E., the Babylonian mathematics had a sophisticated sexagesimal positional numerical system (based on 60 and not on 10) and any positional numerical system needs a placeholder when the positional value is absent, for example, as in the number 203. However, at first, there was no individual notation for zero and it was indicated by a space between sexagesimal digits meaning the lack of a positional value. Later Babylonian mathematicians added a special symbol (digit) for zero. For instance, in a tablet unearthed at Kish and dating from about 700 B.C.E., the scribe Bêl-bân-aplu denoted zeros with three hooks. By 300 B.C.E., Babylonians used a special symbol in the form of two slanted wedges as a placeholder. Nevertheless, despite the invention of zero as a placeholder, the Babylonians never quite discovered zero as a number.

Looking at other countries we can find that some historians speculate late Olmecs had already begun to use the number zero in the 3rd century B.C.E. However other historians assume that only the Mayans in Central America independently invented zero in the fourth century C.E. Their priest-astronomers used a snail-shell-like symbol to fill gaps in the (almost) base-20 positional "long-count" system they used to calculate their calendar. Mayans had a highly developed civilization with skilled mathematicians, astronomers, artists, and architects. However, they failed to make other key discoveries and inventions that might have helped their culture to survive. The Mayan culture collapsed mysteriously around 900 C.E. Both the Babylonians and the Mayans found zero as the symbol playing the role of a place-holder, yet missed zero the number.

Thus, it took long time for the idea of a fully operational zero, filling the empty spaces of missing units and at the same time having the meaning of a null number, to become acceptable to mathematicians.

As historical documents show, the concept of nothing is important in its early Indian religion and philosophy and so for Indian mathematicians, it was much more natural to have a symbol for zero than for mathematicians from the Roman or Greek civilizations. That is why many historians agree that the rules for the operations with zero were written down first by Brahmagupta, in his book *Brahma Sphutha Siddhanta* (*The Opening of the Universe*) which appeared around 628 C.E.

Brahmagupta (ca. 598–668 C.E.) was an Indian mathematician and astronomer. He was the author of two early works on mathematics and astronomy: the *Brāhma Sphuṭa siddhānta*, a theoretical treatise on arithmetic, and the *Khaṇḍakhādyaka*, a more practical text.

He is considered as the one severely neglected, least translated, often maligned, and barely known mathematician in Indian history. Possibly, it was because he used harsh language himself against Aryabhata, Srishena, Varahamihira, and Vishnuchandra and other mathematicians and astronomers.

In *Brahma Sphutha Siddhanta*, Brahmagupta considered not only zero but also negative numbers, and the algebraic rules for the elementary operations of arithmetic with such numbers. He called positive numbers *fortunes*, zero a *cipher*, and negative numbers *debts*. That is why, the concept of zero as a number and not merely a placeholder is attributed to India where by the 9th century C.E. practical calculations were carried out using zero, which was treated like any other number.

Some researchers think that "the earliest recorded use of 0 in human culture was in India in the year 458 C.E." (Byers, 2007)

However, some time ago researchers obtained new information about zero. It was known that a symbol for zero, a large dot, was used throughout the *Bakhshali Manuscript*, a practical manual on arithmetic for merchants, while the date of which was uncertain (Kaye, 1927/2004; Hayashi, 2008). In 2017, samples from the manuscript were tested by radiocarbon technology dating the *Bakhshali Manuscript* between the 2nd century B.C.E. and the 3rd century C.E. and making it the world's oldest recorded use of the zero symbol (Devlin, 2017). It it would imply that the concept of the number zero was known several centuries earlier than it was used in the works of Brahmagupta.

Another great civilization developed in China. It contributed a lot to the development of number systems and arithmetic as practical tools for various calculations.

The simple but efficient ancient Chinese numbering system, which dates back to at least the 2nd millennium B.C.E., used small bamboo rods to represent the numbers 1 to 9, which were then places in columns representing units, tens, hundreds, thousands, etc. It was therefore a decimal place value system, very similar to the one we use today — indeed it was the first such number system, adopted by the Chinese over a thousand years before it was adopted in the West — and it made even quite complex calculations very quick and easy.

Written numbers, however, employed the slightly less efficient system of using a different symbol for tens, hundreds, thousands, etc. This was largely because there was no concept or symbol of zero, and it had the effect of limiting the usefulness of the written number in Chinese.

From the 4th century B.C.E., Chinese counting rods system enabled one to perform decimal calculations giving the decimal representation of a number, with an empty space denoting zero. Thus, while Chinese independently invented a place value for a quantity that was neither positive nor negative, they did not discover the number zero until it was introduced to them by a Buddhist astronomer from India at the beginning of the 8th century.

Documents also illustrate that the ancient Greeks seemed dubious about the status of zero as a number. They asked themselves, "How can nothing be something?" although solution of many practical problems brought mathematicians to the necessity of zero.

With passing time, the Indian decimal zero and related new mathematics were adapted by Muslim mathematicians and astronomers, works of which came from the Arab world to Europe in the Middle Ages. This brought forth philosophical and, by the Medieval period, religious discussions about the nature and existence of zero and the vacuum. The number zero was exceptionally regarded with suspicion in Europe, to such extent that the word cipher for zero became a word for secret code in modern usage. It is very likely a linguistic memory of the time when using decimal arithmetic was deemed evidence of dabbling in the occult, which was potentially punishable by the all-powerful Catholic Church with death (Ifrah, 2000). This tendency is also reflected in the fact that the words derived from the names of zero *sifr* and *zephyrus* were referred to calculation, as well as to privileged (esoteric) knowledge and secret codes.

However, it is necessary to explain that understanding zero as nothing is completely incorrect. Actually, zero as a number does not designates nothing but indicates absence of something. In other words, as the number one indicates one object, the number zero indicates absence of some objects. In the same way, zero as a placeholder in a positional numerical system points to the absence of the corresponding positional value.

In a similar way, in contemporary set theory, the set that has no elements, or in other words, the set with zero elements, is called the *empty set*, and denoted by the special symbol $\emptyset$ (cf., for example, Abian, 1965). Alike in computer science, the word that has no letters, or in other words, with zero letters, is called the *empty word* and denoted by the Greek letter ε, and absence of a symbol is called the *empty symbol* and denoted by another Greek letter Λ (cf., for example, Burgin, 2005). Thus, $\emptyset$ specifies absence of elements in a set, ε shows absence of symbols in a word, and

Λ signifies absence of individual symbol. All these concepts — an empty set, empty symbol, and empty word — are named zeros while zero indicates absence of any thing.

In comparison with other numbers, zero has very specific properties. Adding zero to any number or subtracting zero from any number does not change this number. In contrast to this, any number multiplied by zero becomes zero. The latter property resulted in the prohibition of division by zero in all conventional number systems such as rational or real numbers.

However, mathematicians always wanted to overcome this prohibition making zero similar to other real numbers. Brahmagupta is believed to be the first mathematician who treated zero as a full-fledged number defining operations with zero. However, his definition of division by zero violates the logic of mathematics (Suppes, 1957). For instance, he claimed that zero divided by zero is zero. Later the Indian mathematician Mahāvīra (ca. 800–870) tried to correct what Brahmagupta wrote about division by zero but was not successful also making incorrect statements, such as "any number divided by zero is the same number". Another Indian mathematician Bhaskara II (ca. 1114–1185) suggested a more realistic definition assuming that a number divided by zero gives infinity as the result. English mathematicians John Wallis (1616–1703), Issaac Newton (1642–1726), and Swiss/Russian/Prussian mathematician Leonhard Euler (1707–1783) had the same opinion writing about division by zero (Cajori, 1929).

However, division by zero is still an undefined operation and it is known that in some cases attempts to perform this operation caused problems with computers. For instance, on September 21, 1997, when the program tried to perform division by zero, all the machines on the network aboard the United States Navy *Yorktown* stopped working causing failure of the ship's propulsion system and paralyzing the cruiser for more than a day (Stutz, 1998).

Several approaches to division by zero have been suggested. For instance, the American philosopher, scientist and logician Patrick Colonel Suppes (1922–2014) described five elementary techniques that have been adopted at arriving at "division by zero" problem and various constraints associated with it pointing that there was no uniformly satisfactory solution (Suppes, 1957).

Recently, Mohammed Abubakr from India suggested one approach to this problem based on the introduction of the new kind of numbers called Calpanic numbers where the word *Calpanic* meant "un-real" in Sanskrit (Abubakr, 2011). He also developed the Calpanic arithmetic.

However, many other arithmetics appeared much earlier. While counting different things, people saw that in some situations counting was periodic involving repetition of the same numbers, for example, as it happened in counting of days in a year or hours in a day when after 12 o'clock hours start repeating. This observation resulted in appearance of another system of numbers and operations called *modular arithmetics.* They were the first arithmetics different from the conventional (Diophantine) arithmetic of natural numbers. Namely, all modular arithmetics are finite.

Modular arithmetic, which is also called *residue arithmetic* or *clock arithmetic,* is studied in mathematics and used in physics and computing. In modular arithmetic, operations of addition and multiplication form a cycle upon reaching a certain value, which called the *modulus.* Examples of the use of modular arithmetic occur in ancient Chinese, Indian, and Islamic cultures. For instance, written more than 1800 years ago the Chinese text Sun Zi Suan Jing described the solutions to a polynomial equation in terms of modular arithmetics but, naturally, without using the term *modular arithmetic.* The rigorous approach to the theory of modular arithmetic was worked out by the great German mathematician Carl Friedrich Gauss (1777–1855).

Similar to the history of zero, the chronicle of negative numbers is full of adventures and misunderstanding. Historical sources showed that counting rods were invented in China around 200 B.C.E. Red rods (the auspicious color) were used for positive numbers and black rods (very inauspicious) were used for negative numbers. Negative numbers and operations with them appear in the Chinese text *The Nine Chapters on the Mathematical Art* dated around 150 B.C.E.

From the 7th century C.E., negative numbers were used in India to represent debts. The Indian mathematician Brahmagupta, in *Brahma-Sphuta-Siddhanta* (written in 628 C.E.), discussed the use of negative numbers to produce the general form of the quadratic formula that remains in use today. He also found negative solutions of quadratic equations and gave rules regarding operations involving negative numbers and zero, such as *"A debt cut off from nothingness becomes a credit; a credit cut off from nothingness becomes a debt."*

Thus, historical and anthropological studies show that number 0, negative numbers and thus, integers together with their arithmetic came from financial operations when it was necessary to find profit and loss, credit, debit and debts.

However, when the European mathematicians encountered negative numbers from the books coming from Muslim countries, critics dismissed their sensibility. For instance, the German mathematician Michael Stifel, also known as Styfel (1487–1567), who was also a monk and Protestant reformer used negative numbers in his book (Stifel, 1544). He found negative numbers useful for mathematics and indispensable but called them absurd numbers.

The celebrated Italian mathematician Gerolamo Cardano (1501–1576) made the systematic use of negative numbers in his book Ars Magna but called them fictitious numbers (Cardano, 1545).

It is interesting that there was no generally accepted theory of the elementary operations with negative numbers until the more general subject of operating with the ordinary complex numbers had been placed on a firm basis (Milier, 1925).

However, the general attitude of mathematicians to negative numbers was even worse. For instance, the famous French mathematician François Viète, or in Latin Franciscus Vieta (1540–1603), who made an important-step towards modern algebra developing efficient algebraic notation, insisted on excluding negative numbers from mathematics. Other notable European mathematicians, such as the notable French mathematician Jean le Rond d'Alembert (1717–1783) or the English mathematician William Frend (1757–1841), did not want to accept negative numbers in the 18th century and referred to them as "absurd" or "meaningless" (Kline, 1980; Mattessich, 1998). Even in the 19th century, it was common practice to ignore any negative results derived from equations, on the assumption that they were meaningless (Martinez, 2006). For instance, the prominent French mathematician, physicist and politician Lazare Nicolas Marguerite, Count Carnot (1753–1823) affirmed that the idea of something being less than nothing is absurd (Mattessich, 1998). Some other outstanding mathematicians such as the excellent Irish mathematician William Rowan Hamilton (1805–1865) and the prominent British mathematician and logician Augustus De Morgan (1806–1871) had similar opinions. In a similar way, irrational numbers and later imaginary and complex numbers were firstly rejected. Today these classes of numbers are fully accepted and applied in numerous scientific and practical fields, such as physics, chemistry, biology, accounting, and finance.

Those who objected to negative numbers were not amateurs in mathematics. Their mathematical knowledge was very good but they were closed to new ideas. Here is more information about these people.

Lazare Carnot was born in the village Nolay, France, and studied at the École royal du genie de Mézières (Royal Engineering School of Mézières). After graduation he served in the army continuing his studies of mathematics. After the French Revolution, Carnot entered political life in 1791 and soon was in charge of the military defense of France organizing victory of the French armies. Napoleon made Carnot Minister of War but because of his republican convictions, Carnot resigned when Napoleon crowned himself emperor. As a mathematician, Carnot made important contributions to geometry and to fortification as a military engineer.

William Rowan Hamilton was born in Dublin and graduated from Trinity College, Dublin. He taught at Trinity College, Dublin, and was the Royal Astronomer of Ireland making important contributions to algebra, optics, and classical mechanics. His approach to Newtonian mechanics is called Hamiltonian mechanics and plays the fundamental role in classical and quantum field theories and in quantum mechanics.

Augustus De Morgan was born in Madurai, India, where his father worked with the East India Company. He graduated from Trinity College, Cambridge, and started working as the professor of mathematics at newly founded London University where he made pioneering applications of algebra to logic. De Morgan was also one of the founders of the London Mathematical Society. Interesting, De Morgan wrote the book *"Budget of Paradoxes"*, in which he analyzed many books describing paradoxical way of thinking aimed at such goals as squiring of the circle or perpetual motion (De Morgan, 1972). De Morgan also published *Elements of Arithmetic* (1830), *Elements of Algebra* (1837a), *Elements of Trigonometry* (1837), and *Formal Logic* (1847).

Returning to the history of arithmetic in general we find that after the prominent development in Ancient Greece and preservation in Ancient Rome, mathematical expertise in general and knowledge of arithmetic, in particular, essentially declined in Europe during the Dark Ages. What Greeks and Romans knew about numbers was lost although people in Europe continued using numbers in many practical areas. There were also mathematicians at that time. Although they did not introduced new ideas or obtained new results in mathematics, their work was important for preserving mathematics in Europe. For instance, the Roman philosopher, senator, and consul Anicius Manlius Severinus Boëthius (ca. 475–524), commonly called simply Boethius, wrote textbooks *De Institutione Arithmetica*, that is, *On Arithmetic* (Boethius, 1983), and *Accedit Geometria* (*Geometry*), which remained standard manuals in monastic schools for many centuries (Eves, 1990). Another European medieval scholar Gerbert of Aurillac (ca. 950–1004), who became Pope Sylvester II at the end of his life, also wrote on arithmetic treating it as an essential field of the necessary knowledge.

Boethius was born in Rome after the last Western Roman Emperor was deposed and served as a senator, and consul becoming *magister officiorum*, i.e., the head of the government and court services, under the Ostrogothic King of Italy Theodoric the Great. In the past, the family of Boethius included Roman Emperors and consuls. Boethius achieved high position at a young age but at the end of his life was accused of treason, imprisoned and executed.

Gerbert of Aurillac was born in the town of Belliac, France, and studied mathematics in a monastery in Catalonia. Being at a pilgrimage to Rome, Gerbert met Pope John XIII and Emperor Otto I becoming a tutor of the Emperor's son. Later the Emperor appointed Gerbert the abbot of a monastery in Italy. He became archbishop of Ravenna in 998 and pope in 999. Gerbert was a noted scholar of his time writing on arithmetic, geometry, astronomy, and theology, and building some sophisticated mechanisms such as a hydraulic-powered organ.

However, according to historical data, during the Dark Ages in Europe, knowledge of arithmetic was preserved and further developed in India, China and later among the Arabs although people in Europe continued using numbers. European mathematics revived with the development of trade and overseas exploration, which brought Eastern mathematical knowledge to Europe. As a result, Hindu-Arabic numerals replaced Roman numerals and other non-positional numerical systems, allowing calculations to be more efficiently made on paper, instead of by the abacus.

However, the process of acceptance and adaptation of the mathematical knowledge brought from the East was hard and controversial. For instance, in Britain, a controversy raged during the 18th century about the acceptability of negative numbers while negative numbers were incorporated into counting boards from around 200 B.C.E. in China. Likewise, the rules for the arithmetic of negatives were explicitly stated by Brahmagupta in India ca. 650 C.E. In other words, there was a deep cultural division between the East where negative numbers were accepted from the beginning and the West, where, under Euclid's influence, arithmetic remained the calculus of lengths and areas, both automatically positive quantities (Mumford, 2008).

As we know, arithmetic as a mathematical structure consists of three basic components: (1) numbers, (2) number representations and

(3) operations with and relations between numbers. Consequently, the development of arithmetic essentially involved formation of representation systems for numbers. This process was exceptionally complicated. Only the last part of this process can be described with definite authenticity. There have been many known systems of representation. In Ancient Egypt there were several systems. In one of them there were special symbols for 1, 10, 100, and 1000. Other numbers were represented by means of combinations of these symbols. The basic arithmetic operation in Ancient Egypt was addition. More than four millennia ago, ancient Babylonians used a positional representation system with the base 60. However, they used only two symbols. Alphabetic representation systems were used by ancient Greeks, Jews and Slavs. At the beginning of the new era in India, there was a wide-spread oral positional decimal representation system, with several synonyms for zero and other digits. By the 8th century, this system had spread as far as the Middle East because from the 8th century, the development of arithmetic and other fields of mathematics moved to the Islamic world where it was highly developed for several centuries, especially during the 9th and 10th centuries. This period got underway at the end of the 8th century with the Caliph Harun al-Rashid, the fifth Caliph of the Abbasid Dynasty, who ruled, from the capital city of Baghdad, over the Islam Empire, which stretched from the Mediterranean to India. He encouraged scholarship and translations of Greek texts into Arabic. The next Caliph, al-Ma'mun, supported learning even more strongly than his father al-Rashid, setting up the House of Wisdom in Baghdad, which became the centre for both the work of translating and research.

In particular, Islamic mathematicians learned about negative numbers from Arabic translations of Brahmagupta's works, and by 1000, they were using negative numbers for debts. Arabian scholars courteously gave Indian scientists credit for their number system. For instance, in one of their works, it is written:

"We also inherited a treatise on calculation with numbers from the sciences of India, which Abu Djafar Mohammed Ibn Musa al-Chowarismi developed further. It is the most comprehensive, most practical, and requires the least effort to learn; it testifies for the thorough intellect of the Indians, their creative talent, their superior ability to discriminate and their inventiveness." (Woepcke, 1863)

The most famous Muslim scholar was the prominent medieval intellectual Muhammad ibn Musa al-Khowārizmī (ca. 780–850), who wrote works in mathematics, astronomy, and geography. His last name suggests that his birthplace was in the Middle Asia, somewhere in the territory of Khwarezmia, which was a big state during the High Middle Ages and included large parts of Central Asia and Iran. Working as the astronomer and head of the library of the House of Wisdom in Baghdad, al-Khowārizmī wrote his main work *Hisab al-jabr w'al-muqabala*, which can be rendered in English as *The Compendious Book on Calculation by Completion and Balancing*, as well as a treatise on Hindu-Arabic numerals around the year 825. From the title of this book, our modern term *algebra* was derived becoming one of the major fields in mathematics and was also used in names for a variety of mathematical structures such as a linear algebra, associative algebra, Lie algebra, normed algebra or universal algebra (cf. Kurosh, 1963). Algebra as a mathematical field grew up by generalization from arithmetic as a mathematical field. That is why all arithmetics are at the same time algebras, namely, such algebras elements of which are numbers.

In addition to the rules of arithmetical operations with numbers represented in the decimal positional numerical system, the book *Al-jabr wa'l muqabala* also contained some elements of algebra such as methods of solving linear and quadratic equations with one unknown variable, and the elementary arithmetic of relative binomials and trinomials.

According to the American historian of mathematics Carl Solomon Gandz (1883–1954),

"Al-Khwarizmi's algebra is regarded as the foundation and cornerstone of the sciences. In a sense, al-Khwarizmi is more entitled to be called "the father of algebra" than Diophantus because al-Khwarizmi is the first to teach algebra in an elementary form and for its own sake, while Diophantus is primarily concerned with the theory of numbers." (Gandz, 1936)

In a similar vein, the American historian of mathematics Carl Benjamin Boyer (1906–1976) writes,

"Diophantus sometimes is called the father of algebra, but this title more appropriately belongs to al-Khowarizmi..." because the book *"...the Al-jabr comes closer to the elementary algebra of today than the works of either Diophantus or Brahmagupta..."* (Boyer, 1985)

However, the goal of the book *Al-jabr wa'l muqabala* was not a theoretical development of mathematics but fulfill the practical needs of people. Indeed, al-Khowārizmī wrote that in his book he presented

"... what is the easiest and most useful in arithmetic, such as men constantly require in cases of inheritance, legacies, partition, lawsuits, and

trade, and in all their dealings with one another, or where the measuring of lands, the digging of canals, geometrical computations, and other objects of various sorts and kinds are concerned." (Chester, 1915)

The Europeans were introduced to the decimal positional numerical system based on the digits 1, 2, 3, 4, 5, 6, 7, 8, 9, and 0 as well as advanced for that time methods for operating with numbers represented by such numerals in the 12th century through a treatise on Hindu-Arabic numerals by al-Khowarizmi.

The Arabic text of this treatise was lost but its 12th century Latin translation, *Algoritmi de numero Indorum*, meaning in English *Al-Khowarizmi on the Hindu Art of Reckoning* was preserved and used until the 16th century as the principal mathematical text-book of European universities.

It is also interesting to know that the word *Algoritmi*, which was the translator's Latinization of Al-Khwarizmi's name, gave birth to the term *algorithm*, which at first, started meaning any rules for people for making calculations. In the 20th century, it became one of the most popular terms in general and in computer science and information processing technology, in particular. For instance, all mathematical operations are based on algorithms of their performance.

Serge Abiteboul and Gilles Dowek write:

"*Algorithms are probably the most sophisticated tools that people have had at their disposal since the beginnings of human history. They have transformed science, industry, society. They change the concepts of work, property, government, private life, even humanity. Going easily from one extreme to the other, we rejoice that they make life easier for us, but fear that they will enslave us.*" (Abiteboul and Dowek, 2020)

Note that although some historians think that the concept of algorithm has existed since classical antiquity, this is not true because algorithms were used much earlier in ancient Babylonia, Egypt, India and China millennia ago. However, the concept of algorithm became a rigorous scientific concept only at the beginning of the 20th century.

Another renowned Muslim scholar Shujā' ibn Aslam ibn Muḥammad Ibn Shujā' AbūKāmil [Latinized as *Auoquamel*] (ca. 850–930) was an Egyptian mathematician, who was considered the first mathematician to systematically utilize and acknowledge *irrational numbers* as solutions to and coefficients in algebraic equations.

In the early 11th century, the celebrated Muslim scholar and polymath Abu Rayhan al-Biruni (973–1048), who also came from Khwarezmia as

Al-Khwarizmi, made significant contributions to mathematics, specifically in the areas of theoretical and practical arithmetic, summation of number series, the rules of operation with irrational numbers, method of solving algebraic equations, combinatorial analysis, ratio theory, and geometry. In essence, 95 of 146 books attributed to al-Biruni were devoted to mathematics, astronomy, and related fields. In addition to mathematics, al-Biruni conducted studies in physics, astronomy, and other natural sciences, distinguishing himself as a historian, anthropologist, chronologist, and linguist (Rozenfel'd *et al.*, 1973).

The Muslim mathematician Al-Samawal (1130–1180) used decimal fractions and extended the arithmetic operations to handle polynomials.

One more Muslim astronomer and mathematician Ghiyath al-Din Jamshid Mas'ud al-Kashi (1380–1429) wrote *Treatise on the Circumference* in 1424 where he calculated the number π to nine sexagesimal places, which corresponded to sixteen decimal digits. However, the main mathematical treatise of Al-Kashi was *The Key to Arithmetic* which he completed in 1427. It was intended for teaching arithmetic with applications to astronomy, surveying, architecture, accounting, and trading (Dold-Samplonius, 1992). This work was very popular over several hundred years in mathematics teaching in Medieval Islam throughout the time of the Ottoman Empire. Al-Kashi also used decimal fractions before Stevin and established connections between sexagesimal and decimal numerical systems (Rashed, 1994).

Writing about the history of arithmetic, it is necessary to explain how arithmetic was transformed into algebra. The basic difference between arithmetic and algebra is that arithmetic operates with numbers while algebra operates with symbols. This is explicitly expressed in Indian mathematics as one of the Sanskrit names of arithmetic is *vyacta ganita*, which means the *art of computing with known quantities* (Brahmagupta and Bhaskara, 1817; Volodarsky, 1977). At the same time, one of the Sanskrit names of algebra is *avyacta ganita*, which means the *art of computing with unknown quantities*.

The process of transformation of arithmetic into algebra took many centuries going through three distinct stages determined by the used notation (Katz and Barton, 2007). Namely, we have:

- *Rhetorical algebra* used equations represented by sentences in natural languages. For instance, the rhetorical form of the equation $x + 10 = 12$ is "The thing plus ten is equal to twelve." Rhetorical algebra was first developed by the ancient Babylonians and Egyptians remaining dominant up to the 16th century.

- *Syncopated algebra* can be called partially symbolic because some symbolism is used but it does not contain all of the characteristics of symbolic algebra. For instance, there may be a restriction that subtraction may be used only once within one side of an equation while this is not the case with symbolic algebra.
- *Symbolic algebra* is based on complete symbolism.

Rhetorical algebra also included *geometric algebra*, in which geometric constructions were used for solving algebraic equations or problems involving such equations. Geometric algebra appeared when mathematicians found that numbers, which were multitudes of units equal to 1, were insufficient in solving some practical arithmetical problems and it was necessary to utilize more general objects.

The move away from geometric algebra dates back to Diophantus and Brahmagupta, but was completely changed for arithmetic algebra when Al-Khwarizmi introduced generalized algorithmic processes using only arithmetical operations for solving algebraic problems (Katz and Barton, 2007).

Syncopated algebraic expressions first appeared in Diophantus' *Arithmetic* (3rd century C.E.) and later were used in Brahmagupta's *Brahma Sphuta Siddhanta* (7th century).

The transition to symbolic algebra was made by Islamic mathematicians such as the Moroccan-Arab mathematician, astronomer, Islamic scholar, Sufi, and a one-time astrologer Ibn al-Bannā al-Marrākushī (ca. 1256–1321), also known as Abu'l-Abbas Ahmad ibn Muhammad ibn Uthman al-Azdi Ibn al-Banna, and the Muslim mathematician Abū al-Ḥasan ibn 'Alī ibn Muḥammad ibn 'Alī al-Qurashī al-Qalaṣādī (1412–1486) from Al-Andalus, the area of the Iberian Peninsula ruled by Muslims. However, the first fully symbolic algebra was built by the well-known French mathematician and lawyer François Viète, Seigneur de la Bigotière (1540–1603). Later, the great French mathematician, philosopher, and scientist René Descartes introduced the modern notation into algebra and reduced geometry to arithmetic and algebra.

It is necessary to remark that for a long time, the difference between arithmetic and algebra was only in the peculiarity that algebra studied and used equations. Thus, it is also possible to call this mathematical field by the name *equational arithmetic* when equations were solved by operating with numbers and by the name *equational geometry* when equations were solved by operating with geometrical objects.

Symbolic algebra, in turn, went through the following stages, which form three parts of the contemporary algebra:

- *arithmetic algebra*, in which symbols (letters) represent numbers;
- *conceptual algebra*, in which symbols (letters) represent some entities that are not necessarily numbers;
- *abstract algebra*, in which symbols (letters) represent only elements of abstract algebraic systems, which are defined by axioms.

The first two parts were delineated by George Peacock (1791–1858) in his *Treatise on Algebra* (Peacock, 1842). He called conceptual algebra by the name symbolic algebra and did not write about abstract algebra because at that time it did not exist.

To see the difference between conceptual algebra and abstract algebra, we indicate that conceptual algebra studies groups of transformations, rings of functions or rings of polynomials while abstract algebra studies abstract groups and rings, which are defined only by axioms and rules of operation.

In the historic process, arithmetic algebra emerged with the first algebraic equations, such as $3x - 9 = 0$, which described some practical problems. Different kinds of such and more advanced equations implicitly appeared in the most ancient mathematical sources — Sumerian tablets and Egyptian papyruses dated almost four millennia ago. Mathematicians started solving algebraic equations in a general form around two millennia ago. For instance, because of this, Diophantus was often called the "father of algebra." Utilization of equations made solutions of arithmetical problems much simpler and allowed to find general formulas for such solutions.

Another source of arithmetic algebra was the necessity to described laws of arithmetic for infinitely many numbers, such as natural numbers. For instance, the commutative law $a + b = b + a$, which is valid for all natural numbers utilizes and has to utilize symbols, which denote arbitrary natural numbers.

Much later (in the 18th and 19th centuries) mathematicians transformed arithmetic algebra into conceptual algebra while abstract algebra was fully developed only in the 20th century.

It is interesting how relations between arithmetic and algebra were changing in the history of mathematics. When algebra was emerging, it worked with numbers, or more exactly, with their names. This feature resulted in the conclusion that algebra is a subfield of arithmetic. The reason was that problems that were later treated as algebraic actually demanded operating only with numbers and thus, were solved in arithmetic. For instance, even in the 18th century, the *Encyclopédie* by the French mathematician, scientist, philosopher, and writer Jean le Rond d'Alembert (1717–1783) and French philosopher, art critic and writer Denis Diderot (1713–1784) included a diagram entitled "Figurative System of Human Knowledge", which was developed by d'Alembert. In this diagram, arithmetic included two parts — numeric arithmetic and algebra. In turn, algebra consisted of four parts — elementary algebra, infinitesimal algebra, differential algebra, and integral algebra (Diderot and d'Alembert, 1993).

An interesting peculiarity of the contemporary mathematics is that higher arithmetic (number theory) being a part of arithmetic contains many parts of algebra. For instance, one of the classical books on number theory by the renowned French-American mathematician André Weil (1906–1998) includes sections on finite fields, locally compact fields, local fields, algebraic number fields, simple algebras, and the Brauer group (Weil, 1995).

Exploring history of numbers and arithmetic, it is necessary not only to show the development of this area as well as its transformations and applications but also to describe the role of arithmetic in education. And this role has been always very important. Arithmetic was an important part of the education system in most ancient civilizations, including Vedic society and ancient Egypt, Ancient Greece, and the Roman Empire. For instance, in the dialogues of Plato, Socrates explains:

"We must endeavor to persuade those who are to be the principal men of our state to go and learn arithmetic, not as amateurs, but they must carry on the study until they see the nature of numbers with the mind ... for the sake of the soul." (Plato, 1961)

Thus, we explore how arithmetic was taught at different periods of human history and what sources were used for this purpose. Learning arithmetic has been very important because arithmetic always has been an essential intellectual tool of information acquisition, processing and management during the whole history of human civilization. Consequently, people needed to acquire knowledge about this intellectual tool and its utilization.

In ancient Greece, studies of arithmetic formed two areas: calculation techniques were taught to artisans and other people of middle class status while those who had time, money and inclination could pursue further studies into the science of numbers. Many philosophers including such prominent as Pythagoras and Plato treated arithmetic as the base of all knowledge while the word mathematics originated from the Greek word $\mu\alpha\theta\eta\mu\alpha$ (*mathema*), which meant knowledge, study, and learning. Writing about his ideal state in his work *Republic*, Plato explained that arithmetic, geometry, music and astronomy must become the base of education. Much later these disciplines were called the *Quadrivium*, or *Arts of the Number and Magnitude*, forming the higher level of education in medieval Europe. Note that actually the whole *Quadrivium* was related to arithmetic, which could be considered as the art of number and as it was proven later, of magnitude. Indeed, magnitudes were formalized as real numbers in the 19th century.

Returning to the antiquity, we see that as did the Greeks, so did the Romans: in ancient Rome, arithmetic teaching was not only wholly welcome or accepted, but seen as necessary in many cases. We can see this judging by the following opinions of some prominent individuals of that epoch.

Thus, an eminent Roman author, architect and engineer Marcus Vetruvius Pollio (ca. 75–15 B.C.E.), mostly known as Vetruvius, suggested students of architecture to study arithmetic, geometry, optics, astronomy, and law.

The prominent Roman physician, philosopher and writer Aelius Galenus (129–210 C.E.), better known as Galen of Pergamon, recommended future physicians to study medicine, arithmetic, rhetoric, music, geometry, and dialectics.

One of the greatest ancient Roman scholars Marcus Terentius Varro (116–27 B.C.E.), mostly known as Varro Reatinus, and the famous ancient Roman Stoic philosopher Lucius Annaeus Seneca (4 B.C.E. – 65 C.E.), also known as Seneca the Younger, recommended the study of Arithmetic and Geometry to further develop logical reasoning abilities. It is also believed that Anicius Manlius Severinus Boethius (ca. 475–524), mostly known as Boethius, and renowned writer and scholar of antiquity Flavius Magnus Aurelius Cassiodorus Senator (ca. 485–585), mostly known as Cassiodorus, were the first to use the term *Quadrivium*, which consisted of the four disciplines: arithmetic, geometry, astronomy, and music. Note that Senator was part of Cassiodorus' surname, not his rank.

Later during the period between the end of the 8th century and the first part of the 9th century, the term *Trivium* was coined to denote three disciplines: grammar, logic and rhetoric. They were called *Arts of the Word* constituting *seven liberal arts* together with the *Quadrivium*.

Importance of arithmetic as the basic educational discipline resulted in the situation where the majority of books written and published on arithmetic have been treatises (textbooks) because from ancient times to our days more and more people needed arithmetic.

Such classical books as *Zhoubi Suanjing*, (*Arithmetical Classic of the Gnomon and the Circular Paths of Heaven*) written in China between 1046 B.C.E. and 256 B.C.E. with compilation and addition of materials during the period 202 B.C.E. – 220 C.E. and an important book of mathematics *Sthananga Sutra*, written by Jaina in India between the 2nd century B.C.E. and the 4th century C.E. were composed to teach people arithmetic.

The book *De arithmetica* (*On Arithmetic*) by Boethius was used in Europe as the principal source for most math textbooks till the twelfth century, when the works of Arab mathematicians were brought to Europe and rendered into Latin by different translators such as Robert of Chester, Gerard of Cremona and Adelard of Bath.

The most popular of these books was the treatise on Hindu-Arabic numerals *Hisab al-jabr w'al-muqabala* (*The Compendious Book on Calculation by Completion and Balancing*) written around the year 825 by the famous mathematician Muhammad ibn Musa al-Khowarizmi. It was used until the 16th century as the principal mathematical text book of European universities.

With the advent of printing, arithmetic textbooks appeared in print. This process is described in the next chapter.

Chapter 2

The Chronicle of Numbers and Arithmetic from the Beginning of the Second Millennium to the Beginning of the 20th century

> *One cannot understand ... the universality of laws of nature,*
> *the relationship of things, without an understanding of mathematics.*
> *There is no other way to do it.*
>
> Richard P. Feynman

In the second millennium, the development of society was accompanied by the development of arithmetics. It is possible to separate three approaches in the emergence/construction of new arithmetics:

(1) Creation of new kinds of numbers and extension of existing arithmetical operations to the new domain has been the most popular approach. Examples are the Diophantine arithmetic R of real numbers and the Diophantine arithmetic C of complex numbers.

(2) Preservation of existing kinds of numbers but endowing their sets with new arithmetical operations was the approach that emerged in the 20th century. Examples are different non-Diophantine arithmetics (Burgin, 1977, 1997, 2007).

(3) Finally, in some cases, new arithmetics appropriated both new numbers and new operations. Modular arithmetics give an example of this technique (Kurosh, 1963).

Thus, for a long time, the most popular approach to the development of arithmetic was generation or creation of new classes (types) of numbers. It is possible to separate three directions in this process:

(1) *Intensive generation,* or *inner direction,* gives birth to new classes of numbers taking specific subsets of an existing class of numbers. Intensive generation also brought forth some important individual numbers such e or i.

(2) *Extensive generation,* or *outer direction,* gives birth to new classes of numbers, which contain an existing class of numbers.

(3) *Formative generation,* or *operational direction,* gives birth to new classes of numbers by performing operations with existing classes of numbers.

Let us consider some examples.

(1) Odd numbers, even numbers, prime numbers as classes of natural numbers, as well as algebraic numbers, irrational numbers, and transcendental numbers as classes of real numbers were obtained by intensive generation, i.e., these classes were delineated in the already existing classes of numbers.

(2) Integer numbers, rational numbers, real numbers, complex numbers, transfinite numbers, p-adic numbers were obtained by extensive generation, i.e., these classes were constructed by extensive generation as essentially new classes of numbers, which contained existing classes of numbers. Integer numbers contain whole numbers. Rational numbers contain integer numbers. Real numbers contain rational numbers and so on.

(3) Modular arithmetics, computer arithmetics, and non-Diophantine arithmetics were created by formative generation, i.e., these arithmetics were formed by transformation of existing arithmetics.

In turn, the outer direction can be separated into the linear, dimensional and structural tracks. In the *linear track*, a number system and its arithmetic are developed in one dimension. For instance, in such a way, whole numbers extended natural numbers, integer numbers extended whole numbers, rational numbers extended integer numbers and real numbers extended rational numbers.

In the *dimensional track*, a number system and its arithmetic are extended to additional dimension. For instance, in such a way, complex numbers extended real numbers while quaternions extended complex numbers.

In the *structural track*, a number system and its arithmetic are developed based on specific mathematical structures. For instance, multinumbers are based on multisets (Monro, 1987), named numbers are based on named sets (Burgin, 2011), while fuzzy numbers are based on fuzzy sets (Zadeh, 1975, 1976).

Analyzing the development of arithmetics, it is important to understand that in contrast to modular arithmetics or the arithmetic of integer numbers, non-Diophantine arithmetics do not use new classes of numbers. They only introduce novel operations with the already existing numbers. For instance, any non-Diophantine arithmetic of natural numbers contains the same natural numbers as the Diophantine arithmetic of natural numbers. These numbers satisfy the famous five axioms of Peano (Peano, 1889). At the same time, addition and/or multiplication of natural numbers can be very different in some non-Diophantine arithmetic (Burgin, 1977, 1997, 2001).

However, at the first half of the second millennium, attention of mathematicians was not aimed at the development of arithmetic. The goal was advancement of the methods of calculations.

Arithmetic was important in many areas, especially, in commerce. As a result, different books on arithmetic, the majority of which was textbooks, were written in the Middle Ages and Renaissance in Europe. From the thirteenth century, a new kind of textbooks appeared — the commercial arithmetic or *Abacus Book* written not in academic Latin but in the common language and aimed at the education of merchants and trades people. To achieve this goal, these books contained many practical mathematical problems related to the contemporary marketplace and their solutions. This process was caused by the expansion of international trade and banking. Abacus books taught how to solve practical problems performing arithmetical operations using the abacus and were used in the so-called abacco schools, which flourished from the 14th to the 16th centuries.

As Howard Eves writes, the way around mental and physical difficulties in calculations was the invention of the abacus, which can be considered the earliest mechanical computing device used by people (Eves, 1990). It appeared in various forms in many parts of the ancient and medieval world. The teachers in the abaci schools taught calculations by means of abacus and were called abacists. With the coming of Hindu-Arabic numerals to Europe, direct calculations with numbers became much simpler, and the proponents of this approach were called algorists because it was the book of the prominent Muslim mathematician al-Khowārizmī that introduced these numerals to Europe. The Latin translation of his name was *Algoritmi*, hence algorists. For several centuries algorists competed with abacists, and only around 1500 C.E. computations with Hindu–Arabic numerals won and arithmetic was taught as methods of calculations in the mind and on the paper.

The most important mathematician of the Middle Ages was Leonardo Fibonacci (ca. 1170–1250), also known as Leonardo of Pisa. His name, which was made up only in 1838, was a contraction of the Italian *filius Bonaccio*, or in English, *son of Bonaccio*. His most famous work is *Liber Abaci* (a book on calculations) published in 1202 (Boncompagni Ludovisi, 1862). In spite of its name, which associated it with abacus, it was one of the first Western books to describe the Hindu–Arabic numeral system and operations with natural numbers and fractions based on this system. There was computation of square and cubic roots, the solution of linear and quadratic equations and many applications (Eves, 1990). In particular, the book essentially contributed to financial mathematics.

Using a recurrence relation, Fibonacci defined a sequence of natural numbers, which acquired the name Fibonacci numbers. In this sequence, any number starting with the third is the sum of two previous numbers. Fibonacci numbers occur frequently in mathematics, biology, and computer science.

Fibonacci was the first to develop present value analysis for comparing the economic value of alternative contractual cash flows and a general method for expressing investment returns solving a wide range of complex interest rate problems (Goetzmann, 2003; Goetzmann and Rouwenhorst, 2005). This considerably influenced the progress of capitalist enterprise and public finance in Europe in the centuries that followed.

Fibonacci also wrote another book *Liber Quadratorum* (Book of Squares), in which he presented the theory of Diophantine equations, i.e., equations with whole numbers as coefficients and values of the unknown variables. This is a remarkable area in higher arithmetic (number theory) due to the famous (or as some think, notorious) Fermat's Last Theorem, which was actually a conjecture made by the famous French mathematician Pierre de Fermat around 1637 (Edwards, 1996; Singh, 2012).

It became very famous because the statement was very simple and many mathematicians and non-mathematicians tried to prove it without success until the end of the 20th century. Although in some cases, their efforts resulted in interesting and important mathematical theories. For instance, the principal results of one of the prominent German mathematician Ernst Eduard Kummer (1810–1895), whose works, according to E.T. Bell, set up modern arithmetic (Bell, 1937), came from his attempts to prove Fermat's Last Theorem.

However, some mathematicians questioned importance of this result. For instance, the great German mathematician Gauss wrote in 1816:

"I confess that Fermat's Last Theorem, as an isolated proposition, has very little interest for me, because I could easily lay down a multitude of such propositions, which one can neither prove nor dispose of." (cf., for example, Dunnington, 2004).

However, centuries passed between Fibonacci and Fermat and during that time different authors continued writing treatises on arithmetic. For instance, the French mathematician Nicolas Chuquet (ca. 1445–1488), who was a Parisian and a bachelor of medicine, wrote the book *Triparty en la science des nombres* in 1484. The first part of this work describes computation with rational numbers, including an explanation of the Hindu-Arabic numerals. Chuquet gave a *"règle des nombres moyens"* according to which it was possible to find a fraction between any two given fractions by taking the sum of their numerators and dividing by the sum of their denominators. In addition, he introduced names for powers of 10 writing that the first group was marked *"... million, the second mark byllion, the third mark tryllion, the fourth quadrillion, the fifth quyillion, the sixth sixlion, the seventh septyllion, the eighth ottyllion, the ninth nonyllion and so on with others as far as you wish to go."* (Marre, 1880) That is why, the system the number names *million, billion, trillion*, etc. is sometimes referred to as the Chuquet system.

The second part of Chuquet's book explains computation with irrational numbers obtained as roots of rational numbers, while the third part is dedicated to the theory of equations and thus, can be accredited to algebra. Chuquet considered negative solutions of equations giving a justification of these solutions in terms of debts. Besides, he introduced his own notation for algebraic concepts and exponentiation treating zero as an exponent.

It is considered the earliest French algebra book. However, it was unpublished in his lifetime. Only in 1870s, the French scholar Aristide Marre discovered Chuquet's manuscript and published it in 1880 (Marre, 1880).

On the other hand, the content of the book *Triparty en la science des nombres* was included by another French mathematician Estienne de La Roche (ca. 1470–1530) in his highly successful commercial arithmetic textbook *L'arismetique nouellement composée* (La Roche, 1520). Although La Roche wrote that this work was *"... the flower of several masters, experts in the art, ... Nicolas Chuquet from Paris, Philippe Friscobaldi from Florence, and Luca Pacioli from Burgo"*, many accused him in plagiarism because he did not exactly indicate which part of his work was taken from the works of his predecessors (cf., for example, La Roche, 2019). However, at that time, the authors still did not feel the necessity of the exact attribution of their materials. Besides, it was possible to come to the opposite conclusion that the La Roche's book was the compendium of works of other mathematicians and thus, all accusations in plagiarism were ungrounded. For instance, nobody accuses Euclid in plagiarism although it is known that many result from his famous treatise were obtained by other mathematicians.

With the invention of printing in Europe, the first printed books on arithmetic appeared. Naturally they were textbooks with many practical problems and their solutions.

However, as we know now, printing was first invented in China between 618 and 907 C.E. during the Tang Dynasty. The oldest extant printed book is the *Diamond Sutra*, which was printed around 868 C.E. There are also earlier fragments of a printed sutra, which date two hundred years earlier and are printed on hemp paper. The first technology was woodblock printing while the proliferation of Buddhist texts was a main impetus to large-scale printing.

The first printed book using movable metal print technology was also the Korean Buddhist text *Jikji*, which was the abbreviation for *Selected Teachings of Buddhist Sages and Seon Masters*. It dates 1377.

The first book printed in Europe was the *Gutenberg Bible*, which was printed with movable metal type by Johannes Gutenberg (ca. 1400–1468) in 1455 in Mainz, Germany. Historians suggest that this event started revolution in book publishing.

Very soon after this, people started printing arithmetic textbooks. According to contemporary knowledge, the earliest printed arithmetic was *Treviso Arithmetic*, or *Arte dell' abbaco* (*The Art of Calculations*) published in 1478 in the town of Treviso, just a little way north of Venice (Smith, 1924). Its author is unknown while the printer was either Gerardus de Lisa or Michael Manzolus. The book contained is diverse mathematical problems for merchants, farmers, tailors, builders, and other professionals.

This was the first dated printed book on arithmetic although some historians supposed that it was possible that some undated pamphlets titled *Algorithmus* might predate this work. Note that later the word *algorithmus* became the well known term *algorithm*.

At that time there were two kinds of books on practical arithmetic — Algorithmus books and Abaci books. Algorithmus books presented the rules, i.e., algorithms, of calculations in basic arithmetic with many examples. In contrast to this, Abaci books described solutions of arithmetic problems using algorithms as a tool.

The next two printed books on arithmetic appeared in 1482 in Spain and Germany. At first, historians of mathematics thought that the second printed book on arithmetic was the so-called *Bamberger Rechenbuch* (Wagner, 1482) but later they found the book on arithmetic (Santcliment, 1482) printed in Spain also in 1482 (Escobedo, 2010; Ausejo, 2015). *Bamberger Rechenbuch* was reprinted in 1483 demonstrating popularity of arithmetic and necessity of arithmetical knowledge at that time.

The next known book on commercial arithmetic *Qui comenza la nobel opera de arithmetica*, or simply *Aritmetica mercantile* by the Venetian mathematician Pietro Borghi was printed in Italy by the printer Erhard Ratdolt in Venice, in August 1484. It became the 15th-century's best-selling arithmetic, which ran to seventeen editions (Borghi, 1484).

The third notable printed in Italy book on arithmetic was *Aritmetica* by Filippo Calandri. It was printed in 1491 in Florence by the Italian–German printers Lorenzo Morgiani and Johannes Petri being the first arithmetic to accompany problems with pictorial illustrations (Calandri, 1491). One more its peculiarity in comparison with other printed arithmetics was inclusion of the rules for long division in the form we are now familiar with.

Fra Luca Bartolomeo de Pacioli (ca. 1447–1517), who was also called Paccioli, Paciolo or Lucas di Burgo, was an Italian mathematician and prominent figure of Renaissance humanism. He was a friend of the great Leonardo da Vinci (1452–1519), who illustrated the book *Summa de Arithmetica, Geometria, Proportioni et Proportionalita* (Everything About Arithmetic, Geometry and Proportions) by Pacioli published in 1494 (Pacioli, 1494). There was no another mathematician whose book was illustrated by such a remarkable artist.

It was a big book with more than 600 pages, which were distributed in the following way: 222 pages gave an exposition of arithmetic, 78 pages presented algebra, topics related to business covered 150 pages, and practical geometry including basic trigonometry was explained on 151 pages.

Today Pacioli is considered the father of modern accounting and his book *Summa de Arithmetica, Geometria, Proportioni et Proportionalita* is described as a foundational text in the history of capitalism because it contained the first printed description of double-entry bookkeeping and marks the birth of modern business (Taylor, 1942). Pacioli's system describes most of the accounting cycle as we know it today. However, Pacioli did not invent this system but simply exhibited the scheme used by merchants in Venice throughout the Italian Renaissance.

This was a general tendency at that time because the interest in education that accompanied the Renaissance and the tremendous increase in commercial activity caused appearance of a multitude of popular textbooks in arithmetic. The attention that was paid to teaching arithmetic during at that time was reflected in the large number of published books on arithmetic. Three hundred such books were published in the 15th and 16th centuries (Eves, 1990). Only in Spain, 35 authors published 43 different books on arithmetic, which had a total of 77 editions (Madrid *et al.*, 2020). Their authors can be divided into two groups: classical scholars who wrote in Latin and practical teachers who wrote in the vernaculars and often had other professions such as notaries, gaugers and town surveyors.

Very influential in Germany were arithmetics by Johannes Widmann (ca. 1460 – after 1498) published in Leipzig in 1489 and Jacob Köbel (ca. 1460–1533). According to the history of mathematics, the symbols of arithmetical operations $+$ and $-$ first appeared in print in the book (Widmann, 1489).

However, the most popular arithmetic at that time was written by Adam Riese (1492–1559), who was also called Adam Reis. His books on arithmetic (Ries, 1518, 1522) was so authoritative that the phrase "nach Adam Riese" ("according to Adam Riese" in English) was used in Germany to indicate correct calculations.

The American mathematician and historian of mathematics Howard Whitley Eves (1911–2004) relates the following story about Adam Riese (Eves, 1990). It happened so that Riese and a draftsman entered into a friendly competition to find who of them could, with straightedge and compass, draw more right triangles in one minute. The draftsman used the standard procedure drawing a straight line, erecting perpendiculars to the line and then connecting a point on the line with points on those perpendiculars. Contrary to this, Riese drew a semicircle on a straight line and very fast drew a large number of inscribed right angles. In such a way, he easily won the contest. It is necessary to remark that the draftsman could do much better if he would drew only one perpendicular and the connected one point on the line with a sequence of points on the perpendicular. Riese also could do better if he would drew only one inscribed right angle and then connected one point on the one side of that angle with a sequence of points on the other side of that angle.

Other countries also started printing books on arithmetics. Researchers deem that the first printed in France book on arithmetic appeared in 1512 (Swetz, 1987).

The first printed textbook in Swedish on arithmetic written by Aegidius Aurelius (ca. 1580–1648) appeared in 1614 (Hatami and Pejlare, 2019), but already in 1601, the oldest known manuscript in Swedish on arithmetic was written by Hans Larsson Rizanesander (1574–1646).

Researchers also suppose that the first printed in Portugal book on arithmetic appeared in 1519 (Swetz, 1987).

In England, regardless of its early venture into vernacular printing, the first book on arithmetic in English did not appear until the 16th century. Although some assume that printing of arithmetics in England started in 1537 (Swetz, 1987), the earliest arithmetic was printed in 1525 by Rychard Fakes in London. Fortuitously, the sole surviving page is its colophon, which explains that it is a translation of a French arithmetic.

In spite of the fact that the book printing was invented in Europe in the middle of the 15th century, the first Russian printed books appeared only a century after that owing to the creation of the tsar's typography in 1552 or 1553 in Moscow with the help of a printer from Copenhagen. At first, books were published in Russia without indication of the date of issue.

It might be interesting to know that for a long time, it was taught in Russian (Soviet) schools that printing in general was invented in Russia by Ivan Fyodorov, who actually printed the first book in 1563, that is, long after Gutenberg printed the first book in Europe in 1455 not speaking about printing in China and Korea many centuries before.

Continuing our history of numbers and arithmetic as a mathematical field, we come to *real numbers*, which constitute an extension of and the next major class after rational numbers. Their history started in ancient Greece when a student of Pythagoras discovered that the length of the diagonal of a unit square is not a rational number. This event is described above in more details. It is possible to assume that this was the first irrational number encountered by mathematicians.

For a long time after Pythagoras, mathematicians encountered irrational numbers solving algebraic equations but did not treat them as numbers not speaking about building a theory of real numbers, which consist or rational and irrational numbers.

In a similar way, imaginary and other complex numbers were encountered at various times, beginning with ancient Indian mathematicians, in solving quadratic equations. However, imaginary solutions were discarded as non-existent while intermediate expressions involving complex numbers were treated only as artificial tools but not as full-fledged numbers. According to historians, several mathematicians, including famous Italian mathematicians Cardano, Tartaglia, and Ferrari, demonstrated the first uses of complex numbers in the process of solving algebraic equations. They were famous Italian mathematicians of their time.

Gerolamo Cardano (1501–1576) was a celebrated Italian polymath, whose interests and proficiencies ranged from being a mathematician, physician, biologist, physicist, chemist, astrologer, astronomer, philosopher, and prolific writer, who wrote more than 200 works on science and other topics. Such a breadth of talents and tumult of his personal life make Cardano one of the most amazing characters in the history of mathematics. As a mathematician, he is credited with the first systematic use of negative numbers in Europe and introducing complex numbers in his attempts to find solutions to cubic equations. However, Cardano did not treat negative and complex numbers as numbers but only used them as formal expressions for obtaining positive real solutions of the considered equations. Cardano also wrote the first methodical treatment of probability *Liber de ludoaleae* (*Book on Games of Chance*), which was published posthumously.

Being famous for his mathematical talent, Cardano got the most recognition as a physician. For instance, he was invited and travelled to Edinburg where he cured Scotland's archbishop John Hamilton from asthma. Besides, Pope Paul III, Christian III king of Denmark, Henry II king of France, and Maria de Lorena queen of Scotland offered him the position of the physician of the court, which Cardano declined (Toscano, 2020).

Niccolò Fontana Tartaglia (ca. 1500–1557) was a renowned Venetian mathematician, engineer (as he designed fortifications and wrote on ballistics and on retrieving sunken ships), a surveyor and an official bookkeeper of the Republic of Venice, which became later a part of Italy. He published various books, including the first Italian translations of the works of Archimedes and Euclid.

Lodovico de Ferrari (1522–1565) was a recognized Italian mathematician, who was an assistant of Cardano at the beginning of his carrier finding a solution to equations of the fourth degree in 1541. It involved operations with negative numbers. Later Ferrari became the Director of the Duchy of Milan's land registry.

As the history of mathematics tells us, there was an awkward controversy between Tartaglia and Cardano, who was supported by his student Ferrari (Toscano, 2020). The controversy concerned the formula for solving cubic equations. Tartaglia discovered such formulas for some special cases of these equations using his discovery for winning mathematical debates, as Fabio Toscano calls them, mathematical duels, which were popular at that time.

Considering his discovery as a secret weapon for mathematical duels, Tartaglia did not want to disclose them to anybody else. At the same time, Cardano assumed that discoveries in science and mathematics belong to the whole mankind and the researcher had to share his achievements with other people. As Cardano also searched solutions for cubic equations, he invited Tartaglia to Milan, where he lived, and tried to persuade Tartaglia to share with him those formulas, or more exactly, the rule for solving cubic equations. At first, Tartaglia did not want do this but then when Cardano gave the oath not to disclose the discovery of Tartaglia, the latter explained the rule he had discovered.

For several years, Cardano did not include those rules in his works although he continued writing his numerous books. However, when Cardano and his apprentice Ferrari travelled to Bologna, he found that another Italian mathematician Scipione Dal Ferro found the solutions to cubic equations much earlier than Tartaglia. Cardano felt that this released him from any obligations toward Tartaglia and published those rules in his famous book "*Artis magnae, sive de regulis algebraicis*", which is better known as *Ars magna*.

Although Cardano gave reference both to Tartaglia and Dal Ferro, this caused fury of Tartaglia, who attacked Cardano in his pamphlet starting the famous controversy between the three mathematicians because Ferrari was answering the attacks of Tartaglia instead of Cardano.

It is interesting to know that according to John McLeish, Chinese mathematicians solved arithmetical problem that could be represented by algebraic equations of the fourth and even higher degrees centuries before the controversy between Tartaglia and Cardano about algebraic equations of the third degree (McLeish, 1994).

In Europe, the German mathematician Michael Stifel (1487–1567) was the first to use the term *negative numbers* defining them as numbers that are less than zero and the term *positive numbers* defining them as those that are greater than zero in his book *Arithmetica Integra* (Stifel, 1544). He also explained that zero, rational quantities represented by fractions, and irrational quantities known at that time are actually numbers explicating in what way whole, rational and irrational numbers are related to each other.

The eminent Italian mathematician Rafael Bombelli (1526–1572) was the first to explicitly build the arithmetic of complex numbers. This was a big achievement because even many mathematicians who lived much later were exceedingly confused on this topic.

In his book, Bombelli explained complex arithmetic as follows calling the imaginary numbers i by the name *"plus of minus"* and $-i$ by the name *"minus of minus"*:

> *"Plus by plus of minus, makes plus of minus.*
> *Minus by plus of minus, makes minus of minus.*
> *Plus by minus of minus, makes minus of minus.*
> *Minus by minus of minus, makes plus of minus.*
> *Plus of minus by plus of minus, makes minus.*
> *Plus of minus by minus of minus, makes plus.*
> *Minus of minus by plus of minus, makes plus.*
> *Minus of minus by minus of minus makes minus."*

After explaining multiplication of real and imaginary numbers, Bombelli formulated the rules of addition and subtraction pointing out that real components were added to real components, and imaginary components were added to imaginary components. Bombelli had the understanding that imaginary numbers were decisive and necessary for solving equations of the third and fourth degree. This enabled him to get solutions for many equations where other mathematicians such as Cardano had given up. At the same time, Bombelli also did not treat negative and complex magnitudes as numbers but only used them as formal tools for finding positive real solutions of the considered equations.

Irrational numbers were called, e.g., by Newton, *surdi* (that is, *deaf*) or *false*. Imaginary numbers when they first appeared in Cardano's works were called *sophistic numbers*.

Thus, we can see that numbers utilized before the 18th century were natural, rational and algebraic irrational numbers, such as $\sqrt{2}$ or $\sqrt{3}$. The number π was very different, and the conjecture that the number π is irrational was suggested by Arab mathematicians starting from the 11th century. In the 18th century, mathematicians started the process of unification of all these numbers. The first step in this direction is usually attributed to the Flemish mathematician Simon Stevin (1548–1620), sometimes called Stevinus. He used decimal representation of whole numbers and fractions for merging all types of known at that time numbers, such as whole numbers, positive rational numbers, negative numbers, and known at that time irrational numbers. Besides, Stevin was active in a many fields of science, mathematics and engineering, both theoretical and practical.

However, rigorous construction of *real numbers* was done much later by Meray, Weierstrass, Dedekind and Cantor in the 19th century as it will be described further.

The first use of the term *real* to describe Stevin numbers seems to date back to René Descartes, who distinguished between real and imaginary roots of polynomials (Katz and Katz, 2012). Thus, this term *real number* was used as a way of contrasting real numbers, which were thought of, ever since Stevin, as measuring numbers, on the one hand, and imaginary ones, on the other. Thus, the origin of the term *imaginary number* is often ascribed to Descartes. Gradually the meaning of the term real number has shifted, to a point where today it is used in the sense of "genuine, objective, true number".

The set of all real numbers consists of two parts: rational numbers and irrational numbers. In turn, the set of all irrational numbers also consists of two parts: algebraic numbers, which are solutions of algebraic equations with rational coefficients, and transcendental numbers, which are not solutions of algebraic equation with rational coefficients. As the result, transcendental numbers are more complex than algebraic numbers.

It is assumed that Wilhelm Leibniz was the first mathematician who used the term *transcendental number* (cf. Ribenboim, 2000). Leonhard Euler used another term *too irrational* to denote transcendental numbers (Euler, 1988).

It is also worth noting that later studies in higher arithmetic (number theory) brought mathematicians to the creation of the theory of algebraic numbers.

Parallel to the expansion of the scope of numbers mathematicians continued the development of their arithmetic. As we saw, even in ancient and pure arithmetics. While the first was considered only as an educational and practical field, pure arithmetic has preserved its research status till our days acquiring the name *higher arithmetic* or as it was called later, *number theory.* For instance, the British mathematician and historian Thomas Little Heath (1861–1940) explained:

"By arithmetic, Plato meant, not arithmetic in our sense, but the science which considers numbers in themselves, in other words, what we mean by the Theory of Numbers." (Heath, 1921)

The main concern of number theory is properties of whole numbers although mathematicians who work in number theory also study properties of integer, rational and algebraic real numbers, that is, real numbers, which are solutions of polynomial equations with whole coefficients.

Because the goal of number theory (higher arithmetic) is the acquisition of information and knowledge about numbers, its beginning can be traced to Pythagoras, who studied different types of natural numbers: odd and even numbers, prime and compound numbers, triangular numbers, and perfect numbers. After Pythagoras, Greek mathematicians continued developing number theory. For instance, books 7, 8 and 9 from the famous *Elements* of Euclid contain elements of number theory represented by 102 propositions. These results include: the demonstration of infinity of prime numbers, the description of what is now called the Euclidean algorithm, and the Fundamental Theorem of Arithmetic, which was formulated and proved in the archaic terms of the ancient Greek mathematics. In terms of modern number theory, this theorem was proved by Gauss (1801). Later this theorem was extended to rings and fields of algebraic numbers such as the ring of Gaussian integers or the ring of Eisenstein integers (Gauss, 1831; Conway and Guy, 1996) as well as to some classes of prearithmetics (Burgin, 2018c).

One more example of the number theoretical results in Euclid's *Elements* is Euclid's theorem that the number $2^n(2^{n+l} - 1)$ is perfect if the second factor is a prime number.

Later many outstanding mathematicians contributed to the vast area of number theory while arithmetics of other classes of numbers, such as complex numbers, were mostly treated as applied areas. Indeed, all numbers have very diverse applications. For instance, even imaginary and complex numbers found utilization in varied fields of physics.

The notable French mathematician Francois Viète (1541–1603), whose main field was algebra and who coined the term coefficient, also contributed to arithmetic by extending Diophantus's results to continuous quantities. However, the main contribution of Viète to mathematics was the development of the mathematical notation. In particular, he suggested using letters as variables, or more exactly, as names of variables.

Francois Viète was born at Fontenay-le-Comte, France, and studied law at the University of Poitiers. He started his career as a counselor at Rennes and later the privy counselor to the King of France. However, due to religious differences his political enemies banished him from the court. Viète returned to Fontenay-le-Comte and dedicated his time to mathematical studies. After some time, the political's Viète situation changed and continued his work for French Kings. From time to time, gave lectures on mathematics but his main application of his mathematical knowledge was deciphering letters of the Spanish King.

However, the origin of the contemporary number theory is usually associated with the famous French mathematician Pierre de Fermat (1607–1665), who was a lawyer by his profession. He never published his mathematical works and in particular, his results in number theory are contained almost completely in his letters to other mathematicians as well as in his personal notes. Fermat was not a professional mathematician either by his education or by his jobs as he was a commissioner of requests and later a King's counselor in the parliament of Toulouse. Fermat lived temperately and quietly all his life avoiding profitless disputes (Bell, 1937). This made possible to use his talent for the development of mathematics in general and number theory (arithmetic) in particular.

The most famous (or as some think, notorious) contribution of Fermat to number theory was Fermat's Last Theorem, also called Fermat's conjecture. Namely, around 1637, he wrote in the margin of Diophantus' Arithmetica that the equation $x^n = y^n + z^n$ cannot have positive integer solutions for any natural number n greater than 2. Although Fermat claimed to have a general proof of his conjecture, no proof by him has ever been found and his conjecture stood unsolved in mathematics for the following three and a half centuries.

It is interesting that Gauss wrote about Fermat's conjecture that *"with luck, its solution might perhaps turn up as a by-product of a wide extension of the higher arithmetic..."* (cf., Ribenboim, 1985). Indeed, almost two hundred years passed and Fermat's Last Theorem was proved by the English mathematician Andrew Wiles with the help of his student Richard Taylor using the theory of modular forms and elliptic curves (Taylor and Wiles, 1995; Wiles, 1995).

Other important results of Fermat in number theory include:

- Fermat's theorem stating that for any natural number n and any prime number p, the number $n^p - n$ is divisible by p; the first proof of Fermat's theorem was given by Leinbniz.

- Every prime number of the form $4m + 1$ is the unique sum of two squares, while no prime number of the form $4m - 1$ has such form; the first proof of this theorem was given by Euler in 1749.
- Every whole number can be expressed as the sum of four or fewer squares; the proof of this theorem was given by the prominent German mathematician Carl Gustav Jacob Jacobi (1804–1851).

The next important mathematician who contributed a lot to number theory was the great Leonhard Euler (1707–1783), who was born in Basel, Switzerland and worked both in Russian and Prussian Academies of Sciences. He was also a physicist, geographer, astronomer, logician and engineer writing mathematical works in all these areas. He was the most prolific mathematician of all times. More than 800 of his papers and books have been published. According to André Weil,

"No mathematician ever attained such a position of undisputed leadership in all branches of mathematics, pure and applied, as Euler did for the best part of the eighteenth century." (Weil, 1984)

Being a talented boy, Leonhard Euler entered the University of Basel at the age of 14. In two year, he completed his Master's degree in philosophy writing about philosophical ideas of Descartes and Newton.

Leonard's father was a Protestant minister and wanted his son to follow him into the church. However, although he was a devout Christian, Euler could not resist his great mathematical talent and switched to studying mathematics at the University of Basel under the guidance of the famous Swiss mathematician Johann Bernoulli (1667–1748).

After graduation, Euler was offered a position in the St Petersburg Academy of Sciences in Russia due to the recommendation of Bernoulli. Very soon, he became the professor of physics and three years later, was appointed the senior chair of mathematics at the Academy. Euler was working all time writing one paper after another and becoming famous after he won the Grand Prize of the Paris Academy in 1738 and 1740.

When the situation in Russia deteriorated, Euler accepted the invitation of Frederick the Great, the King of Prussia, and went to Berlin to work in the Academy of Science as the Director of mathematics. Even there, Euler continued publishing his works in St Petersburg and receiving part of his salary from Russia. In Berlin, Euler spent 25 years writing several books and around 380 articles and serving as an advisor to the government on insurance, annuities, pensions, artillery and state lotteries.

In 1766, Euler returned to Russia where he continued his highly productive work in the St Petersburg Academy of Sciences and writing almost half of his entire works in spite of becoming almost entirely blind.

Let us look at some of Euler's results in number theory

Euler determined the number of natural numbers less than a given natural number n and prime to n defining in such a way the function, which is now called Euler's φ–function.

Euler also proved some particular cases of the of Fermat's Last Theorem, which was considered one of the most attractive and at the same time, hard results of numbers theory. Being the greatest mathematician of his time, Euler wrote to his colleague — the notable German–Russian mathematician Christian Goldbach (1690–1764) that "the general case [of Fermat's Last Theorem, MB] still seemed quite unapproachable" (Edwards, 1977). In addition, Euler proved that every prime number of the form $4n+1$ is a sum of two squares.

Mathematicians assume there are five the most important numbers:

Number 1 was used as the generator of all natural numbers and through them, of all other numbers.

Number 0 has very specific properties (for example, it is neither positive nor negative), plays the role of a place-holder in positional numerical system, and mathematicians did not discover it for a long time.

Number π (pi) represents an invariant property of circles, namely, that the ratio of the length of any circle to its diameter is always equal to π. It is interesting that π represents other invariant properties. For instance, the ratio of the area of any circle to the square of its radius is also equal to π.

Number i is the generator of all complex numbers with the help of real numbers. It was Euler who introduced the notation $i = \sqrt{-1}$.

The last of these regal numbers with eminent importance is number e, which was discovered by the famous Swiss mathematician Jacob Bernoulli (1655–1705) around 1683 when he examined compound interest. However, e was implicitly used as the base for natural logarithms by the renowned Scottish scholar John Napier (1550–1617) at the beginning of the 17th century (Napier, 1614).

Euler suggested the name e to this number and found many important applications for it (McCartin, 2006). That is why e is often called Euler's number. It is approximately equal to 2.71828.

In addition, Euler derived one of the most beautiful formulas in mathematics

$$e^{i\pi} + 1 = 0$$

Now it is called Euler's identity.

In his book on the history of number theory, André Weil describes the main topics in Euler's work on numbers as follows (Weil, 1984):

(1) Fermat's Last Theorem and the multiplicative group of integers modulo n;
(2) sums of squares and quadratic forms;
(3) Diophantine equations of degree 2, of genus 1, and others;
(4) elliptic integrals;
(5) continued fractions and Pell's equation, which has the form $x^2 - ny^2 = 1$, where n is a given positive non-square integer and for which integer solutions are sought for x and y;
(6) summation of series related to prime numbers;
(7) "partitio numerorum" and formal power series;
(8) prime divisors of quadratic forms;
(9) large primes.

Important contributions to arithmetic in general and higher arithmetic, in particular, were made by the prominent Italian–French–Prussian mathematician Joseph-Louis Lagrange (1736–1813) and important French mathematician Adrien-Marie Legendre (1752–1833).

Louis Lagrange was born in Turin, Sardinia-Piedmont (now Italy), and studied at the College of Turin. Being 19 years old, Lagrange was appointed professor of mathematics at the Royal Artillery School in Turin and later succeeded Leonard Euler as the Director of Mathematics at the Berlin Academy. When he 51 years old, Lagrange left Berlin to be a member of the Académie des Sciences in Paris and work later as a professor of analysis at the École Polytechnique becoming a Senator of France in 1799.

Lagrange was one of the creators of the calculus of variations, developed foundations of group theory, and transformed Newtonian mechanics into Lagrangian mechanics developing a formalism very useful for classical and quantum field theories.

According to Weil, Lagrange worked in the following areas of arithmetics (Weil, 1984):

(1) The theory of Pell's equation where he proved Fermat's assertion about the existence of infinitely many solutions for this equation, and building an algorithm to obtain them.

(2) The theory of real numbers where he proved that all quadratic irrationals have periodic continued fractions.

(3) Solution in integers of the equations of degree 2 in two unknowns, as well as of the equations $z^2 = Ax^2 + By^2$.

(4) Fermat's Last theorem, which he proved for the sums of four squares;

(5) A study of binary quadratic forms.

(6) The study of the Diophantine equations originated with a problem of Fermat on Pythagorean triangles.

The main publication of Lagrange in number theory (higher arithmetic) was the monograph *Recherches d'Arithmétique* (Lagrange, 1773/1775).

Adrien-Marie Legendre was born in Paris, France, studied at the College Mazarin in Paris and taught at Ècole Militaire. Later he worked at the Académie des Sciences, which was closed after the French Revolution and then reopened as the Institute National des Sciences. Legendre made contributions to the theory of functions, elliptic integrals, and differential equations.

Legendre's work in number theory included (Weil, 1984):

(a) The discovery of the quadratic reciprocity law although he did not prove it.

(b) The discovery of the existence of infinitely many primes in arithmetic progressions, with difference and first term relatively prime.

(c) The discovery that every quadratic form $ax^2 + bxy + cy^2$ represents infinitely many primes, provided a, b, and c have no common divisor.

The main publication of Legendre in number theory (higher arithmetic) was the monograph *Théorie des Nombres*, in which he sought to give a comprehensive account of number theory, as he saw it at the time (Legendre, 1830).

With its further development in the 19th century, higher arithmetic became very popular in the mathematical community acquiring the name number theory. The classical treatise in this area *Disquisitiones Arithmeticae* (Arithmetical Investigations) was written by the great German mathematician Johan Carl Friedrich Gauss (1777–1855) when he was only 21 years old. In this formidable work, he presented numerous brilliant and at the same time, precise methods that, while not always obtained by him, formed and systematized number theory as the field in arithmetic. In this work, Gauss established this field by not only giving his new results but also formalizing previously scattered and mostly informal methods and results, providing original answers to some important problems in arithmetic, developing the theory of factorizations, and creating the landscape for future researchers. The cornerstone result of Disquistiones is the rigorous proof of the everlasting Fundamental Theorem of Arithmetic, which states that any greater than 1 natural number is either prime or is the product of prime numbers, which is unique up to the order of prime numbers.

Besides, Gauss made the first methodical study of modular arithmetic, which had been utilized before without rigorous foundation and which later was successfully applied in number theory, abstract algebra, computer science, cryptography, and even in visual and musical art.

Leopold Kronecker, mentioning Gauss's celebrated results, pointed out that many of the most fundamental constants of geometry and physics had, in their basic definitions, purely arithmetical properties (Ueno, 2003).

In his book, Gauss studied the integer solutions of quadratic polynomials in two variables giving the first proof of the law of quadratic reciprocity, which described important properties of prime numbers and which before was only conjectured by Leonard Euler and the Adrien-Marie Legendre. These and other results of Gauss formed algebraic number theory, the roots of which can be traced to the famous Greek mathematician Diophantus (ca. 3rd century C.E.).

Before this, in 1797, Gauss proved the so-called Fundamental Theorem of Algebra, which describes an important property of the arithmetic of complex numbers. Namely, this theorem states that any polynomial with one variable has as many zeroes (roots) in complex numbers as it is the degree of this polynomial. This means that the field of complex numbers is algebraically closed.

Historians of Mathematics discovered that much earlier, the statement (but without a proof) of the Fundamental Theorem of Algebra was published by the Holland mathematician Albert Girard (1595–1632) in Girard (1629). Later attempts to prove this theorem were made by d'Alembert (1746), Euler (1749), de Foncenex (1759), Lagrange (1772), and Laplace (1795), who suggested incomplete proofs.

Important results in number theory were obtained by the talented German mathematician Ferdinand Gotthold Max Eisenstein (1823–1852). He did research in a variety of fields including quadratic forms and cubic forms, the reciprocity theorem for cubic residues, quadratic partition of prime numbers and reciprocity laws. He proved several results that avoided even the great Gauss, who was of a very high opinion of Eisenstein saying that the three most brilliant mathematicians of all time were Archimedes, Newton, and Eisenstein (cf., Bell, 1937).

The history of higher arithmetic (number theory) shows that for a long time, it was not separated from arithmetic as the entire discipline in its study of properties of natural numbers. Even when the separation from the elementary arithmetic took place, number theorists continued their studies of properties of natural numbers. However, in the 19th century mathematicians started studying arithmetics (fields) of algebraic numbers creating *algebraic number theory*. In a similar way, the new area called *non-Diophantine number theory* emerged in the 21st century (Burgin, 2018c).

The principal goal of higher arithmetic (number theory) is the study of the properties of whole and in some cases integer numbers. At the same time, there were much more numbers, and mathematicians continued exploring properties of existing classes of numbers, of individual numbers, such as π and e, and introducing new types of numbers.

An important result in arithmetic was obtained by the Swiss polymath Johann Heinrich Lambert (1728–1777), who proved that numbers π and e are irrational (Lambert, 1770). Extending these results, the French mathematician Charles Hermite (1822–1901) proved the transcendence of number e (Hermite, 1873) and the German mathematician Carl Louis Ferdinand von Lindeman (1852–1939) proved transcendence of number π (Lindeman, 1882).

However, till the middle of the 19th century, there was no rigorous mathematical theory of real numbers. The principal step in this direction was made by the famous French mathematician, engineer, and physicist Augustin-Louis Cauchy (1789–1857), who represented irrational numbers by limits of specific sequences of rational numbers (Cauchy, 1821). However, Cauchy did not define how to perform operations with these representations of numbers and thus, he did not create a rigorous foundation for the arithmetic of real numbers. However, without this arithmetic, the theory of real numbers was essentially incomplete.

Cauchy was born in Paris, France, and studied at the École Polytechnique and the engineering school École des Ponts et Chaussées. After graduation, he worked as an engineer. Later Cauchy taught at the École Polytechnique, the Collège de France, and the University of Turin, Italy, also conducting research at the Académie des Sciences in Paris.

In 1830s, an approach to building a theory of real numbers based on logical foundation was suggested by the Bohemian mathematician, logician, philosopher and theologian Bernard Placidus Johann Nepomuk Bolzano (1781–1848), who began setting up rigor into mathematics in general and mathematical analysis, in particular (Bolzano, 1975). He applied the exhaustions method and the concepts of the exact least upper bound, and the sequence convergence criterion, which were introduced in Bolzano (1817), in his attempt to develop a theory of a real number in his manuscript entitled *Theory of Values* (*Einleitung zur Größenlehre*) with the goal of putting the calculus on purely arithmetical foundations. In this work, Bolzano introduced the concepts of a measurable number, equality and inequality relations, infinitely great and infinitely small numbers asserting completeness of a space of real numbers (Bolzano, 1975). Because of this, Felix Klein called Bolzano "the father of arithmetization" assuming that arithmetic was the foundation of mathematics (Boyer, 1985).

Bolzano was born in Prague, Austro-Hungary (now Czech Republic) and studied at the University of Prague becoming an ordained priest and the professor of religion and philosophy at this university. He was even elected Dean of the Philosophical Faculty but his progressive ideas alienated other faculty and church leaders. As a result, he was removed from his positions and exiled to the neighborhood of Prague where he continued writing his mathematical, social, philosophical and religious works but prohibited to teach, preach, and publish.

If we understand arithmetization as the process of extending utilization of numbers and arithmetic to various mathematical and non-mathematical areas essentially transforming those areas, it becomes clear that "Arithmetization is doubtlessly the most dominating principle in the history of mathematics . . . " (Milier, 1925)

An important case in this process was the arithmetization of trigonometry by changing the definitions of the elementary trigonometric functions. With possibly a few isolated exceptions these functions were dealt with in the form of line segments (Milier, 1925). Arithmetization of trigonometry converted them to numerical functions of a single variable.

Arithmetization of the calculus demanded utilization of the arithmetic of real numbers, and it took several centuries for mathematicians to construct rigorous mathematical representation of real numbers developing a complete and consistent theory. It was done in the 19th century by several mathematicians. The first was the French mathematician Hugues Charles Robert Méray (1835–1911), who in 1869 published the theory of a real numbers and operations with them based on their representation by Cauchy sequences, i.e., converging in some ideal sense sequences of rational numbers (Méray, 1869). Charles Méray studied at École Normale Supérieure and taught at the Lycée of St Quentin, University of Lyon, and University of Dijon. However, such issues as his complicated language, inelegant terms as well as his detachment from the mathematical community were the cause of the situation in which his theory was ignored by other mathematicians. In addition, French mathematicians were not very interested in rigorous definition preferring intuitive constructions.

That is why rigorous definitions and constructions of real numbers and operations with them are usually attributed to other mathematicians whose theories were published in 1872. Note that all of them lived and worked in Germany. There were three main approaches to this problem. One was suggested by the outstanding German mathematician Karl Theodor Wilhelm Weierstrass (1815–1897), who represented real numbers by converging sequences of decimal or binary digits. For instance, the number one and one half was represented by the finite sequence 1.5 and the number one-third was represented by the infinite sequence 0.33333... However, these ideas of Weierstrass were published by another German mathematician Heinrich Eduard Heine (1821–1881), who gave their first systematic presentation attributing their origin to Weierstrass (Heine, 1872).

Weierstrass also made contributions to higher arithmetic (number theory) and other mathematical fields. For instance, he proved what is now called Lindemann–Weierstrass theorem, which allows finding interesting classes of transcendental numbers (Weierstrass, 1885). In his 1863 lectures, Weierstrass proved that complex numbers form the only commutative algebraic extension of real numbers.

Karl Weierstrass was born in Ostenfelde, Westphalia, and studied at the University of Bonn to prepare for the government position. As this direction contradicted his interests in mathematics, which he studied privately, he left the university without a degree. That is why he taught mathematics, physics, botany, and gymnastics at schools in West Prussia and Braunsberg until the University of Königsberg gave him an honorary doctor's degree. Soon after this Weierstrass became a chair at the Technical University of Berlin becoming later a professor at the Friedrich-Wilhelms-Universität Berlin.

Another approach to real numbers was developed by the outstanding German mathematician Julius Wilhelm Richard Dedekind (1831–1916), who represented real numbers by cuts of rational numbers (Dedekind, 1872). Although Dedekind's definition is still used in courses of analysis as logically and categorically flawless, Georg Cantor was doubtful about its worth for analysis. Nevertheless, in the 20th century, Dedekind's approach was used by John Conway for creation of surreal numbers, which was a far reaching generalization of Cantor's theory of transfinite numbers comprising a variety of different classes of numbers (Knuth, 1974; Conway, 1976).

Richard Dedekind was born in Braunschweig, Germany, and studied at the University of Göttingen under the supervision of Gauss. Later Dedekind worked at this university as well as at Zurich Polytechnic and Technical High School in Braunschweig. There he developed his approach to the definition of real numbers by cuts.

The third approach, which was similar to what Meray did, was elaborated by the great German mathematician Georg Cantor (1845–1918), who represented real numbers by Cauchy sequences, i.e., contracting sequences of rational numbers (Cantor, 1872). This construction provided solid foundations for the "arithmetization of analysis", which is described further in this chapter. Georg Cantor also proved that the set of algebraic irrational numbers is countable while the set of real numbers and, therefore, the set of all transcendental numbers is uncountable (Cantor, 1874).

The rigorous formation of the totality of all real numbers brought forth the concept of a *number line*. Although it was formed in works of Cantor and Dedekind, the term itself emerged much later. At first, it was known as a *number scale* and only later (since 1912) it acquired the name a *number line*.

Real numbers had such an affluent structure that their generalizations bifurcated into several directions:

- transition from real numbers to complex numbers and further to hypercomplex numbers by Herman Grassmann, William Hamilton, John Graves, Arthur Cayley, James Cockle, Richard Dedekind, Karl Weierstrass, and Corrado Segre;
- introduction of infinitesimals in the form of magnitudes by Otto Stolz, Rodolfo Bettazzi, Giuseppe Veronese, and Paul Du Bois-Reymond;
- construction of transfinite real numbers by Georg Cantor;
- transition from real numbers to hyperreal numbers by Abraham Robinson and further to surreal numbers by John Conway;
- transition from real numbers to real hypernumbers by Mark Burgin.

All these kinds of numbers are considered further while now we once more discuss the development of arithmetic.

There have been many inventions and developments to make the performance of the arithmetic operations easier. One of the most important was the invention of logarithms by the Scottish mathematician, physicist, and astronomer 8th Laird of Merchiston John Napier of Merchiston (1550–1617) in 1614 (Napier, 1614). His main argument was that addition of numbers in the decimal representation is much simpler than multiplication of the same numbers. To achieve this goal, Napier introduced the *logarithmic function*, which taking a number, assigns to it its *logarithm* allowing changing multiplication to addition and by applying the exponential function, obtaining with sufficient accuracy the result of multiplication. Logarithms also provided substitution of division by subtraction, which is easier than division. The word logarithm was formed from the Greek words *arithmos* meaning *number* and *logos* meaning *reason* or *word*.

An interesting peculiarity of the logarithmic approach is that it uses a special case of the basic relation from the theory of non-Diophantine arithmetics, which is called projectivity (Burgin, 1977, 1997, 2019).

Based on this discovery, William Oughted (1574–1660) invented a *logarithmic slide rule* around 1622. His invention was based on calculating device, which was elaborated in 1620 by English mathematician, astronomer and clergyman Edmund Gunter (1581–1626), which consisted of a single logarithmic scale. It was a mechanical analog calculator, which suggested easy multiplication and division of decimal numbers. Since then, many forms of ready reckoners, mechanical and electronic calculators, and computers have been invented. However, a logarithmic slide rule remained the basic computing device for engineers till the middle of the 20th century.

In addition to logarithms, Napier introduced and studied the so-called *location arithmetic* as an additive (non-positional) binary numeral system aimed at amplification of computation techniques (Napier, 1990).

It is necessary to remark that the Swiss clockmaker, maker of astronomical instruments and mathematician Joost Bürgi (1552–1632) independently invented a system of logarithms between 1603 and 1611 although he did not develop a clear notion of the logarithmic function.

Writing even a brief history of numbers and arithmetic, it is necessary to indicate that different mathematical fields were reduced to arithmetic. Let us look how this happened. In essence, numbers and arithmetic have been and will be applied in mathematics and beyond through the whole history of human civilization. However, some of these applications had a great impact on mathematics and science. One of such cardinal transformations of mathematical knowledge was invention/discovery of *analytic geometry* by the great French mathematician and philosopher René Descartes (1596–1650) and the great French mathematician and lawyer Pierre de Fermat (1607–1665). Descartes published his work *La Géométrie* on analytic geometry as an appendix to *Discours de la méthode* (Descartes, 1637). Contrary to this, Fermat did not publish in his lifetime his manuscript *Ad locos planos et solidos isagoge* (*Introduction to Plane and Solid Loci*) on analytic geometry although it was circulating in Paris in 1637 and later.

René Descartes was born in La Haye en Touraine, France. He began his education at a Jesuit College going after this to the University of Poitiers to become a lawyer according to his father's wishes. However, after university Descartes went to serve in the army. Still in army, he started research in mathematics and philosophy. At the age of 24, Descartes returned to Paris and having enough means from his property was able to continue his philosophical and mathematical explorations in comfortable conditions. Later he can to the Netherlands and for some time studied at the University of Franeker and Leiden University. He wrote his main works in mathematics, epistemology, metaphysics, physics, and cosmology in the Netherlands becoming one of the founders of the modern philosophy.

The essence of analytic geometry is the reduction of geometry to arithmetic and its further development — algebra, which grew up from arithmetic. This outstanding result has a long history starting with Pythagoras, who, according to legends, proclaimed "Everything is number" implying that it was possible to reduce everything including geometry to arithmetic. However, as we already know, it was demonstrated that it was impossible to represent some simple geometrical objects by numbers known at that time. As the result, mathematicians started to treat geometry as a more fundamental area in comparison with arithmetic using geometrical techniques for solving arithmetical and algebraic problems. However, two millennia later, Pythagoras was vindicated when Descartes and Fermat reduced geometry to arithmetic constructing arithmetization of geometry.

Geometrical objects from Euclidean geometry are represented by objects in Euclidean spaces, elements of which are finite ordered sets of real numbers. This representation was inherited by the more abstract mathematical field called *topology*. Namely, the important topological structures called *manifolds* are constructed from parts of Euclidean spaces and thus, are also based on numbers and arithmetic.

Negative coordinates in Euclidean spaces were correctly employed by John Wallis in 1656 although not all mathematicians accepted this innovation being opposed in general to negative numbers. Besides, an additional transformation of a geometrical theory into arithmetic was done by John Wallis, who constructed a numerical form of the geometrical theory of indivisibles created by Bonaventura Cavalieri.

Bonaventura Francesco Cavalieri (ca. 1598–1647) was an Italian mathematician and a Jesuate monk. Starting with the works of Archimedes, Cavalieri developed a new geometrical approach to finding volumes and areas of geometrical figures (Cavalieri, 1635). It was called the method of indivisibles and was an important step towards the calculus. To find the area, Cavalieri used "all the segments" of the considered figure, and to find the volume, Cavalieri employed "all the parts of the planes" in the considered figure. Those segments and parts of the planes were called indivisibles. During his life, Cavalieri was the chair of mathematics at Bologna and the prior of the Jesuati convent in Bologna. In addition to mathematics, Cavalieri worked on the problems of optics and motion.

John Wallis (1616–1703) was an English clergyman and the most influential English mathematician prior to Newton. His work on arithmetization of the indivisibles contributed to the development of the calculus. He also served as chief cryptographer for Parliament and the royal court.

However, the most radical approach was suggested by the reputed German mathematician Leopold Kronecker (1825–1891), who started the program *arithmetization of analysis*. He called his program "*General Arithmetic*" (*Allgemeine Arithmetik*) assuming that arithmetic was the building brick of the entire edifice of mathematics and included algebra and analysis (Kronecker, 1887/1889; 1901; Ueno, 2003). Note that at that time, all algebra was arithmetic algebra while abstract algebra came into being only later.

The initial reason for Kronecker was related to the calculus because until the second half of the 19th century, there were two pillars of the calculus — the discrete tools of arithmetic and the continuous apparatus of geometry. The primary goal of arithmetization was to develop the calculus exclusively in the context of the natural numbers. Important steps relevant to this program were the definitions of real numbers in terms of rational numbers as the latter are built from integers, rigorous utilization of real numbers instead of geometrical magnitudes, and the constructive epsilon-delta definition of limits in the spaces of numbers and functions.

Leopold Kronecker was born Liegnitz, Prussia (now Legnica, Poland) and studied at the University of Berlin, University of Bonn, and University of Breslau. However, after graduation, he focused on business as member of wealthy family. Making enough money, Kronecker returned to Berlin to do research in mathematics as an independent scholar becoming a member of the Berlin Academy of Sciences due to his important results in such mathematical fields as number theory, elliptic functions, and algebra. At the end of his life, he became the chair at the University of Berlin.

The program of arithmetization was supported by Wilhelm Weierstrass, who argued the geometric foundations of calculus were not solid enough for rigorous work. To achieve this goal, the German mathematician Hermann Hankel (1839–1873) suggested creation of a *universal arithmetic* such that belonged to the purely intellectual mathematics detached from all perceptions. In this context, real numbers became "intellectual structures" rather than "intuitively given magnitudes inherited from Euclid's geometry" (Hankel, 1867).

Hermann Hankel was born in Halle, Germany, and studied at Leipzig University, the University of Göttingen and University of Berlin. He taught at Leipzig University, the University of Erlangen-Nurenberg and University of Tübingen. He obtained new properties of real numbers, complex numbers and quaternions. For instance, he proved the theorem, which said that the only distributive multiplication in the set of all real numbers as the extension of the multiplication of positive real numbers had to satisfy the law of signs.

The program of arithmetization of analysis had essential influence on the work of many mathematicians and later was transformed into the more comprehensive program of arithmetization of the whole mathematics (Hatcher, 2000; Ueno, 2003). It implied the study of various branches of mathematics by methods that utilize only the basic concepts and operations of arithmetic.

In the 20th century, arithmetization of analysis brought forth the new field in mathematics called *reverse mathematics*, which analyzes what axioms are needed to prove this or that theorem (Stillwell, 2018).

Let us continue the history of emergence and adaptation of new kinds of numbers in arithmetic. Although the most advanced mathematicians continued using complex numbers more and more, the Norwegian surveyor Caspar Wessel (1745–1818) was the first one to obtain and publish the earliest accurate representation for complex numbers as points in a plane, or two-dimensional vectors in the contemporary mathematical lexicon. In 1797, he presented his result to the Royal Danish Academy of Sciences publishing it in the Academy's Memoires of 1799 (Wessel, 1799). However, this paper, written in Danish, went unnoticed until 1897.

Later than Wessel, the Parisian bookkeeper Jean-Robert Argand (1768–1822) independently developed an accurate representation for complex numbers. Historians of mathematics do not know whether he had mathematical training but he produced a mathematical pamphlet, run by a private press in small print, although he did not include his name in the title page (Argand, 1806). One copy ended up in the hands of the well-known mathematician Adrien-Marie Legendre (1752–1833) who in turn mentioned it in a letter to the professor of mathematics Joseph François Français (1768–1810). After François Français died, his brother Jaques Frédéric Français (1775–1833), who also was a mathematician working as a professor of military art, inherited his papers with the Legendre's letter describing Argand's mathematical results without the name of Argand. Based on this letter, Jaques Francais published an article in 1813 in the *Annales de Mathémathiques* describing the basics of complex numbers and urging the unknown author to come forward. Argand learned of this and his reply appeared in the next issue of the journal.

It is interesting that in 1847, the famous French mathematician Augustin-Louis Cauchy (1789–1857) suggested an essentially different representation of complex numbers as elements of the algebraic factorization of the field of real number polynomials with one variable. In this representation, he did not need the imaginary unit i suggesting its elimination from mathematics.

Gauss knew the geometric representation of complex numbers as vectors since 1796. However, it went unpublished until 1831, when he explained that complex numbers form points in the plane in the same way as real numbers form points in the line. Besides, Gauss introduced the term *complex number* trying to change the terminology in that area. He wrote:

"If this subject has hitherto been considered from the wrong viewpoint and thus enveloped in mystery and surrounded by darkness, it is largely an unsuitable terminology which should be blamed. Had +1, −1 and $\sqrt{-1}$, instead of being called positive, negative and imaginary (or worse still, impossible) unity, been given the names say, of direct, inverse and lateral unity, there would hardly have been any scope for such obscurity." (Gauss, 1831)

Carl Friedrich Gauss was born in Brunswick, Duchy of Brunswick (now Germany) being a mathematical prodigy from the very young age. He studied at Göttingen University although he left without a diploma. Gauss preferred doing research to teaching that is why he did not worked at universities although he had a few students. At first, he had a stipend from the Duke of Brunswick and was able only to do research in mathematics, which was very successful. After losing this stipend, Gauss accepted the position of the director of the Göttingen observatory.

In addition to work in the observatory and his mathematical investigations, Gauss carried out a geodesic survey of the state of Hanover inventing for this work the heliotrope — an instrument that worked by reflecting the Sun's rays with a system of mirrors and a small telescope.

Being a hard worker, Gauss essentially contributed to many fields in mathematics and science. At the same time, being an ardent perfectionist, Gauss often refused publishing his works that did not accord with his high standards of being complete and above any criticism.

The notable German mathematician Felix Klein (1849–1925) described the changes in the attitude of the European mathematicians to complex numbers by the demonstration of what was written about them in three mathematical books at those times (Klein, 1924). Note that at that time, the highest level of mathematics was achieved in Europe.

Felix Klein was born in Düsseldorf, Prussia (now Germany) and studied at the University of Bonn. He taught at the University of Erlangen-Nurenberg, Technische Hochschule Munchen, Leipzig University, and the University of Göttingen. Klein made essential contribution to function theory and non-Euclidean geometry establishing connections between geometry and group theory.

In the first textbook *Mathematische Anfangsgrunde* analyzed by Klein, which was written by the German mathematician and epigrammatist, i.e., a coiner of epigrams, Abraham Gotthelf Kästner (1719–1800) and published in 1786, it is stated:

"*Whoever demands the extraction of an even root of a 'denied' quantity (one said 'denied', then, instead of 'negative'), demands an impossibility, for there is no 'denied' quantity which would be such a power.*" (cf., Klein, 1924).

However, as Klein writes, after a dozen of pages, "one finds: '*Such roots are called impossible or imaginary*', and, without much investigation as to justification, one proceeds quietly to operate with them as with ordinary numbers, notwithstanding their existence has just been disputed as though, so to speak, the meaningless became suddenly usable through receiving a name. You recognize here a reflex of Leibniz's point of view, according to which, imaginary numbers were really something quite foolish but they led, nevertheless, in some incomprehensible way, to useful results." (Klein, 1924).

In the second book analyzed by Klein, which was published by the German mathematician Martin Ohm (1792–1872) in 1822, the clear extension of the number system is given. The author explains that, just like with negative numbers, square root of negative 1 must be added to the real numbers as a new number. However, even his book lacked the geometric interpretation of complex numbers (Ohm, 1822).

Finally, in the third textbook analyzed by Klein, which was written by Bardeys, the principle of extension of real numbers to complex numbers came to the fore and the geometric interpretation was properly explained (Bardeys, 1907).

For a long time, only basic classes of numbers, such as natural, whole or positive rational numbers, were delineated, while their primary subclasses, such as prime numbers or even numbers as subclasses of the class of natural numbers, were differentiated. Then, as we already know, it took centuries to construct and accept all integer numbers, all rational numbers, real and complex numbers. However, the 19th century witnessed an explosion of diverse number systems. Moreover, this process persisted in the 20th century and continues now.

It is possible to discern two trends in this process. In the *internal trend*, different subclasses of real numbers were introduced and studied. The main of these subclasses are as follows:

- *irrational numbers*, which are real numbers that are not rational, i.e., all real numbers consist of rational and irrational numbers, for example, $\sqrt{2}$ and $\sqrt{3}$ are irrational numbers;
- *algebraic numbers*, which are real numbers that are roots of polynomials with rational coefficients, for example, $\sqrt{6}$ and $\sqrt{7}$ are algebraic numbers;
- *algebraic integer numbers*, or simply, *algebraic integers*, which are real numbers that are roots of polynomials with integer coefficients and the leading coefficient equal to 1, for example, $\sqrt{5}$ and $\sqrt{10}$ are algebraic integers;
- *transcendental numbers*, which are irrational numbers that are not algebraic, for example, it is proved that and e are transcendental numbers.

The *external trend* was aimed at creation of new kinds of numbers. Moreover, in addition to the existing diversity of number systems, mathematicians introduced and studied various generalizations of numbers and their arithmetics.

It is possible to discern three directions in these developments:

- The *algebraic direction* brought forth a variety of algebraic systems by more and more abstracting number systems, which existed at that time.
- The *geometric direction* advanced from numbers to magnitudes developing theories of magnitudes, treating numbers as some kinds of magnitudes, and finally coming to numerical representation of magnitudes.
- The *transfinite direction* moved on from finite numbers, such as natural and whole numbers, to infinite numbers, such as cardinal and ordinal numbers.

We start with the algebraic direction, which resulted in the birth of modern algebra. Different kinds of numbers introduced in the framework of the algebraic direction are usually unified under the name *hypercomplex numbers* because these numbers are considered as the further development of complex numbers. For instance, according to the Dutch mathematician and historian of mathematics Bartel Leendert van der Waerden (1903–1996), hypercomplex numbers are numbers with properties departing from those of the real and complex numbers (van der Waerden, 1985). However, in some cases, properties of what was called hypercomplex numbers went so far from conventional numbers that they were more adequately considered not as numbers but as elements of algebraic systems. In spite of this, some mathematicians continue calling these elements by the name *number*. For instance, Kantor and Solodovnikov define a hypercomplex number as an element of a finite-dimensional linear algebra over the real numbers that is unital and distributive but not necessarily associative (Kantor and Solodovnikov, 1989).

The first specific class of hypercomplex numbers consisted of the *quaternions* introduced by the famous Irish mathematician William Rowan Hamilton (1805–1865) in an attempt to find generalizations of complex numbers (Hamilton, 1843, 1844, 1866). Quaternions have different applications in mathematics and physics, such as calculations involving three-dimensional rotations used in three-dimensional computer graphics, computer vision, robotics, navigation, flight dynamics, crystallographic texture analysis, and orbital mechanics (Girard, 1984).

Another generalization of complex numbers, which were called the *octonions*, came next. Octonions form a non-associative extension of the quaternions. They were discovered in 1843 by an Irish jurist and mathematician John Thomas Graves (1806–1870), who was inspired by discovery of quaternions made by his friend Hamilton. Graves called these new numbers the *octaves*, and mentioned them in a letter to Hamilton dated 16 December 1843. However, the first publication of his result in (Graves, 1845) was slightly later than Arthur Cayley's article on the same type of numbers, which were called *octonions* by Cayley, who discovered them independently. That is why octonions are sometimes referred to as *Cayley numbers* while their arithmetic is called the *Cayley algebra*. The octonions form 8-dimensional algebra over real numbers. Their multiplication is neither commutative nor associative but alternative.

In a series of articles in Philosophical Magazine, English lawyer and mathematician James Cockle (1819–1895) introduced and studied one more generalization of complex numbers called the *tessarines* (Cockle, 1848; 1849; 1849a).

The Italian mathematician Corrado Segre (1863–1924), who is remembered today as a major contributor to the early development of algebraic geometry, introduced the *bicomplex numbers*, which form an algebra isomorphic to the *tessarines* (Segre, 1892).

There is one more generalization called *sedenions*. They form a 16-dimensional algebra over real numbers, in which multiplication is neither commutative, nor associative, nor alternative but only power associative.

In general, quaternions, octonions, sedenions, etc. are defined only for dimensions $2m$ for $m \geq 2$ and are obtained using the Cayley–Dickson construction.

At the same time, the outstanding German mathematician and polymath Hermann Günther Grassmann (1809–1877) developed a theory of extensive, or in contemporary terms, multidimensional magnitudes introducing in such a way the concept of an n-dimensional vector space (Grassmann, 1844). Before only magnitudes of dimensions one (linear magnitudes), two and three were studied reflecting the number of spatial dimensions. In contrast to this, Grassmann asserted that the number of possible dimensions is in fact unbounded building systems of the so-called *hypercomplex numbers*.

Hermann Grassmann was born in Stettin, Prussia (now Szczecin, Poland), and studied at the University of Berlin where he studied theology. Even without university training in mathematics, he made important discoveries in mathematics.

Hermann Grassmann was an extraordinary researcher. In addition to mathematics, he worked in physics and linguistics. In optics, Grassmann's law describes human color perception. In linguistics, Grassmann's law explains phonological processes in Ancient Greek and Sanskrit.

However, in spite of all his achievements, Hermann Grassmann worked all his adult life as a gymnasium teacher teaching mathematics, physics, German, Latin, chemistry, mineralogy, and religious studies. At the same time, his application for becoming a mathematics university professor was rejected. The main reason was that his innovative ideas and results were much more advanced than the mediocre mathematicians could understand and accept.

As the American author James Gleick wrote, *"Shallow ideas can be assimilated; ideas that require people to reorganize their picture of the world provoke hostility."* (Gleick, 1989).

Even outstanding mathematicians have problems with understanding the ideas of Hermann Grassmann. For instance, the great mathematician Carl Friedrich Gauss (1777–1855) wrote to Hermann Grassmann in December 1844:

"...in order to discern the essential core of your work it is first necessary to become familiar with your special terminology. ...however that will require of me a time free of other duties ..."

As a result, Hermann Grassmann was outside the mathematical establishment all his life while his revolutionary results were ignored.

Only by the late 1860s, his work slowly was being recognized by other mathematicians. One of the first was the German mathematician Hermann Hankel (1839–1873), a student of the great German mathematician Bernhard Riemann (1826–1866). The renowned German mathematician Felix Klein (1849–1925) learned about Grassmann from Hankel. He referred to Grassmann in his famous "Erlanger Programm" (Klein, 1872). The prominent Italian mathematician Giuseppe Peano (1858–1932) was deeply influenced by Grassmann when he elaborated axioms for vector spaces. Although Peano's axiomatic definition of a vector space (Peano, 1888) remained largely unknown, the outstanding mathematician and physicist Hermann Klaus Hugo Weyl (1885–1955) used it for the mathematical base of relativity theory. He was referring to Grassmann's results as an "epoch making work" (Weyl, 1918).

At the same time with Herman Grassmann, the well-known English mathematician Arthur Cayley (1821–1895) wrote about "analytical geometry of n dimensions" implicitly bringing in the concept of an n-dimensional vector space.

Returning to hypercomplex numbers, we find that they include various extensions of complex numbers, which were considered above such as quaternions (H), octonions (O), sedenions (S), tessarines, coquaternions, and biquaternions. All these generalized numbers form algebraic systems with specific properties. However, not all of them have sufficiently good properties as later it was proved that there were only four finite-dimensional real division algebras: the algebra of real numbers R, the algebra of complex numbers C, the algebra of the quaternions Q, and the algebra of the octonions O (Adams, 1960). It is necessary to remark that any extension of the concept of a complex number is possible only at the cost of some of the customary properties of numbers such as commutativity or associativity. For instance, multiplication of quaternions is not associative. Hypercomplex numbers of higher dimensions, such as sedenions, contain divisors of zero.

Hypercomplex numbers in the general form were also studied by the venerated German mathematicians Karl Theodor Wilhelm Weierstrass (1815–1897) and Richard Julius Wilhelm Dedekind (1831–1916), who explored properties of such numbers paying special attention to divisors of zero (Weierstrass, 1884; Dedekind, 1885, 1887).

The *dual numbers* were introduced by the English mathematician and philosopher William Kingdon Clifford (1845–1879). The dual numbers extend the real numbers by adjoining one new element ε with the property $\varepsilon^2 = 0$. In mathematics, it means that the element ε is nilpotent. Thus, a dual number has the form $z = a + b\varepsilon$ where a and b are uniquely determined real numbers. Going one step further, dual numbers form coefficients of *dual quaternions*.

Clifford was born in Exeter, England and studied at Trinity College, Cambridge. He taught at University College London. In his mathematical works, he developed the theory of geometric algebras and conjectured that gravitation might a manifestation of the underlying geometry of the space.

Another thread in the development of dual numbers brought mathematicians to *supernumbers*, also called *anticommuting numbers* or *Grassmann numbers*, which extended the concept of dual numbers to n distinct anti-commuting generators, possibly taking n to infinity.

Adding multiple commuting dimensions to supernumbers, mathematicians and physicists came to the concepts of *superspace* and *superalgebra*, which play an important role in contemporary physics as a mathematical base for supersymmetry (Frappat *et al.*, 2000; Rogers, 2007).

One more kind of numbers called the *supernatural numbers* was used by the German mathematician Ernst Steinitz (1871–1928) in his work on field theory (Steinitz, 1910). In the same paper, he was the first to introduce the standard definition of rational numbers as equivalent classes of fractions.

Recently, Redouane Bouhennache introduced spherical and hyperspherical hypercomplex numbers, which form a kind of prearithmetics in terms of the theory of non-Diophantine arithmetics (Bouhennache, 2014; Burgin, 2019).

An important peculiarity of the studies in arithmetic in the second half of the 19th century was the treatment of different classes of numbers as algebraic systems such as fields or rings. For instance, *numbers rings* are systems of numbers with three arithmetical operations — addition, multiplication and subtraction, which satisfy additional identities. In a similar way, *numbers fields* are systems of numbers with four arithmetical operations — addition, multiplication, subtraction and division, which satisfy additional identities. Number fields and rings are constructed extending rational numbers with irrational algebraic numbers and are studied in algebraic number theory. In general, it is possible to call the study of arithmetics as specific algebraic systems by the name *algebraic arithmetic*.

Coming to the transfinite and geometrical directions in arithmetic, we find that they are intrinsically interwoven. As we will see, the mathematical study of magnitudes naturally overlaps with the exploration of infinite numbers. Now it is generally assumed that Georg Cantor was the first who developed a mathematically rigorous theory of infinity. However, even before Cantor started publication of his groundbreaking works on set theory and transfinite numbers (cf., Cantor, 1878; 1883), two other German mathematicians Paul David Gustav du Bois-Reymond (1831–1889) and Carl Johannes Thomae (1840–1921) elaborated consistent and relatively sophisticated theories of infinite numbers in the form of quantities or magnitudes. However, while Cantor developed theories of infinitely big (transcendent) numbers, du Bois-Reymond and Thomae concentrated on infinitely small numbers.

However, even before this was done, the Dutch philosaopher and mathematician Bernard Nieuwentijt (1654–1718) wrote a book *Analysis Infinitorum* proposing a system containing the infinite quantity (number), as the largest quantity, while the infinitesimal quantity was formed by dividing finite numbers by the infinite one (Nieuwentijt, 1695). Nieuwentijt postulated that the product of two infinitesimals should be exactly equal to zero.

Infinitely small numbers appeared in the form of quantities or magnitudes because ancient Greeks did not know irrational numbers. So, instead they operated with magnitudes when they found that in contrast to the Pythagorean program, not all geometrical objects (even very simple) were presentable by numbers and by ratios, which substituted positive rational numbers. The discovery of incommensurable magnitudes was a shock to the Pythagorean School and other mathematicians of that time. However, as time passed by, the existence of incommensurable magnitudes was no longer something strange and in the period of Plato and Aristotle, their existence was common place up to the point of being considered absurd that all magnitudes were commensurable (Fernandes, 2017). Nevertheless, being unable to accept ratios as numbers, ancient Greeks made a strict distinction between numbers and magnitudes. For instance, Aristotle explained that, although numbers and magnitudes belonged to the same encompassing concept (category) *quantity*, they are essentially different because *number* is discrete coming from counting and *magnitude* is continuous coming from measurement (Aristotle, 1984).

Even though the concept of number was amplified all the way through the centuries until real numbers came into being including the notion of continuity, magnitudes were still studied and advanced in the 19th century. There were two reasons for this situation. On the one hand, as more and more new kinds and types of numbers were introduced, it was hard to treat them as numbers at the beginning. So, they were called magnitudes or quantities. Herman Grassmann following Aristotle wrote, *"magnitude is anything that may be said to be equal to or not equal to another"* and adding *"Two things are said to be equal, if in each statement you can substitute the one for the other"* (Grassmann, 1861). Mathematicians of the 19th century, who studied magnitudes, drew inspiration from these words (Ehrlich, 2006).

The goal of du Bois-Reymond was to represent the relative size of real functions as the rate of their growth, which can be called the order of greatness or order of smallness (Hardy, 1910). Du Bois-Reymond treated classes of functions with the same rate of growth as a kind of quantities, which he called *infinities* (du Bois-Reymond, 1870/1871, 1875, 1877, 1882). Hardy demonstrated importance of this work for several mathematical fields such as the theory of Fourier's series, the theory of integral functions, and the theory of singular points of analytic functions in general (Hardy, 1910). In addition, growth rates became important tools in the analysis of algorithms (Kleinberg and Tardos, 2006; Goodrich and Tamassia, 2015).

Based on his general approach, du Bois-Reymond also developed a theory of infinitesimals, about which he wrote:

"The infinitely small is a mathematical quantity and has all its properties in common with the finite ... " (du Bois-Reymond, 1877).

More exactly, an infinitesimal is a quantity, magnitude or number, which is less than any conventional quantity, magnitude or number but larger than zero.

Du Bois-Reymond did not pay much attention to the arithmetical operations and did not study the arithmetic of his infinities but established a number of order-theoretic results for them. Doing this, du Bois-Reymond maintained that his system of infinities constitutes a "broader idea of the numerical continuum" than the Cantor–Dedekind continuum of real numbers (du Bois-Reymond, 1877).

Du Bois-Reymond other mathematical results included an example published in 1875 of a continuous function which was nowhere differentiable. It was inspired by a function with similar properties constructed by Weierstrass in 1872 but not published by him until much later.

It is interesting to remark that almost hundred years before du Bois-Reymond, Leonhard Euler undertook an analogous study of what he called "degrees" of infinity and arrived at results analogous to some of those of du Bois-Reymond (Euler, 1780; Ehrlich, 2006).

Based on the rate of change of functions, Thomae also developed a system similar to the system of du Bois-Reymond calling his quantities *measures of the orders of functions* and not orders of infinity (Thomae, 1870, 1872, 1880). In his work, Thomae was primarily concerned with the *orders of vanishing* of functions or, the *orders of infinite smallness* of functions, as other mathematicians preferred to call them later.

Thomae declared that his system of "measures of the orders of vanishing of functions constitutes a "number domain" that is "infinitely more dense than ... the ordinary real numbers" demonstrating a member of the domain as "*a number which while not zero is smaller than each specifiable number, i.e., a number in the ordinary number domain*" (Thomae, 1870).

This shows that in the study of infinitesimals, Thomae preceded nonstandard analysis for almost a century. His results brought infinity in the form of infinitesimals, i.e., infinitely small magnitudes or quantities, into mathematics in general and arithmetic, in particular. Later systems of infinitesimals were equipped with arithmetical operations.

In his work, Thomae realized that the orders of infinity associated with the members of a particular class of functions can be represented by members of a lexicographically ordered system of complex numbers. The weight of these structures for non-Archimedean systems was demonstrated by the Austrian mathematician and philosopher Hans Hahn (1879–1934), who proved that it is possible to embed every ordered Abelian group in an appropriate lexicographically ordered system of complex numbers (Hahn, 1907).

Hans Hahn was born in Vienna and studied at the universities of Vienna, Strasbourg, Munich and Göttingen. After graduation, he worked the universities in Vienna, Innsbruck, Czernowitz, and Bonn. The famous logician Kurt Gödel was a student of Hahn in Vienna. Hans Hahn made important contributions to functional analysis, topology, set theory, the calculus of variation, and ordered algebraic systems. Many important mathematical theorems bear his name, such as the famous Hahn-Banach theorem in functional analysis, Hahn embedding theorem for ordered algebraic systems, Hahn-Kolmogorov theorem, and Hahn-Mazurkiewicz theorem.

The noted Austrian mathematician Otto Stolz (1842–1905) laid the groundwork for the modern theory of magnitudes aimed at a rigorous theory of non-Archimedean systems (Stolz, 1881–1884, 1888). Developing this theory, Stolz introduced two number systems and a related theory thereof. The first system is based on du Bois-Reymond system, while the second system describes moments of functions, which are named after Newton's fluents and fluxions, with the goal to represent some of the more contentious practices of the early differential calculus using algebraic and analytic tools.

Magnitudes were also studied by Italian mathematicians. In his theory of magnitudes, the Italian mathematician Rodolfo Bettazzi (1867–1941) studied the properties of order, addition and continuity of magnitudes. He used an abstract approach based on the construction of structures without an empirical counterpart (Bettazzi, 1890). As a result, magnitudes were principally defined as structures without any reference to the physical operation of measurement. At the same time, providing a deeper analysis than Stolz, Bettazzi introduced and studied a wider range of classes of magnitudes such as classes of *one-directional* and *two-directional* magnitudes.

The well-known Italian mathematician Giuseppe Veronese (1854–1917) also came to infinitesimal magnitudes developing his non-Archimedean geometry and taking inspiration in the approach of Stolz (Veronese, 1889, 1896, 1898, 1909).

In the context of contemporary mathematics, it is possible to assume that magnitudes are measures of physical objects, which can be mathematically represented by numbers. In turn, numbers can be physically represented by magnitudes, which are intrinsically related to infinitesimals.

However, infinitesimals have not been treated as numbers for a long time. Being useful in mathematics, infinitesimals could scarcely withstand logical inquiry. They were severely criticized through the history being derided by the famous English philosopher George Berkeley (1685–1753) in the eighteenth century as "ghosts of departed quantities," execrated by Georg Cantor (1845–1918) as "cholerabacilli" infecting mathematics in the nineteenth century, and roundly condemned by the famous British philosopher, logician, mathematician, essayist and social critic Bertrand Russell (1872–1970) as "unnecessary, erroneous, and self-contradictory" in the twentieth century. After the works of the prominent French mathematician Augustin-Louis Cauchy (1789–1857), mathematicians assumed that the use of infinitesimals in the calculus and mathematical analysis was finally supplanted by the limit concept which took rigorous and final form in the latter half of the 19th century. By the beginning of the twentieth century, the concept of infinitesimal had become, in analysis at least, a virtual "unconcept" (Bell, 2019).

In spite of this, the outlawing of infinitesimals by some mathematicians was not successful in excluding them from mathematics and science. It would be better to say they were, rather, pushed underground. Physicists and engineers, for example, never deserted their utilization as heuristic devices for the obtaining correct results in the application of the calculus to physical problems. Besides, infinitesimals also lived on in theories of magnitudes and non-Archimedean fields in a mathematically rigorous sense. Later infinitesimals were resurrected in nonstandard analysis (Robinson, 1966) and the theory of surreal numbers (Knuth, 1974; Conway, 1976).

As it was described above, people used named numbers, such as five apples or ten meters, long before they came to the general concept *number*. Named numbers represented quantities of objects, e.g., apples, or essences, e.g., meters, of the same kind. These objects (essences) form units from which numbers are built. Practical applications of named numbers demonstrated existence of two types of units: *simple units*, such as distance in meters, time in hours or weight in grams, and *compound units* formed from simple units, such as square meters (m^2) for area, kilometers per hour for velocity (km/hr), grams per liter (g/L) for density or distance in meters and centimeters, e.g., 5 m 15 cm.

The necessity to operate with different types of units brought forth the *compound unit arithmetic* (Walkingame, 1849). In this arithmetic in addition to the basic arithmetic operations with real numbers, that is, addition, multiplication, subtraction, and division — there are three more operations:

(1) *Reduction* changes a compound unit quantity to a single unit quantity, i.e., quantity with one unit (name). For instance, conversion of a time expressed in hours and minutes to the time expressed in minutes, e.g., changing "1 h 90 min" to "150 min".

(2) *Expansion* is the inverse operation to reduction, i.e., it is the conversion of a quantity that is expressed with a single unit of measure to a quantity that is expressed with a compound unit, such as converting "105 cm" to "1 m 5 cm".

(3) *Normalization* is the conversion of compound units to a standard form. For instance, changing "1 h 90 min" to "2 h 30 min".

Knowledge of the relations between the diverse units of the same or different quantitative measures, their multiples and their submultiples is an important for the compound unit arithmetic because operations are performed according to these relations.

In addition to the standard techniques of operating with abstract numbers, the compound unit arithmetic contains two schemas of operation.

The first one is the *reduction–expansion method*, in which all the compound unit variables are reduced to single unit variables, then the calculation performed and the result is converted back to compound units. This schema works well in automated calculations.

The second schema is the *normalization method*, in which each unit is treated separately and the problem is continuously normalized as the solution develops. This schema works well for manual calculations.

Formalization of the compound unit arithmetic is achieved by using named numbers and multinumbers (Burgin, 2011).

However, mathematicians did not only extend the scope of number systems and arithmetics but also worked on their rigorous representation. We already considered how rigorous constructions for real and complex numbers were introduced. In contrast to this, the rigorous approach to the most basic natural numbers based on axiomatization was developed later in the 19th century. This analysis started in the 19th century in works of Hermann Günther Grassmann (1809–1877), Charles Sanders Peirce (1839–1914), Richard Julius Wilhelm Dedekind (1831–1916) and Giuseppe Peano (1858–1932), who suggested and explored systems of axioms for the Diophantine arithmetic of natural numbers (Grassmann, 1861; Peirce, 1881; Dedekind, 1888; Peano, 1889). However, only in the 20th century, these studies, as it is described in the next chapter, came to the forefront of mathematics.

The outstanding German mathematician and polymath Hermann Günther Grassmann was the first to build an axiomatics for the Diophantine arithmetic (Grassmann, 1861). He was also a linguist, physicist, neohumanist, general scholar, and publisher. In his work, Hermann Grassmann introduced mathematical systems with operations, which later were called vector spaces. He studied the concepts of linear maps, scalar products, linear independence and dimension although Peano was the first to give the modern definition of vector spaces and linear maps (Peano, 1888). Grassmann even exceeded the framework of vector spaces as he used multiplication bringing him to another new concept of a linear algebra. Nevertheless, in spite of all his achievement and innovations, the mathematical work of Hermann Grassmann was little noted until he was in his sixties.

Hermann's brother Robert Grassmann (1815–1901) aimed to further develop this approach to arithmetic (Grassmann, 1872, 1891). According to Robert Grassmann, arithmetic, and more generally, mathematics as a whole, did not need syllogistic logic, and, even more strongly, no logical theory at all as its foundation. On the contrary, logic as a discipline had to be a chapter of mathematics.

In this context, Robert Grassmann treated mathematics as a science of forms (Formenlehre). If according to the general theory of structures (Burgin, 2012), we interpret forms as structures, it is possible to see that Hermann and Robert Grassmann were the first mathematicians to suggest the structural account of mathematics. In the 20th century, this approach was promoted by the very influential group of mathematicians who published books under the name Bourbaki (Lautman, 1938; Bourbaki, 1957, 1960; Dieudonné, 1979) and supported by philosophers of mathematics (Resnick, 1997; Shapiro, 1997) becoming very popular under the name *structuralism in mathematics* (Parsons, 1990; Hellman, 2001; Carter, 2008). Interestingly, arithmetic is the central example in most discussions of structuralism in contemporary philosophy of mathematics (Reck and Price, 2000).

One more logical description of the Diophantine arithmetic of natural numbers was suggested by the outstanding American philosopher, logician, mathematician, and versatile scientist Charles Sanders Peirce (1839–1914), who published his axiomatic system in (Peirce, 1881). According to Shields, it is not generally known that his paper provided the first abstract formulation of the notions of partial and total linear order introducing recursive definitions for arithmetical operations independently of Grassmann (Shields, 1997).

Charles Peirce was born in Cambridge, Massachusetts, USA, and studied at Harvard University graduating with the bachelor of science degree in chemistry. He worked at the United States Coast Survey, as an assistant in Harvard's astronomical observatory, and a lecturer in logic at John Hopkins University, which was the only academic appointment he ever held because of the covert opposition of some North American scientists.

In addition to works in logic and mathematics, Peirce also originated semiotics as a scientific discipline and pragmatism as a philosophical direction. In his scientific and philosophical works, Peirce was an enthusiastic classificator developing a host of triadic typologies.

One more axiomatization of the Diophantine arithmetic of natural numbers was developed by the prominent German mathematician Richard Julius Wilhelm Dedekind (1831–1916) in (Dedekind, 1888). However, as in other his works, he was ahead of his time aiming at generalization of existing structures and making important contributions to abstract algebra, axiomatic foundation for natural numbers, algebraic number theory and the definition of the real numbers.

However, the most popular logical foundations for the Diophantine arithmetic of natural numbers was developed by the Italian mathematician Giuseppe Peano (1858–1932). He directly gave axioms for natural numbers, which constituted the substance of the Diophantine arithmetic in a clear and formalized form (Peano, 1889). Comparing the works of Hermann Grassmann, Charles Peirce, Richard Dedekind, and Giuseppe Peano, we can see that often the attempt to build a more general system results in more hardships for the reader. In turn, this explains why the system of Peano was better accepted and became so popular in the mathematical community.

Giuseppe Peano was born in Piedmont, Italy, and studied at the University of Turin. He taught at the University of Turin and at the same time, a the Military Academy in Turin. In general, Giuseppe Peano was a very active researcher authorizing more than 200 papers and books, creating his own school of mathematics, becoming one of the founders of mathematical logic and essentially developing logical notation. As a part of his research, Peano made significant contributions to the contemporary precise and systematic treatment of mathematical induction. The traditional axiomatization of natural numbers is named the Peano axioms in his honor and consists of five axioms (Peano, 1989).

Another crucial achievement in the domain of numbers and arithmetic was the introduction of transfinite, i.e., infinite, numbers and elaboration of their arithmetics based on set theory. Thus, the first step to transfinite numbers was elaboration of an informal set theory, which is a mathematical theory describing finite and infinite sets in a unified form.

The first who wrote about properties of infinite numbers and infinite quantities was the great Italian physicist Galileo Galilei (1564–1642). The next was the Bohemian mathematician and theologian Bernard Placidus Johann Nepomuk Bolzano (1781–1848). In his book *Paradoxien des Unendlichen*, which was published due to the efforts of his friend F. Prihonsky, Bolzano used a one-to-one correspondence to investigate infinite sets, making it clear that a possibility of a one-to-one correspondence between a set and a proper subset of itself was fundamental to the nature of infinite sets, rather than a justification for avoiding actual infinite sets (Bolzano, 1851). Hence, he maintained, the infinite did actually exist, and was the first mathematician to explicitly define a set. In addition, Bolzano, contended that, if a quantity was infinite, then the measure of that quantity in the form of a number also must be infinite (Bolzano, 1851).

However, to do this in a detailed comprehensive form, it was necessary to have a set theory and such a theory was created by the great German mathematician Georg Ferdinand Ludwig Philipp Cantor (1845–1918). In his works, Cantor rigorously defined infinite numbers, or as he called them, *transfinite numbers* and their arithmetics opening a new page in the whole mathematics (Cantor, 1878, 1883).

Georg Cantor was born in St Petersburg, Russia, and moved to Germany with his family when he was eleven years old. His father wanted Cantor to become an engineer and Cantor started his studies at Polytechnic of Zurich. However, after his father's death, Cantor moved to the University of Berlin and began studying mathematics. After graduation, he moved to Halle to work at the Martin Luther University of Halle-Wittenberg. Cantor obtained important results for trigonometric series and construction of a rigorous definition of real numbers.

However, his main achievement was creation of set theory, which met active opposition of several important mathematicians including Leopold Kronecker and Henry Poincaré. This created problems in publication of his works and advancement in his career. In spite of this, set theory became the most popular foundation of the whole mathematics unfortunately only after the death of its creator.

Indeed, reflecting the impact of modern technology on society, the American writer and historian of science James Gleick wrote, *"Shallow ideas can be assimilated; ideas that require people to reorganize their picture of the world provoke hostility."* (Gleick, 1989).

The size of sets is measure by cardinal numbers while the size of ordered sets is assessed by ordinal numbers. Consequently, there are two basic types of transfinite numbers: cardinal numbers and ordinal numbers. Informally, cardinal numbers are numbers of elements in sets, which can be either finite or infinite. That is why finite cardinal numbers are well-known whole numbers. Ordinal numbers are measures of ordered discrete sets, which also can be either finite or infinite. That is why finite ordinal numbers also coincide with whole numbers.

By its nature, Cantor's set theory was informal, or as it called now, naive, and soon it was found that without some restrictions, this theory led to contradictions, which were called paradoxes. For instance, the Cantor paradox demonstrated that it was impossible to consider the set of all sets. Indeed, as Cantor proved the set of all subsets of any set X is larger than X. Then taking the set U of all sets, we see that on the one hand, it is the largest possible set while on the other hand, the set of all subsets of U must be even larger.

As we know, contradictions are prohibited in mathematics. That is why mathematicians suggested different ways for eliminating contradictions from mathematical theories. The most radical way was simply to disregard set theory as an inconsistent system. For instance, this was suggested by the great French mathematician Henri Poincaré. However, other mathematicians understood importance of set theory and aimed at improving set theory by its formalization. For instance, another great mathematician David Hilbert defended Cantor's set theory from its critics writing, *"No one shall expel us from the paradise that Cantor has created."* (Hilbert, 1967)

Based on their belief in the necessity and usefulness of set theory, mathematicians and logicians started to rebuild set theory on axiomatic foundations. The first were Alfred North Whitehead (1861–1947) and

Bertrand Russell (1872–1970), who proposed an axiomatic theory of sets called the theory of types (Whitehead and Russell, 1910–1913).

Alfred North Whitehead was a famous English mathematician, logician and philosopher. He was born in Ramsgate, Kent, England, and studied mathematics in Trinity College, Cambridge. He worked at Trinity College, University College London, Imperial College London, and the University of London. At the end of his life he moved to the United States and worked at Harvard University. At the beginning of his career, Whitehead did research and published papers and books on mathematics. Later his interests moved to philosophy of science and metaphysics becoming one of the leading figure in the direction known as process philosophy.

Bertrand Arthur William Russell, 3rd Earl Russell, was born in Monmouthshire, England, and studied mathematics in Trinity College, Cambridge. Having enough means for comfortable living, Russell did research in mathematics creating the theory of types to eliminate paradoxes of set theory writing together with Whitehead *Principia Mathematica*, which became a milestone in mathematics, foundations of mathematics, and mathematical logic. He also published different papers on mathematics and logic. However, at the age of 38 years, he started teaching at Trinity College. For a long time, Russell was a pacifist and for this he was even dismissed from Trinity College during World War I.

Later Russell went from mathematics to philosophy becoming one of the founders of modern analytic philosophy. He also taught philosophy in America at the University of Chicago and UCLA.

During his long career, Trinity College, Russell also made contributions to ethics, politics, and educational theory. For his writings, Russell received the Nobel Prize for Literature.

However, mathematicians found the theory of types too complex for utilization because they never encountered set types in their mathematical theories. Another axiomatic set theory, which was closer to the mathematical practice being, at the same time, highly formalized, was elaborated in 1908 by Ernst Zermello (1871–1953). Later this axiomatic system was improved by Dmitry Semionovich Miramanoff (1861–1945), Abraham A. Fraenkel (1891–1965) and Thoralf Albert Skolem (1887–1963), who independently added one more important axiom called the Replacement Axiom (Miramanoff, 1917; Fraenkel, 1922; Skolem, 1922/1923). However, as the paper of Fraenkel was the most accessible and thus, the most influential, this axiom is often credited to Fraenkel and the whole system is now called the Zermello-Frankel (ZF) axiomatic set theory. Later many other axiomatic systems for set theory have been constructed. These axiomatics allowed getting rid of all contradictions found before.

As a result, the theory of transfinite numbers, which includes their arithmetics, became rigorously grounded and accepted by the majority of mathematicians.

As it has been explained, natural numbers have two functions — ordering and measuring discrete finite sets. However, when Cantor started exploring infinite sets, it became clear that these two functions demand two different types of infinite numbers. As a result, Cantor introduced transfinite numbers of two types — *cardinal numbers* or *cardinals* and *ordinal numbers* or *ordinals* (Cantor, 1895/1897). Ordinal numbers carry out the function of measuring ordered infinite sets, which corresponds to counting their elements, while cardinal numbers perform measuring infinite sets without additional structures. In other words, ordinal numbers organize information about well-ordered sets giving them mathematical names and allowing representation of operations with ordered sets by operation with ordinal numbers. At the same time, cardinal numbers provide information about the quantity of elements in a set or the magnitude of this set. Thus, it is interesting that while natural numbers are the same as counting numbers and as measures of discrete quantities, their extension to the realm of infinity brings forth two different classes of transfinite numbers

Defining transfinite numbers, Cantor described corresponding arithmetics with addition, multiplication and powers. Both arithmetics of transfinite numbers contain the conventional arithmetic of natural numbers as their subarithmetic because for finite sets their cardinal and ordinal numbers are the same coinciding with natural numbers.

Exploring properties of operations in transfinite arithmetics, Cantor found that they were often very different from properties of operations with natural numbers, which are both finite cardinal and ordinal numbers. For instance, the sum of any two natural numbers is larger than any of them. At the same time, it is possible that the sum of two cardinal numbers is equal to one of them. In particular, if we add 1 to a natural number, we obtain a larger number. At the same time, if we add 1 to a transfinite cardinal number, we obtain the same number. Interestingly, some non-Diophantine arithmetics have these properties.

It is important to understand that in their theories Bolzano and Cantor introduced and studied only infinitely big numbers. However, as we already know, infinity came in rigorous form to mathematics in the form of transcendent numbers and infinitesimal magnitudes before Cantor published his works on set theory introducing infinite numbers and calling them *transfinite numbers*. In spite of this, the works of Cantor are still considered being the first rigorous study of infinity in mathematics. Even now, some researchers believe that Cantor's theory of transfinite numbers is the only known mathematical or more exactly arithmetical theory of infinities (cf., for example, Swamy, 1999).

Chapter 3

Acceleration of the Development in Arithmetic and Numerical Systems in the 20th and 21st Centuries

Obvious is the most dangerous word in mathematics.

E. T. Bell

At the very end of the 19th century, the German mathematician Kurt Wilhelm Sebastian Hensel (1861–1941) introduced an important class of numbers called *p-adic numbers* where p is an arbitrary prime number (Hensel, 1997). We write about p-adic numbers in this chapter because the main studies of these numbers and their application were developed in the 20th century.

Similar to real numbers, p-adic numbers are obtained by extending the field of rational numbers. Both real and p-adic numbers make rational numbers complete with respect to limits of certain numerical sequences but they use different types of convergence.

Hensel and those mathematicians who worked after him essentially developed the theory of p-adic numbers finding various applications in mathematics and beyond (Hensel, 1908; Volovich, 1987; Brekke and Freund, 1993, 1995, Vladimirov *et al.*, 1994).

Later Hensel also introduced *g-adic numbers* where g was an arbitrary natural number larger than 1 (Hensel, 1913). His construction was similar to the construction of p-adic numbers.

Kurt Hensel was born in Königsberg, Prussia (now Kaliningrad, Russia) and studied at the University of Berlin and University of Bonn. He worked as a professor at the University of Marburg.

The German mathematician Ernst Steinitz (1871–1928) introduced *supernatural numbers*, sometimes called *Steinitz numbers*, in his research in field theory (Steinitz, 1910). A supernatural number is a product of the powers of prime numbers. When this product is finite, that is, the number of finite powers of prime numbers that constitute the product is finite, we obtain a natural number. However, it is possible that the product contains an infinite number the powers of prime numbers or/and some powers are infinite, i.e., equal to ∞. In this case, we have a proper supernatural number. It is possible to multiply supernatural numbers but mathematicians do not know how to define natural addition for them.

The Diophantine, i.e., conventional, arithmetic existed as a unique arithmetic of natural numbers for millennia before other, i.e., non-Diophantine, arithmetics of natural numbers were discovered/invented at the end of the 20th century. In contrast to this, other arithmetics of cardinal and ordinal numbers were discovered/invented relatively soon after the Cantorian arithmetic of ordinal numbers appeared. Namely, in 1906, the German mathematician Gerhard Hessenberg (1874–1925), defined the natural addition and natural multiplication of ordinals constructing in such a way a non-Cantorian arithmetic of ordinal numbers (Hessenberg, 1906). When in the 20th century, surreal numbers were introduced by John Conway, it was discovered that these operations coincide with the addition and multiplication of ordinals as surreal numbers (Hickman, 1983).

One more arithmetic of transfinite numbers was constructed by the German mathematician Ernst Eric Jacobsthal (1882–1965), who found a third, intermediate way of multiplying ordinals by transfinite iteration of natural addition and defined exponentiation by transfinite iteration of this multiplication (Jacobsthal, 1909). Much later, Jacobsthal's operations were rediscovered by John Conway (1976, 1994) and by Abraham and Bonnet (1999). Jacobsthal's arithmetic was also studied by Gonshor (1980) and by Hickman (1983). Using a recurrence relation, Jacobsthal defined a sequence of natural numbers, which acquired the name *Jacobsthal numbers*.

Transfinite numbers of Cantor extended natural numbers into the realm of the infinite. Later other classes of ordinary numbers were extended in a similar way. At first, the Polish mathematician Roman Sikorski (1920–1983) constructed transfinite integer numbers and transfinite rational numbers (Sikorski, 1948). Then the German mathematician Dieter Klaua (1930–2014) built transfinite real numbers (Klaua, 1959/1960, 1960).

There were other approaches to measuring infinite sets with infinite, or transfinite, numbers. For instance, in Cantor's theory, the set of natural numbers and all its infinite subsets have the same size (cardinality) while intuition suggests that the size of all even numbers is less than the size of all natural numbers. To support this intuition by rigorous mathematical structures, Fred Katz introduced *class size* as a logical theory of set cardinalities (Katz, 1981) while Vieri Benci defined and later generalized with Mauro Di Nasso the concept of numerosity (Benci, 1995; Benci and Di Nasso, 2003; 2005). Informally, *numerosity* means the quantity of objects in a discrete set. The mathematical theory of numerosities defines how to assign size to infinite sets of natural numbers in such a way as to preserve the part-whole relation, that is, if a set A is properly included in B then the numerosity of A is strictly less than the numerosity of B. Paolo Mancosu discussed philosophical issues of these alternatives to Cantor's theory (Mancosu, 2009).

Another alternative to Cantor's theory of transfinite numbers is developed by Meir Buzaglo (1992), while a general approach to set enumeration with respect to their equivalence is constructed by Mark Burgin and Meir Buzaglo (1994). Cantor's theory of cardinal numbers becomes a special case of these enumerations.

An interesting peculiarity of the 19th century study of infinities in number systems was Cantor's attitude to these studies. One the one hand, he was the creator of the arithmetic of infinitely big numbers. On the other hand, during most of his life, he expressed himself as a vigorous opponent of infinitesimals describing them as the "'infinitesimal Cholera bacillus of mathematics', which had spread from Germany through the work of Thomae, du Bois Reymond and Stolz, to infect Italian mathematics" (cf., Dauben, 1980; Ehrlich, 2006). According to the American historian of science and mathematics Joseph Warren Dauben, recognition of infinitesimals in the works of Thomae, du Bois-Reymond, Stolz and Veronese inevitably meant that Cantor's own theory of real numbers was incomplete renouncing the perfection of Cantor's personal creation. Understandably, Cantor launched a systematic campaign to discredit works on infinitesimals in every possible way (Dauben, 1980).

In addition, the mainstream of mathematicians did not accept infinitesimals as numbers for a long time. Even more, the followers of Cantor, Dedekind, and Weierstrass sought to purge infinitesimals from analysis, while their philosophical allies, such as Bertrand Russell (1872–1970) and Rudolf Carnap (1891–1970), declared infinitesimals to be "pseudoconcepts".

In spite of the conformism of the mathematical establishment, even in the 20th century, different types and kinds of numbers continued multiplying like Fibonacci's rabbits and even faster. It will take too much space even to name all of them. So, we will describe only the most important novel number types and classes, as well as new categories and clusters of number types and classes.

In addition to big classes of numbers, such as real, complex, transcendental, and transfinite numbers, mathematicians isolate and study classes of numbers and individual numbers that bear names of mathematicians. It is natural to call them *personalized numbers*. Let us consider some examples of personalized numbers.

The *Gaussian integers* usually denoted by $Z[i]$ are complex numbers, the real and imaginary parts of which are both integers. The Gaussian integers, with ordinary addition and multiplication of complex numbers, form an arithmetic.

The *Gaussian rationals* usually denoted by $Q[i]$ are complex numbers, the real and imaginary parts of which are both rational numbers. The Gaussian rationals, with ordinary addition and multiplication of complex numbers, form an arithmetic.

The *Fermat numbers*, also called a *Fermat primes*, are prime numbers having the form $2^{2^n} + 1$ where n is a natural number. The reason for this name was that the famous French mathematician Pierre de Fermat proposed the conjecture that all numbers of this form are prime. However, in 1732, Euler discovered when $n = 5$, we get the number $2^{2^5} + 1$, which is divisible by 641 and thus, it is not a prime number.

The *Mersenne numbers*, also called a *Mersenne primes*, are numbers having the form $2^n - 1$ which are prime. The reason for this name was that the French polymath Marin Mersenne (1588–1648) was communicating by mail with many mathematicians encouraging their study.

The *Goldbach numbers* are even natural numbers that can be expressed as the sum of two odd prime numbers. The reason for this name was that the German mathematician Christian Goldbach (1690–1764) posed the following conjecture:

Every even integer greater than 2 can be written as the sum of two primes.

This is one of the oldest and best-known unsolved problems in number theory and the whole mathematics (cf., for example, Chudakov, 1937; Estermann, 1938; Van der Corput, 1938).

The most important of individual personalized numbers is the number e, which is called *Euler's number* after the great Swiss–Russian–Prussian mathematician Leonhard Euler, who often used this number in his mathematical works denoting it by e deriving this notation from the word exponent.

The Eddington number is the number of protons in the observable universe. The reason for this name was that the famous British astrophysicist, astronomer and philosopher of science Arthur Stanley Eddington (1882–1944) was the first to estimate the value of this number and to clarify how this number might be essential for cosmology and foundations of physics.

However, personalized numbers formed systems only of a particular interest, while in addition to the basic classical systems of numbers such as natural or complex numbers, it is possible to delineate the following innovative fundamental number classes and clusters of number classes:

(1) infinite (including transfinite) numbers,
(2) abstract named numbers,
(3) fuzzy numbers,
(4) multinumbers,
(5) hypernumbers (finite, infinite, periodic, integer, positive and negative),
(6) hyperreal numbers and magnitudes (finite, infinite and infinitesimal),
(7) hypercomplex numbers,
(8) surreal numbers,
(9) computable numbers,
(10) feasible numbers,
(11) numbers in categories.

In what follows, we describe these classes from the historical point of view and information perspective.

As we know, numbers emerged in the form of concrete named numbers, e.g., five trees or two mountains. It was a long process before they became abstract numbers, i.e., numbers *per se*, such as 5 or 2. However, in the 21st century, mathematics came back to named numbers with the creation of the theory of abstract named numbers based on the theory of named sets (Burgin, 2011). To transform/abstract concrete named numbers into abstract numbers, mathematicians eliminated names of concrete named numbers for example, three trees, three sticks and three dogs became the abstract number 3. To transform/abstract concrete named numbers into abstract named numbers, mathematicians made names of concrete named numbers abstract, for example, three trees, three sticks and three dogs became the abstract named number $(3, e, X)$ where X is a general name, e.g., trees, dogs, sticks, of the number 3.

An important progress in the development of numbers is related to named numbers. Indeed, people use numbers for counting, representing quantities and measuring. When numbers are used for representing quantities, they are tied to the names of the corresponding objects, e.g., 5 apples, 7 buildings, 9 planets or 11 cars. When numbers are used for measuring, they are tied to the names of the measurement units, e.g., 5 minutes, 7 feet, 9 pounds of oranges or 11 kilograms. The discipline that tells how numbers are represented and how to operate with them is arithmetic. In mathematics, arithmetic comprises numbers *per se*, i.e., abstract numbers, and develops operations with these abstract numbers. However, in real life, people operate not with abstract numbers but with numbers that are related to some things, i.e., with named numbers such as 1 car, 2 books, 10 dollars or 1 gallon of milk.

People use named numbers in virtually every domain, from kitchens and stores to high-tech science laboratories, banks and governments. Experts who work in different areas elaborated specialized classes of named numbers. Examples are 5 meters and 1 light year in physics and 20 Kelvin or 70 Fahrenheit in chemistry. Biologists use the same named numbers as physicists, e.g., 10 kg or 5 m. Historians and archeologists created new names for named numbers. For instance, historians use such named numbers as 1000 B.C.E. or 500 A.C. Archeologists use such named numbers as 3 ka (kilo annum, thousand years) which means 3 thousand calendar years ago, while the named number 10 ka BP stands for 10 thousand years before present.

As Miller writes, "to count is to count something. Every sum is a sum of something. A sum identifies only what is common to all the things counted. You must *identify* a unit which is common to the whole collection of things you wish to count in order to *identify* your sum." (Michael Miller, 1999)

The concept of a named number naturally comes from real life to programming languages, where it plays an important role. For instance, in Ada 95 Reference Manual, the following operation is defined:

A number declaration declares a named number.

In contrast to this, mathematicians, when they study, utilize and develop mathematics, add one and one and get two, multiply two times two and get four or add ten and twenty and get thirty. In real life, people add one car to one car and get two cars, take two books two times and get four books or add ten dollars to twenty dollars and get thirty dollars.

It means that in practice, only named numbers are used representing quantities and measures (Depman, 1959; Borovik, 2010). Moreover, numbers originated as named numbers (Burton, 1997; Cunnington, 1904; Everett, 2017; Flegg, 2002; Ifrah, 2000). In spite of this, till the beginning of the 21st century, there was no mathematical theory of named numbers although people knew how to add the same quantities. For instance, 5 apples plus 7 apples is equal to 12 apples. However, people were perplexed when they had to add objects from different categories because conventional rules of arithmetic apply only to quantities that have the same name, that is, these objects are counted in common units. If the units were not the same, arithmetic was considered useless. This is expressed in the proverb "*You can't add apples and oranges.*" This proverb is often used even as motto. For instance, the Scientific Digital Visions company mission describes their primary quality objective as *To Never Add Apples and Oranges.* As it is written on the website (http://www.dataenabled.com/sdv_quality.html), this mission focuses on measurement units, which are the apples and oranges of data and it extends to all phases of the data life cycle. Mission-critical systems may fail if people or programs inappropriately combine data with different units.

Nevertheless, in spite of persistence of named numbers in the ordinary life, a rigorous mathematical theory of *named numbers* was created only in the 21st century by Mark Burgin (Burgin, 2006). Later he further developed this theory into the theory of *multinamed numbers* (Burgin, 2011). An interesting peculiarity of these theories is the discovered duality between multinamed numbers and multinumbers introduced in multiset theory (Monro, 1987).

Another leap forward in the area of numbers is related to infinite numbers, or more exactly, to numbers that represent infinity. Actually, real numbers already enclosed infinity because the numerical representation of the majority of real numbers, namely, irrational numbers, demanded infinite number of digits. It can be proved that this feature of real numbers does not depend on the numerical system chosen for the representation. That is why some mathematicians and notably Leopold Kronecker (1825–1891) rejected the existence of irrational numbers.

However, real numbers and complex numbers, which inherited implicit infinity from real numbers, represented infinity in the concealed form. At the same time, mathematical practice demanded to develop means of working with overt infinity. Naturally, this entailed elaborating the construction of infinite numbers, which gave important information about infinite mathematical objects. In this domain, we can discern the following areas and directions:

(1) the theories of new and conventional transfinite numbers;
(2) nonstandard analysis and related areas oriented at infinitesimals;
(3) the theory of surreal numbers and related areas;
(4) the theory of hypernumbers and related areas;
(5) the theory of multicardinals;
(6) the theory of cardinal multinumbers;
(7) the theory of numerosities;
(8) the theory of asymptotic numbers.

Different types of numbers have been introduced for infinity quantization. Taking the most popular of them, we see that the transfinite quantization of infinity is based on counting processes (Cantor, 1878, 1893), while the nonstandard quantization of infinity is based on measuring procedures (Robinson, 1966). At the same time, the construction of surreal numbers combines counting and measuring approaches (Conway, 1976). All these approaches to infinity quantization use set-theoretical methods prevalent in mathematics. In contrast to this, infinity quantization by hypernumbers is based on measuring procedures and employs topological techniques inherent in physics (Burgin, 2002; 2004; 2012).

We already discussed some approaches to numerical representation of infinity in the form of infinitesimals and transfinite numbers in the previous chapter. So, why once more are we speaking about infinitesimals? As it often happens in human history, mathematical works on infinitesimal magnitudes, which were published in the 19th century, were forgotten and mathematicians started a new attack to reinstate infinitesimals in mathematics. In the 20th century, the first step was done by the Polish mathematician, logician, philosopher, avant-garde painter, literary critic, and theoretician of modern art Leon Chwistek (1884–1944), who used sequences of real numbers to represent infinitely small numbers (Chwistek, 1935). The German mathematicians Curt Otto Walther Schmieden (1905–1991) and Detlef Laugwitz (1932–2000) further developed this approach (Schmieden and Laugwitz, 1958). In a more advanced form, it was done by Laugwitz (1960; 1961), where sequences of real numbers were used for investigation of distributions and operators. The American mathematician James M. Henle further expanded this sequential technique (Henle, 1999, 2003).

However, the most popular direction in the rigorous mathematical theory of infinitesimals was initiated by the prominent Israeli–British–American mathematician Abraham Robinson (1918–1974), who elaborated nonstandard analysis based on set-theoretical techniques, which determined hyperreal numbers as classes of equivalent sequences of real numbers where two sequences were equivalent if almost all elements of them coincide (Robinson, 1961, 1966).

Abraham Robinson was born in Waldenburg, Germany, which is now Wałbrzych, Poland. With Hitler coming to power, the Robinson family moved to Jerusalem, Palestine. Robinson started studying mathematics at the Hebrew University of Jerusalem, got a scholarship to study at Sorbonne in Paris, went to France but barely escaped German occupation fleeing to England. There he enlisted in the Free French Air Force, which together with the British Air Force fought with Germans. As a mathematician, Robinson became a Scientific Officer working on aerodynamics of delta wings and supersonic flow. However, his main interest was in mathematical logic and after the war, he continued logical explorations working at the University of Toronto, Hebrew University of Jerusalem, University of California, Los Angeles (UCLA), and finally Yale University. As a mathematician, Robinson was a leading expert in different areas of mathematics such as model theory, metamathematics, logic, and analysis as well as in aerodynamics.

His main achievement — nonstandard analysis — provided quantization of infinity in two directions — for infinitely small (contracting) quantities and for infinitely big (expanding) quantities, which found various applications in different areas such as functional analysis, number theory, probability, dynamical systems, and mathematical economics, demonstrating its usefulness for constructing mathematical models for diverse phenomena (Cutland, 1988). However, the initial goal of Robinson in creating nonstandard analysis was providing a rigorous mathematical construction for infinitesimals, which from the time of Newton and Leibniz, had been so successfully used in the calculus.

Nonstandard numbers — hyperreal and hypercomplex numbers — form the mathematical structure called a field, in which it is possible to perform all four basic arithmetical operations — addition, multiplication, subtraction, and division (Robinson, 1965). Real numbers constitute a special case of hyperreal numbers while complex numbers are a special case of hypercomplex numbers.

One more generalization of real numbers (as well as of ordinal numbers) is provided by surreal numbers introduced by John Horton Conway (Knuth, 1974; Conway, 1976, 1994). He is an English–American mathematician, who spent the first part of his long career at the University of Cambridge, in England, and the second part at Princeton University, in the USA. Conway has also made contributions to the theory of finite groups, knot theory, number theory, cellular automata, combinatorial game theory and coding theory as well as to many branches of recreational mathematics including the Game of Life.

Surreal numbers are constructed from real numbers and transfinite numbers using Dedekind cuts similar to the cuts of rational numbers used by Richard Dedekind for building a rigorous mathematical representation of real numbers (cf. Chapter 2).

Surreal numbers are extremely general containing all hyperreal numbers introduced by Abraham Robinson (Robinson, 1966), all superreal numbers introduced by the English philosopher and mathematician David Orme Tall (Tall, 1980), and all ordinal and cardinal numbers introduced by Georg Cantor (Cantor, 1932) as their subclasses. Surreal numbers form a real-closed field including all real numbers, ordinals and many less familiar numbers such as negative ordinals, fractions of ordinals, inverses to ordinals or an ordinal plus a real number. Based on this generality, the American philosopher Philip Ehrlich wrote that the system of surreal numbers contains "all numbers great and small" (Ehrlich, 1994, 2001).

However, this is an exaggeration because surreal numbers do not comprise many types and classes of numbers. For instance, the system of surreal numbers does not contain fuzzy numbers, multinumbers, named numbers, cardinal multinumbers and multicardinal numbers. Moreover, new developments of mathematics brought forth novel kinds of numbers, which represent infinity and go beyond surreal numbers. For instance, to deal with the problem of divergence in theoretical physics, which is called the *infinity puzzle* by the English physicist Francis Edwin Close (Close, 2011), a topologically based *quantization* of expanding infinity has been developed by the Russian–Ukrainian–American mathematician Mark Burgin in the form of real and complex hypernumbers (Burgin, 1990b, 2012, 2017). Construction of real hypernumbers from real numbers is similar to construction of real numbers from rational numbers as it was done by Charles Méray and Georg Cantor. Namely, real hypernumbers are represented as classes of equivalent sequences of real numbers or of rational numbers. Utilization of real numbers or of rational numbers for constructing real hypernumbers gives the same result. Complex hypernumbers are constructed by a similar technique from complex numbers. Hypernumbers essentially increased the capabilities of the calculus allowing mathematicians finding sums of arbitrary series of real or complex numbers, integrating any real function, elaborating a rigorous mathematical construction of the Feynman path integral, solving a variety of operator equations and disentangling different problems in probability theory (Burgin, 2008, 2012, 2017; Burgin and Dantsker, 1995, 2015; Burgin *et al.*, 2012; Burgin and Krinik, 2009, 2010, 2015). All innovative techniques based on hypernumbers well correlate with other extensions of the classical methods of summation, differentiation, and integration. For instance, numerical and functional series summation in hypernumbers comprises such approaches as asymptotic series and formal power series (cf., for example, Estrada and Kanwal, 1994).

Comparing hypernumbers in the sense of (Burgin, 2012) with other approaches to infinity, we can see that Cantor's theory of transfinite numbers is a natural set-theoretical extension of natural numbers, while the theory of real hypernumbers is a natural topological extension of real numbers. As a result, transfinite numbers were created and have been used to measure infinite discrete sets by correspondence and counting, while real hypernumbers were created and have been used to measure infinite geometrical and topological objects (Burgin, 2005a).

It is necessary to remark that some people erroneously perceive the theory of hypernumbers as a new version of nonstandard analysis. This is completely incorrect due to the following reasons (Burgin, 2012, 2017).

First, nonstandard analysis has both infinitely big and infinitely small hyperreal numbers, while the theory of hypernumbers has only infinitely big hypernumbers. As the result, real hypernumbers are essentially different from hyperreal numbers, which form the base of nonstandard analysis.

Second, nonstandard analysis is aimed at providing rigorous calculus foundations. At the same time, the goal of the theory of hypernumbers is giving new efficient mathematical tools for physicists in solving the problem of infinity. Certainly, quantum physics frequently encounters infinitely big values in the form of divergent integrals and series but never comes across infinitely small numbers or infinitesimals, which were called "the ghosts of departed quantities" by the famous British philosopher George Berkeley (1685–1753). As the American philosophers Frederick Kronz and Tracy Lupher write, *"A rigorous foundation was eventually provided for infinitesimals by Robinson during the second half of the* 20th *century, but infinitesimals are rarely used in contemporary physics."* (Kronz and Lupher, 2012).

Third, nonstandard analysis is elaborated using set-theoretical principles and constructions, which are fundamental for mathematics. In contrast to this, the theory of hypernumbers is created employing topological principles and constructions, which are indispensable for physics. Similar dissimilarities exist between real hypernumbers and surreal numbers, which include hyperreal numbers from nonstandard analysis (Burgin, 2012, 2017).

As nonstandard analysis and its extensions, such as the theory of surreal numbers, were initiated by problems in mathematics, it is possible to call this area *"mathematical mathematics"*. At the same time, the theory of hypernumbers was set off by problems in physics and thus belongs to the area *"physical mathematics"* (Heaviside, 1893, 1894; Moore, 2012, 2014).

As a result, hypernumbers are better suited for problems in physics. One more argument for this statement was suggested by one of the greatest physicists of the 20th century Paul Adrien Maurice Dirac (1902–1984) when he explained:

"Sensible mathematics [from the point of view of physics, MB] *involves neglecting a quantity when it turns out to be small — not neglecting it just because it is infinitely great and you do not want it!"* (Dirac, 1978)

Paul Dirac was born in Bristol, England, while his father was an immigrant from Switzerland who worked as a teacher. Dirac graduated from the University of Bristol where he studied electrical engineering. After this, his interest moved to general relativity and quantum physics where he became the leading expert introducing innovative theories and receiving the Nobel Prize for his achievements.

In the above statement, Dirac expressed the common opinion of physicists. In contrast to this, nonstandard analysis employs not only infinitely big numbers but also infinitely small numbers. At the same time, hypernumbers contain infinitely big numbers, which can represent divergent values from physics, but it does not have infinitely small numbers.

In essence, all these mathematical theories of the infinite (Cantor's theory, nonstandard analysis, surreal numbers, and superreal numbers) were initiated by problems of mathematics, i.e., it is possible to call them "mathematical mathematics". At the same time, distribution theory and the theory of hypernumbers and extrafunctions were set off by problems of physics and belong to physical mathematics (Burgin, 2017).

One more kind of infinite numbers — *asymptotic numbers* — were introduced by the Bulgarian–American mathematicians Christo Christov and Todor Dimitrov Todorov, who developed the arithmetic of asymptotic numbers and studied their properties (Christov and Todorov, 1974).

Recently Paolo Giordano and Mikhail Katz used Cantor's renowned formalization of real numbers in terms of Cauchy sequences of rational numbers to build two classes of the number systems, which are Cauchy complete and have infinitesimals (Giordano and Katz, 2011). In one class, infinitesimals are invertible while in the other, infinitesimals are nilpotent, e.g., a nonzero infinitesimal taken to the second power can be equal to 0. Real numbers with nilpotent infinitesimals are called Fermat reals (Giordano, 2010).

One more new area in arithmetic is related to computations in science and engineering. In the 20th century, *interval arithmetic* emerged from approximate computations and *interval numbers* acquired a unblemished status in mathematics. The first system of rules for operation with interval numbers was published in a 1931 (Young, 1931). Later applications of interval arithmetic for improving the reliability of digital systems intervals were used to measure rounding errors associated with floating-point numbers (Dwyer, 1951). One more step in this direction was made by the Japanese mathematician Teruo Sunaga (1929–1995), who published a comprehensive paper on interval analysis of numerical computations (Sunaga, 1958). Interval arithmetic became a fully recognized mathematical field after the publication of the book by the American mathematician Ramon Edgar (Ray) Moore (1929–2015) in 1966 (Moore, 1966).

However, the first known interval number was introduced in the 3rd century B.C.E. by Archimedes, who calculated lower and upper bounds for the number π demonstrating that $223/71 < \pi < 22/7$. In terms of interval arithmetic, Archimedes represented π as the interval number $[223/71, 22/7]$.

An *interval number* is an interval of the form $[a, b]$ where a and b are real numbers. In interval arithmetic, arithmetical operations — addition, subtraction, multiplication, and division — with interval numbers are defined. These numbers are similar to the interval-valued fuzzy numbers considered further.

Interval arithmetic forms the base of *interval analysis*, which is also called *interval mathematics* or *interval computation* providing tools for mathematical techniques used to represent rounding and measurement errors in numerical computations.

Interval arithmetic was extended to *complex interval arithmetic* by defining a *complex interval* (*complex interval number*) as the set of all complex numbers inside a circle (Dawood, 2011). In it, the basic arithmetical operations for real interval numbers (real closed intervals) are extended to complex interval numbers. In a similar way, it is possible to extend interval arithmetic to other multidimensional number systems such as quaternions and octonions.

Researchers also constructed *affine arithmetic* as the further development of interval arithmetic (Comba and Stolfi, 1993; Figueiredo and Stolfi, 1996, 2004). In it, the quantities are represented as linear combinations (affine forms) of primitive variables, which stand for sources of uncertainty in the data or approximations made during the computation. Affine arithmetic is useful for solving many kinds of numeric problems, such as solving differential equations, analyzing dynamical systems, integrating functions, equations, plotting curves, error analysis, and process control.

Many important directions in the development of numbers and arithmetics are connected to the further development of the concept of a set. This development brought forth multisets, fuzzy sets, rough sets and their generalizations.

Multisets are collections of objects which can contain indistinguishable members while in sets all members are distinguishable by definition (cf., Burgin, 2011; Fraenkel *et al.*, 1973)). For instance, $\{1, 1, 2\}$ is a multiset while $\{1, 2, 3\}$ is a set. Natural numbers used by people actually form a multiset and not a set. Indeed, to perform addition $1 + 1$, we need two elements 1, while in the set of natural numbers there is only one element 1. The concept of a multiset was rediscovered many times. The Indian mathematician Bhāskara Acharya (1114–1185), also known as Bhāskarācārya ("Bhāskara, the teacher"), and as Bhāskara II, was the first to study assemblies with indistinguishable elements, that is multisets. He described their permutations in his work written ca. 1150 (Knuth, 1997). Later multisets were rediscovered many times appearing under different names. They were also called *aggregates, heaps, bunches, samples, weighted sets, occurrence sets,* and *firesets* (finitely repeated element sets) (cf., Blizard, 1991; Singh, 1994; Singh *et al.*, 2007)). Now multisets play an important role in modern combinatorics, databases and computer science.

Having a multiset, it is possible to each type of its elements, to assign the number of elements of this type. For instance, in the multiset $\{1, 1, 2, 2, 2\}$, there are two numbers 1 and three numbers 2. Formally, we can write it as a multinumber $\{(1, 2), (2, 3)\}$, which is a measure of the initial multiset $\{1, 1, 2, 2, 2\}$. In a general case, the structure of types with numbers of elements of these types is called a *multinumber*. For a long time, multinumbers were studied as a kind of representations of multisets and only in 1987, the Australian mathematician Gordon Monro suggested treating multinumbers as distinct mathematical objects (Monro, 1987).

Multinumbers are generalizations of natural numbers. It is possible to consider a natural number as the quantity of elements having the same type, e.g., three tigers or eight eggs. Multinumbers correspond to multiplicities with elements that can have different types, e.g., three tigers and eight eggs, showing the number of elements of each type, e.g., (tiger, 3; egg, 8). Multinumbers form an arithmetic, in which it is possible to add and in some cases, subtract multinumbers. For instance, (tiger, 3; egg, 8) plus (cat, 5; tiger, 2) is equal to (tiger, 5; cat, 5; egg, 8).

Different generalizations of multisets have been introduced, studied and applied to solving problems. Those generalizations have given birth to the corresponding classes of generalized multinumbers as measures of such multisets.

The English mathematician Wayne Douglas Blizard (1944–2010) introduced multisets with positive and negative multiplicity bringing forth *integer-valued multinumbers* (Blizard, 1990). Later the American mathematician Peter Loeb introduced *hybrid sets* as a generalization of multisets in which multiplicity of an element is any integer number (Loeb, 1992). Hybrid sets are also integer-valued multinumbers. However, much earlier, the American mathematician Hassler Whitney (1907–1989) introduced and studied generalized characteristic functions with values in integers, i.e., they could be positive, negative or zero. He called a structure defined by these functions by the name a *generalized set* developing an algebra of generalized characteristic functions (Whitney, 1933). Naturally generalized sets are in fact *integer-valued multinumbers*.

As negative numbers in the case of integer numbers, integer-valued multinumbers allow performing subtraction of arbitrary multinumbers extending in this ways the arithmetic of multinumbers. The meaning of negative values for elements of type d is similar to the meaning of negative numbers. For instance, the meaning of -100 dollars is a debt of $100.

Blizard also generalized multisets to *real valued multisets*, i.e., multisets in which multiplicity of an element can be any nonnegative real number (Blizard, 1989). Real-valued multisets are *real-valued multinumbers* by their essence.

Using multirelations, Polish–American mathematician Jerzy Grzymala-Busse introduced the concept of a *rough multiset* what gave *rough multinumbers* as their measures (Grzymala-Busse, 1987).

The Ukrainian researcher Alexander Chunikhin introduced and studied *natural multidimensional numbers* and their arithmetic, which also generalize multinumbers and their arithmetic (Chunikhin, 1997, 2012; 2012a).

Malaysian mathematicians Shawkat Alkhazaleh, Abdul Razak Salleh and Nasruddin Hassan introduced *soft multisets*, giving birth to *soft multinumbers* (Alkhazaleh *et al.*, 2011). Later Alkhazaleh and Salleh introduced *fuzzy soft multisets*, which allowed constructing a new class of generalized multinumbers, namely, *fuzzy soft multinumbers* (Alkhazaleh and Salleh, 2012).

Another sizeable direction in number studies and their utilization originated when the American mathematician, computer scientist, electrical engineer, and artificial intelligence researcher of Azerbaijani–Iranian–Jewish origin Lotfi Aliasker Zadeh (1921–2017) introduced and studied fuzzy numbers and fuzzy arithmetic (Zadeh, 1975, 1976). These constructions were based on his main contribution to mathematics — fuzzy sets. Zadeh also developed the theory of fuzzy sets and fuzzy logic (Zadeh, 1965a, 1974), which formed a big area in mathematics and were very popular in a diversity of applications (Kaufmann, 1973–1977; Hellendoorn, 1991; Chen, 1994; Klir and Yuan, 1995; Chunikhin, 1997; Bector and Chandra, 2005; Hanss, 2005). For instance, fuzzy set theory became one of the most fashionable mathematical approaches to problems of uncertainty and imprecision.

Lotfi Zadeh was born in Baku, Aizerbaijan, but at the age of 10 years, he moved with his family to Tehran, Iran. There he graduated from the University of Tehran and in 1944, he went to the United States becoming a graduate student at the Massachusetts Institute of Technology (MIT) and after this, he started doctoral studies at Columbia University becoming an assistant professor there after receiving PhD. Later he became a professor at the University of California, Berkeley where he started publishing his works on fuzzy set theory and fuzzy logic.

Fuzzy sets reflect an important peculiarity of the real world. Namely, they allow partial membership. While given a set R, an object either completely belongs to this set or wholly does not belong, taking a fuzzy set F, it is possible that an object belongs to F only to some extent. For instance, taking the set of all blue balls, we know that a blue ball belongs to it while a ball that is not completely blue, e.g., half blue and half yellow, does not belong to it. At the same time, a ball that is half blue and half yellow partially belongs to the fuzzy set of all blue balls.

To reflect this aspect more precisely, fuzzy membership is used. It allows showing the extent to which an object belongs to some set. For instance, a ball that is half blue and half yellow belongs to the fuzzy set of all blue balls with the membership 0.5 and to the fuzzy set of all yellow balls also with the membership 0.5. In a similar way, a ball that is three fourth green and one fourth brown belongs to the fuzzy set of all green balls with the membership 0.75 and to the fuzzy set of all brown balls with the membership 0.25.

Fuzzy numbers also have these features. For instance, it is possible to consider the integer fuzzy number α, which is to some extent 5 and to some extent 6. In real life, if an individual is asked when she returned home, she can say 5 or 6 hours ago. In terms of fuzzy numbers, it means that she thinks that she returned home α hours ago.

In general, fuzzy sets allow various definitions of fuzzy numbers, which were introduced and utilized by different researchers. In the most general form a *fuzzy number* is a fuzzy set in a set of numbers (Burgin, 2011). Fuzzy numbers constructed by Zadeh and studied by many researchers usually satisfy additional conditions.

In addition to ordinary fuzzy sets, Zadeh introduced and studied type 2 fuzzy sets (Zadeh, 1975, 1976). In turn, type 2 fuzzy sets were used to define type 2 fuzzy numbers and generalized type 2 fuzzy numbers, which were studied by different researchers (Haven *et al.*, 2010; Hernández *et al.*, 2017; Hesamian, 2016).

With the goal to achieve a more adequate concept for description of real-world information, Zadeh defined Z-numbers, which are ordered pairs of fuzzy numbers (Zadeh, 1984, 2011). Z-numbers were used in decision making (Xiao, 2014), estimation of risk levels (Abiyev *et al.*, 2018) and other real-life areas (Aliyev *et al.*, 2015, 2016).

Introduction of a variety of other generalizations of fuzzy sets also brought forth new classes of fuzzy numbers and fuzzy arithmetics. Here we consider only some of them.

American mathematician Josef Amadee Goguen (1941–2006) introduced *L-fuzzy sets*, which naturally instigated *L-fuzzy numbers* (Goguen, 1967).

Several authors studied *interval-valued fuzzy sets*, which implicitly set off *interval-valued fuzzy numbers* and their arithmetic (Grattan-Guinness, 1975; Jahn, 1975; Sambuc, 1975; Zadeh, 1975).

Bulgarian mathematician Krassimir Todorov Atanassov pioneered intuitionistic fuzzy sets (Atanassov, 1983, 1986), which brought about *intuitionistic fuzzy numbers* and their   arithmetic (Delgado *et al.*, 1998; Mitchell, 2004; Wang, 2008; Mahapatra and Roy, 2013; Kumar and Kaur, 2013; Prakash *et al.*, 2016).

In comparison with fuzzy sets, elements of an intuitionistic fuzzy set have two estimates — one estimate of their membership and another of their non-membership. The necessity to have two characteristics is related to the situations when it is impossible to have all information about membership in a fuzzy set. The same is true for intuitionistic fuzzy numbers.

Similar construction of pseudo fuzzy sets was introduced by Sukanta Nayak and S. Chakraverty (2017) bringing about *pseudo fuzzy numbers*.

The American mathematician Ronald Yager constructed *fuzzy multisets* under the name *fuzzy bags*, which brought forth *fuzzy multinumbers* as their measures (Yager, 1986).

The Belgian mathematicians Chris Cornelis, Martine De Cock and Etienne Kerre brought in intuitionistic fuzzy rough sets instigating *intuitionistic fuzzy rough numbers* (Cornelis *et al.*, 2003).

The Polish mathematician Anna Maria Radzikowska and Belgian mathematician Etienne Kerre introduced *L*-fuzzy rough sets, which brought forth *L-fuzzy rough numbers* (Radzikowska and Kerre, 2004).

The Indian mathematicians K.V. Thomas and Latha S. Nair established rough intuitionistic fuzzy sets instigating *rough intuitionistic fuzzy numbers* (Thomas and Nair, 2011).

The Chinese mathematicians Dan Meng, Xiaohong Zhang and Keyun Qin introduced *soft rough fuzzy sets* and *soft fuzzy rough sets*, which brought about *soft rough fuzzy numbers* and *soft fuzzy rough numbers* (Meng *et al.*, 2011).

In his study of the foundations of mathematics, the Russian-Ukrainian-American mathematician Mark Burgin unified sets, multisets, fuzzy sets and all their generalizations by the concept *named set*, which also comprises all other generalizations of sets (Burgin, 1990, 2004b, 2011). This gave birth to three classes of numbers studied in Burgin (2011):

- a formalized form of *named numbers* and their arithmetic;
- *cardinal multinumbers* and their arithmetic;
- *multicardinal numbers* and their arithmetic.

In turn, the variety of new types of numbers generated the diversity of arithmetics. In contrast to this, at the beginning of mathematics and for a long time of its existence, there was only one arithmetic — the Diophantine arithmetic of natural numbers. It gave birth to and was the foundation of the arithmetics of all basic classes of numbers discovered/created before the 20th century. A lot of applications gave evidence for the uniqueness of this arithmetic. However, as in the case of geometry, this uniqueness was extinguish when non-Diophantine arithmetics appeared (Burgin, 1977).

Creation of new arithmetics with new numbers was also caused by new tools for caculations. Due to the vital role, which computers play in the modern society, an important place in this diversity of arithmetics is occupied by *computer arithmetic*, which includes algorithms, hardware designs, software implementations, number systems, and arithmetic applications in emerging technologies (Parhami, 2002). Although computer arithmetics were created to allow computers to perform conventional operations with numbers, it was impossible to do this to the full extent because there were infinitely many natural and other standard types of numbers while computers were able to store and operate with only finite sets of numbers. As a result, any computer arithmetic contains only a part of numbers from any standard arithmetic, either from the conventional arithmetic of natural numbers or the conventional arithmetic of real numbers.

Computers brought forth a variety of arithmetics because there are different types of numbers in computers, such as fixed-point and floating-point numbers, and every one demands a specific arithmetic. Namely, there are three basic number systems — integers, fixed-point numbers, and floating-point numbers. These kinds of numbers have different representations and for each of them, the corresponding numeric arithmetic, such as the *decimal floating-point arithmetic* or *binary fixed-point arithmetic*, is constructed (Kneusel, 2015).

In addition, the concept of arithmetic was extended to mathematical systems, elements of which are not numbers. Examples are the arithmetic of real vectors, the arithmetic of complex vectors, the arithmetic of real matrices, or the arithmetic of complex matrices. To make terminology more consistent, we call such systems by the name *prearithmetic* (Burgin, 2019).

Thus, we see that the development of number systems and arithmetics has been very active and this process continues now. Looking at the history of numbers and arithmetic, we can observe the following regularity. Operating with numbers, mathematicians, scientists and other researchers come to problems, which cannot be solved using existing classes of numbers. To overcome these limitations, mathematicians and scientists start using formal expressions for intermediate results without accepting these expressions as final results. It happened, for example, with irrational expressions such as square roots of 2, 3 and many other natural numbers. Mathematicians had the same attitude at first to negative numbers and later to square roots of negative numbers. For instance, Cardano wrote about square roots of negative numbers:

"... *This, however, is closest to the quantity which is truly imaginary since operations may not be performed with it as with a pure negative number, nor as in other numbers. ... This subtlety results from arithmetic of which this final point is as I have said as subtle as it is useless.*" (Cardano, 1545)

However, time passes and instead of avoiding and/or rejecting strange objects encountered in their studies, mathematicians find how it is possible to rigorously represent them by new classes of numbers. In this process, they went from integer numbers to real numbers to complex numbers and so on.

One of the most lucid examples of such situations is related to infinite numbers. The great physicist Galileo Galilei, who was the first European researcher explicitly exploring properties of the actual infinity in a mathematical setting, discovered the counterintuitive fact that the whole can be equal to a part of it demonstrating that there were as many squares of natural numbers as all natural numbers (Galileo, 1638). Later based on this discovery, the great German mathematician and philosopher Leibniz asserted that existence on infinite numbers was absurd because they lead to a paradox (Leibniz, 1860–1875). Two centuries passed and in spite of this opinion, the great mathematician Cantor built a rigorous mathematical theory of infinite numbers (Cantor, 1878). This shows how infinite sets and infinite numbers being at first rejected, later achieved the authorized status in a rigorous mathematical theory.

Another example is related to physics. On the one hand, theoretical models of physical processes entailed divergence, that is, they brought in infinities. On the other hand, physicists never encountered infinities in their experiments. As a result, physicists started trying to avoid and eliminate infinities in their theories, while those infinite numbers that were known in mathematics at that time — transfinite numbers or hyperreals — were not helpful. However, mathematicians already created new classes of infinite numbers called hypernumbers providing tools for working with infinities in physics (Burgin, 1990b, 2012, 2017).

Non-Diophantine arithmetics also provide new means for theoretical research in physics (Benioff, 2012, 2016, 2020; Czachor, 2016, 2017, 2017b, 2020, 2020a; Czachor and Posiewnik, 2016; Chung and Hassanabadi, 2019; Chung and Hounkonnou, 2020) but in contrast to the theory of hypernumbers and extrafunctions, they do not employ new classes of numbers. Non-Diophantine arithmetics only introduce novel operations with already existing numbers. For instance, non-Diophantine arithmetics of natural numbers contain the same natural numbers as the Diophantine arithmetic N. These numbers satisfy the famous five axioms of Peano for the Diophantine arithmetic (Peano, 1889). However, addition and/or multiplication of natural numbers can be very different in many non-Diophantine arithmetics.

As we have seen, irrational and complex numbers emerged when mathematicians tried to solve more algebraic equations. Multinumbers come in the process of dealing with collections with groups of indistinguishable elements. To solve the problem of multiplication of distributions, the French mathematician Jean-François Colombeau introduced new generalized functions (Colombeau, 1982; 1984; 1985). The spaces of these generalized functions contained the space of the new kind of numbers, which were later called *Colombeau numbers* (Aragona and Juriaans, 2001; Vernaeve, 2010). Construction of these numbers involves high mathematical abstractions. Not to overload the reader of the first part of this book with abstract mathematical structures, we only mention that Colombeau numbers are highly abstract mathematical structures defined as equivalence classes in the space of mappings of function spaces.

Researchers also identified and taught arithmetics used in different areas. Examples are *financial arithmetic* (Blake, 2006), *physical arithmetic* (MacFarlane, 2010), *chemical arithmetic* (Chauvenet, 2012; Wells, 2018), *medical arithmetic* (Modell and Place, 1957) or *biological arithmetic* (Ledford, 2013).

In 1940s, a new mathematical structure called *category* was introduced in the framework of abstract algebra and topology by the distinguished French mathematician Samuel Eilenberg (1913–1998) and noted American mathematician Saunders Mac Lane (1909–2005) (Eilenberg and Mac Lane, 1942, 1945). This structure reflected the idea that the outer structure of mathematical objects based on the notion of structure-preserving mapping, such as homomorphisms in algebra or continuous mappings in topology, was more important in many situations than the inner structure of mathematical objects, which described their construction from separate elements. As the New Zealand mathematical logician Robert Ian Godlblatt explains, category theory takes an arrow (morphism) as the basic mathematical object instead of the membership concept basic for set theory (Goldblatt, 1979). Note that an arrow is a special kind of an elementary named set (Burgin, 2011).

Eilenberg was born in Warsaw, Russia (now Poland) and studied at the University of Warsaw. His main contribution was to algebraic topology and category theory.

Mac Lane was born in Norwich, Connecticut, and studied at Yale University, the University of Chicago and University of Göttingen. He taught at Yale University, Harvard University, Cornell University, and the University of Chicago. Mac Lane did research, in field theory, valuation theory, algebraic topology, algebra in general and category theory in particular making also contribution to the philosophy of mathematics and writing popular university textbooks.

According to Mac Lane, the term *category* was borrowed from the famous German philosopher Immanuel Kant (Mac Lane, 1998).

At first, categories were used only as a flexible mathematical language and were not considered as a worthy mathematical field in its own right. However, extending the utilization of categories in mathematics, mathematicians demonstrated that it was possible to build the whole mathematics in the framework of categories because such kind of categories as *topoi* allowed reconstructing set theory as a subtheory of category theory (cf., for example, Goldblatt, 1979). In particular, mathematicians elaborated a categorical counterpart of the arithmetic of natural numbers. Namely, the American mathematician Francis William Lawvere introduced a *natural numbers object* in topoi (Lawvere, 1963). Separate numbers were represented as morphisms of this object while the Peano (or Dedekind–Peano) axioms were also adopted in the categorical setting.

Later Robert Paré and Leopoldo Román studied natural numbers objects in a special kind of categories called *monoidal categories* (Paré and Román, 1989). Based on this construction, they developed the theory of primitive recursive functions in monoidal categories. Román also studied natural numbers objects in Cartesian categories (Román, 1989). He demonstrated that taking a natural numbers object N, it is possible to define conventional arithmetical operations — addition, multiplication, and subtraction — making N an arithmetic or a commutative semiring, which is a special case of prearithmetics (Burgin, 2019).

Writing about the history of numbers and arithmetic, it is necessary to emphasize that in the 20th century mathematicians discovered/created not only new classes of numbers but many new arithmetics based on already existing classes of numbers. Before the 20th century, new arithmetics originated together with new classes of numbers. In the 20th century, this process became independent with three main approaches: logical, algebraic and substantial.

In the *logical approach*, new arithmetics were created by changing axioms of the conventional Diophantine arithmetic. Research in this approach was stimulated by logical problems in mathematics and especially, by the problem of finding arithmetics with the decidable logical theory. As we know from the first Gödel theorem, the logical theory of the Diophantine arithmetic, which is called Peano arithmetic, is undecidable (Gödel, 1932).

In the *algebraic approach*, new arithmetics were created by changing operations in the conventional Diophantine arithmetic. Research in this approach was instigated by inability of the Diophantine arithmetic to correctly model some real-life situations.

Thus, the traditional schema of changing the numbers to get new arithmetics can be treated as the third approach, which can be called the *substantial approach*. As we already saw, research in this approach was set off by the efforts of mathematicians to solve more and more mathematical as well as practical problems. For instance, in such a way, arithmetics of integer numbers, rational numbers, and real numbers emerged.

Exploring the algebraic approach, it is possible to see that a key event in the history of arithmetic happened in 1975 when non-Diophantine arithmetics were discovered in the Platonic world of Ideas or as some assume, were constructed in the mental world or/and the physical world, in which people live.

As we know, for a long time, mathematicians and other people thought that there was only one geometry — the Euclidean geometry. Nevertheless, in the 19th century, many non-Euclidean geometries were discovered/constructed. This was the major mathematical discovery and advancement in the 19th century, which changed understanding of mathematics and the work of mathematicians providing new insights and tools for mathematical research.

The same event happened in the history of arithmetic because even longer than in the case with geometry, mathematicians and other people thought that there was only one arithmetic — the Diophantine arithmetic. Nevertheless, in the 20th century, many non-Diophantine arithmetics were discovered/constructed.

This discovery potentially has even more implications than the discovery of new geometries because the impact of the latter was only on mathematics and physics as the majority of people do not utilize geometry in their life. At the same time, all people use arithmetic of whole numbers. Operation with numbers is one of the most important activities of people in the contemporary society. Every woman and every man perform additions, subtractions and counting many times every day. Mathematicians created arithmetic to provide rigorous and efficient rules and algorithms for operation with numbers. Calculators and computers were invented to help people to operate with numbers. Modern science and technology are impossible without operation with numbers. Political, social and economic applications of mathematics are now multiplying with a great speed and the majority of these applications involve numbers and arithmetic operations.

Consequently, non-Diophantine arithmetics found applications in different areas. They were explicitly applied to physics (Czachor, 2016, 2017, 2017b, 2020; 2020a; Czachor and Posiewnik, 2016; Chung and Hassanabadi, 2019; Chung and Hounkonnou, 2020), psychology (Czachor, 2017a), cryptography (Czachor, 2020b), fractal theory (Aerts *et al.*, 2016, 2016a, 2018; Czachor, 2019), economics and finance (Burgin and Meissner, 2017). There are also implicit applications of non-Diophantine arithmetics to physics (Benioff, 2011, 2012, 2012a, 2013, 2015, 2016, 2016a), economics (Tolpygo, 1997), in idempotent analysis (Maslov, 1987; Maslov and Samborskii, 1992; Maslov and Kolokoltsov, 1997; Kolokoltsov and Maslov, 1997), and in tropical geometry, cryptography and analysis (Litvinov, 2007; Itenberg *et al.*, 2009; Speyer and Sturmfels, 2009; Grigoriev and Shpilrain, 2014; Maclagan and Sturmfels, 2015; Zhang *et al.*, 2018), information theory, and geometry (Benioff, 2020, 2020a; Berrone, 2010).

In more detail, we consider non-Diophantine arithmetics in the next chapter while now we consider interrelations between arithmetic and other mathematical structures. In particular, the development of arithmetics included enrichment of their structures. At first, additional operations were introduced. It was a graduate process instigated by the practical demands. Initial operations — addition and multiplication — are applied to pairs of numbers. Such operations are called binary in mathematics. Repetition of these operations gives new operations with three, four, five and more numbers. For instance, it is possible to add three numbers, e.g., $2+3+4 = 9$, or to multiply five numbers, e.g., $23456 = 720$. Then other operations, such as taking the arithmetical average, taking percent, taking maximum or taking minimum, were defined and utilized. Even more, infinite operations with numbers were also introduced and utilized.

In addition, an important property of numbers is that they are ordered. For instance, we know that 5 is larger than 2. This added the order structure to arithmetic. The order structures is defined for all known classes of numbers, only in some classes, such as natural, rational or real numbers, this structure is total, while in others, such as complex numbers, the standard order structure is partial.

One more structure discovered in many arithmetics was metric (distance function) determined by the distance between numbers. For instance, the distance between 5 and 2 is 3. A system with a metric is called a metric space. Metric structures or their generalizations are defined for all known classes of numbers. They bring better knowledge of number systems.

Metric spaces are special cases of topological spaces (Kuratowski, 1966). That is why arithmetic also inherited topological structures from intrinsic properties of numbers. Different number systems, such as real numbers or p-adic numbers, are constructed using the topological structure of rational numbers (Feferman, 1974; Gouvêa, 1997). The construction of hypernumbers is essentially topological (Burgin, 2012; 2017). Topology is a very useful tool in number theory (Hatcher, 2014).

There is even the field *arithmetic topology*, which is a combination of algebraic number theory and topology (Kapranov, 1995; Resnikov, 1997; Deninger, 2002).

Arithmetics were studied and developed not only in mathematics but also in logics. That is why we describe here the history of the logical approach to numbers and arithmetic.

Logical studies of numbers and arithmetic, which had started in the 19th century in works of Hermann Grassmann, Charles Peirce, Richard Dedekind and Giuseppe Peano, flourished in the 20th century.

The most groundbreaking result in this area was obtained by the famous Austrian-American logician Kurt Gödel (1906–1978) that any consistent finite system of axioms cannot allow deduction of all true sentences in the Diophantine arithmetic N (Gödel, 1931/1932). Now it is called the *first Gödel incompleteness theorem*.

This result was very unexpected and surprised all mathematicians and logicians because from the beginning of the 20th century the great German mathematician David Hilbert (1862–1943) assumed that mathematicians can solve any mathematical problem claiming in his address to the Society of German Scientists and Physicians in Königsberg in 1930:

Wir müssen wissen — wir werden wissen.

We must know — we will know.

However, Gödel's constraint result, which delimits logical deduction in arithmetic, is based on utilization of the so-called recursive algorithms, such as conventional logical deduction or Turing machine, for proving theorems in mathematics (Burgin, 2005). At the same time, application super-recursive algorithms, such as inductive Turing machine, allows proving all true sentences in the Diophantine arithmetic N due to the fact that there is, for example, a hierarchy of inductive Turing machines, which can compute and decide the whole arithmetical hierarchy (Burgin, 2003a; 2018d). It means that what is impossible to do with a system of deduction algorithms is possible to achieve using more powerful induction algorithms based on inductive Turing machines.

Arithmetical truth is not only nonprovable by recursive algorithms but it is also not representable in the standard arithmetic according to the undefinability theorem proved by famous Polish-American logician and mathematician Alfred Tarski (1901–1983), which states that arithmetical truth cannot be defined in formal arithmetic (Tarski, 1936). Now this result is also called Gödel–Tarski theorem because Gödel also discovered it in 1930, while proving his incompleteness theorems published in 1931 (Murawski, 1998). Note that this theorem is essentially based on the formal language used in arithmetic. Building a hierarchy of languages and axioms for arithmetic, which form a logical variety in the sense of (Burgin, 1997d; Burgin and de Vey Mestdagh, 2015), it is possible to define and prove arithmetical truth in this system.

Kurt Gödel (1906–1978) was born in Brünn, Austria–Hungary, and studied at the University of Vienna becoming a Privatdozent at this university. After Germany annexed Austria, Gödel immigrated to the United States and worked at the Institute of Advanced Studies in Princeton.

Alfred Tarski was born as Alfred Teitelbaum in Warsaw, Russia (now Poland) and studied biology at the University of Warsaw but then turning to mathematics abandoning biology. He taught logic and mathematics at the Polish Pedagogical Institute and the University of Warsaw becoming a prominent logician. He left Poland for the United States days before the German and Soviet invasion of Poland. There Tarski worked at Harvard University, City College of New York, the Institute of Advanced Studies in Princeton, and the University of California, Berkeley. He published works in set theory, logic, metamathematics, logic application to a variety of mathematical systems and on such topics as relation algebras and cylindric algebras.

The results of Gödel and Tarski were related to the ideas of Hilbert. Namely, to substantiate his claim about solvability of all mathematical problems, David Hilbert also elaborated an advanced program of building axiomatic foundations of the whole mathematics and proving consistency of these foundations by the so-called finitary methods, which meant that logical inference and logical operations had to be based on extralogical concrete objects, which were intuitively present as immediate experience prior to all thought (Hilbert, 1926).

However, Kurt Gödel also proved his *second incompleteness theorem*, which asserted that it was impossible to prove consistency of the Diophantine arithmetic (Peano arithmetic) N using the language of N and traditional deduction (Gödel, 1931/1932). As the Diophantine arithmetic is the most indispensable mathematical structure and traditional deduction was considered being the universal inference mechanism of mathematics, many believed that it was unquestionably impossible to prove consistency of the Diophantine arithmetic of natural numbers and Gödel theorems put absolute boundaries on what is possible to prove in mathematics.

However, as the history of mathematics shows, when mathematicians cannot solve some problems, they invent new structures and innovative techniques to overcome this "impossibility" (Burgin, 2012). The same happened with problem of consistency of the Diophantine arithmetic of natural numbers. The German mathematician and logician Gerhard Karl Erich Gentzen (1909–1945) suggested several proofs of the consistency of the Diophantine arithmetic of natural numbers using transfinite induction (Gentzen, 1936, 1938, 1943).

Gerhard Gentzen was born in Greiswald, Germany and studied at the University of Greiswald, University of Göttingen, University of Munich, and University of Berlin. He worked as an assistant to Hilbert joining the Nazi Party in 1937 and at the Mathematical Institute of the German University of Prague.

The German mathematician Wilhelm Friedrich Ackermann (1896–1962) also proved the consistency of the Diophantine arithmetic of natural numbers using methods from his paper (Ackermann, 1924) and the transfinite induction (Ackermann, 1940).

Wilhelm Ackermann was born Herscheid, Germany, and studied at the University of Göttingen. He taught at the gymnasiums in Burgsteinfurt Ludenscheid becoming a honorary professor at the University of Münster.

There were also other proofs the consistency of the Diophantine arithmetic of natural numbers using the transfinite induction (cf., for example, Lorenzen, 1951; Schütte, 1951, 1960; Hlodovskii, 1959; Gauthier, 2000). It was also found that the possibility of proving consistency of the Diophantine arithmetic depended on the way in which the metamathematical property of consistency is expressed in the language of the considered theory (Murawski, 2001).

It is necessary to explain that all these undecidability and incompleteness results related to arithmetic are about names and incompleteness of definite named sets in the sense of Burgin (2011). Indeed, proving his famous incompleteness theorem, Gödel used enumeration, which is now called Gödel numbering, of all arithmetical formulas (Gödel, 1931/1932; Rogers, 1958). However, enumeration is a specific naming where numbers are names of formulas. Formulas of a logical language are names of relations and sets (Church, 1956). Thus, Gödel–Tarski undefinability theorem asserts that arithmetical truth does not always have a name in the set of arithmetical formulas. In the most explicit form, the naming context of undecidability theorems was presented by Cornelis Huizing, Ruurd Kuiper and Tom Verhoeff, who proved two interesting generalizations of Rice Theorem, which demonstrate undecidability of many problems in the theory of algorithms and abstract automata (Huizing *et al.*, 2012).

The most popular logical formalization for the Diophantine arithmetic of natural (or in some versions, whole) numbers is called the *Peano arithmetic* because it is based on the axioms suggested by Giuseppe Peano or *first-order arithmetic* because it allows quantification only over numbers. In it, addition and multiplication are described by recursive axioms (cf., for example, Rasiowa and Sikorski, 1963; Manin, 1991; Kleene, 2002), which were first suggested by Hermann Grassmann (Wang, 1957).

Logicians started exploring the first-order arithmetic but found that they cannot deduce many important properties from the axioms. As the result, they began building and studying more restricted arithmetics. Later logicians proved that it was impossible to solve many problems in the Diophantine arithmetic using conventional mathematical deduction based on recursive algorithms (Gödel, 1931/1932, 1934; Church, 1936; Turing, 1936).

In 1923, the well-known Norwegian mathematician Thoralf Skolem (1887–1963) introduced the *primitive recursive arithmetic* (PRA), which was later called *Skolem arithmetic* (Skolem, 1923). It is the first-order theory of the natural numbers with multiplication, which is much weaker than Peano arithmetic because the latter includes not only multiplication but also addition. That is why, in contrast to Peano arithmetic, Skolem arithmetic is a decidable theory. This means that for any sentence in the language of Skolem arithmetic, it is possible to effectively determine, whether that sentence is provable from the axioms of Skolem arithmetic.

In 1929, the Polish mathematician Mojżesz Presburger (1904–1943) introduced another arithmetic, which was later called *Presburger arithmetic* (Presburger, 1929). It is the first-order theory of the natural numbers with addition, which is much weaker than Peano arithmetic because the latter includes not only addition but also multiplication. That is why, in contrast to Peano arithmetic, Presburger arithmetic is a decidable theory. It means that it is possible to find by a recursive, that is, conventional, algorithm whether a statement in this arithmetic is true or false. However, later it was proved that such algorithms will be extremely complex (Fischer and Rabin, 1974).

In 1950, the American mathematician Raphael Mitchel Robinson (1911–1995) introduced one more arithmetic, which was later called *Robinson arithmetic* (Robinson, 1950). It satisfies all axioms of Peano arithmetic but the axiom schema of induction. Although it is weaker than Peano arithmetic, Robinson arithmetic is also incomplete and essentially undecidable.

Predicative arithmetic is the arithmetic without the induction principle. It is mathematically sufficiently strong, but logically very weak working in special kind of theories (Nelson, 1986).

One more logically instigated mathematical structure is the *second-order arithmetic*. It allows quantification over sets of natural numbers while first-order arithmetics such as Peano arithmetic allow quantification only over numbers. Second-order arithmetic includes, but is significantly stronger than, its first-order counterpart — Peano arithmetic. In second-order arithmetic, it is possible to formalize not only the natural, whole or integer numbers but also the rational and real numbers because real numbers can be represented as (infinite) sets of natural numbers.

It is possible to treat second-order arithmetic as a weak version of set theory in which every element is either a natural number or a set of natural numbers (Simpson, 2009). Although it is much weaker than Zermelo–Fraenkel set theory, second-order arithmetic is sufficient for proving almost all of the results of classical mathematics expressible in its language. As a subfield of mathematical logic, second-order arithmetic is a collection of axiomatic systems that formalize the natural numbers and their subsets (Friedman, 1976).

Logicians also studied third-order arithmetic (Skelley, 2004; Hachtman, 2017) and higher-order arithmetics (Simpson, 2009; Cheng, 2019). Higher-order arithmetics are arithmetics of natural numbers formalized in higher-order logics (Väänänen, 2021). First-order logic uses quantifiers only on variables that range over individuals. Second-order logic, in addition, employs quantifiers over sets. Third-order logic, additionally, exploits quantifiers over sets of sets, and so on.

The concept *general arithmetic* is essentially related to *Peano arithmetic*, which is determined by the system of Peano axioms. One of these axioms tells that each numbers has the unique successor defining in such a way the successor function. In Peano arithmetic, there are additional demands on the successor function. First, it is total. Second, it is an injection of natural numbers into themselves, i.e., different numbers have different successors. Third, there is the first number which is not the successor of any other number. General arithmetic is the theory of natural numbers based on induction on the successor function but abandoning all these additional restrictions on this function. In spite of this, it is still possible to prove many properties of arithmetical operations such as commutativity and associativity of addition and multiplication, as well as Lagrange's Four-Square Theorem. Adding one more axiom, the injectivity of the succession function, allows proving many more interesting theorems such as quadratic reciprocity and Fermat's Theorem.

It is interesting that not only logic was employed in the studies of arithmetic but the process was also going in the opposite direction where arithmetic was used for exploration of logical systems. This process is called *arithmetization of logic* and *arithmetization of metamathematics* (Feferman, 1960; Gauthier, 2015).

Such important area of contemporary mathematical logic as model theory was also arithmeticized (Hajek and Pudlak, 1993).

Arithmetization of logic means replacing the reasoning with logical expressions by operating with natural numbers. This is achieved by enumeration of logical expressions using natural numbers. In this process, natural numbers become names of the expressions. The method of arithmetization provides a convenient system of naming sentences (Woleński, 2005/2004). This allows transformation of relations between and operations on expressions into relations between and operations on natural numbers.

It is assumed that the first arithmetization by enumeration was constructed by Kurt Gödel for the proof of the incompleteness of formal arithmetic (Gödel, 1931/1932). Now this enumeration is called a *Gödel numbering* or *Gödel enumeration* while the number assigned to an expression is called its *Gödel number*.

Gödel numbering as a tool of arithmetization has been used by different researchers in logic and computer science. For instance, the American logician Alonzo Church (1903–1995) used arithmetization in logic to obtain the first example of an undecidable algorithmic problem of arithmetic (Church, 1936).

Alonzo Church was born in Washington, DC, USA, and studied at Princeton University. He taught logic, mathematics, and philosophy at Princeton University and the University of California, Los Angeles. Among other contribution to logic, Church invented lambda-calculus.

Arithmetization is also an important process in computer science where computation with symbols is replaced by computing with natural numbers. Important results of computer science demonstrate equivalence of different models of algorithms and computation. One of these results proves the equivalence between Turing machines and partial recursive functions. Turing machines operate with words in some alphabet while partial recursive functions map natural numbers and their sets into the set of all natural numbers. Demonstrating that it is possible to change computing with words by computing with natural numbers uses a Gödel numbering (cf., for example, Martin, 1991).

However, influence of arithmetic on computer science has much deeper roots. Indeed, the central for computer science term *algorithm*, as we already discussed, originated from a book on arithmetic, which described basic computational algorithms working with numbers. The first mathematical model of algorithm was a special kind of recursive functions, which worked with natural numbers (cf. (Burgin, 2005)). Now models of algorithms that work with real numbers are elaborated (cf., for example, Blum, *et al.*, 1989; 1998). Many important characteristics of algorithms, such as computational complexity or algorithmic complexity, have numerical values (Balcazar, *et al.*, 1988; Blum, *et al.*, 1989; Burgin, 2010d; 2016b; Li and Vitanyi, 1997).

It is also necessary to understand that even the most abstract mathematical structures have numerical characteristics. For instance, fields in algebra have characteristics with values in whole numbers, e.g., the characteristic of the field of all real numbers is 0 (Kurosh, 1963). Vector spaces and linear algebras have dimensions, which are natural numbers for finite-dimensional vector spaces and linear algebras and are transfinite numbers for finite-dimensional vector spaces and linear algebras (Kurosh, 1963). Any algebraic or topological structure, e.g., a ring, torus or group, has the number of its elements. Numerical characteristics provide information about mathematical structures.

However, not only mathematicians, philosophers, and logicians studied numbers and arithmetic. In the 20th century, psychologists and educators started exploring how people work numbers and arithmetic (cf., for example, Piaget, 1964; Baird, 1975a; Baird and Noma, 1975; Noma and Baird, 1975; Weissman *et al.*, 1975; Dehaene, 1997; 2003; Dehaene and

Cohen, 1994; Slovic and Slovic, 2015). They explored such topics as the innate comprehension of numbers in individual mentality (cf., for example, Núñez, 2009; Dehaene, 1997; 2002; Dehaene *et al.*, 2003), how the brain operates with numbers performing arithmetical operations (cf., for example, Lakoff and Núñez, 2000; Núñez and Lakoff, 2005), and how children learn arithmetic (cf., for example, Piaget, 1952; Wynn, 1992a; Voigt, 1994; Dehaene, 1997a; Marzocchi *et al.*, 2002).

Moreover, at the end of the 20th century and the beginning of the 21st century, psychologists and biologists studied numerical information processing by insects, fish, birds and animals, such as bees, monkeys, lions, cats and rats (cf., for example, Agrillo *et al.*, 2006; Brannon and Terrace, 1998; Breukelaar and Dalrymple-Alford, 1998; Cantlon and Brannon, 2007; Dehaene *et al.*, 1998; Hauser *et al.*, 2003; Howard *et al.*, 2018; Merritt *et al.*, 2009; Nieder *et al.*, 2002; Nieder and Miller, 2004; Roberts, 1995; Roberts and Boisvert, 1998; Sawamura *et al.*, 2002; 2010; Sulkowski and Hauser, 2001; Thompson *et al.*, 1970; Vasas and Chittka, 2018; Verguts and Fias, 2004).

Psychologists found that "from the neural viewpoint, adult humans, nonhuman primates and young children activate the same brain regions involved in approximate representation of magnitudes, and these regions are also related to the development of mathematical intelligence quotient from infancy to adulthood in educated human beings" (d'Errico *et al.*, 2018).

Archeologists also started to study when and how people had begun using numbers and arithmetic. For instance, using comparative data from modern societies to form an interpretive framework, it is demonstrated that the Abri Blanchard bone, the Blombos Cave beads (ca. 75,000 years ago), Abri Cellier artifact (ca. 28,000 years ago), and Taï plaque (ca. 14,000 years ago) may signify the use of cognitive technologies, concepts like numbers and time that transform the cognitive and physical environments through mechanisms of imposed structure (Overmann, 2013, 2019). Archeologists also explored when hominins moved from the numerical cognition that we share with the rest of the animal world to number symbols (d'Errico *et al.*, 2018).

In addition, for over 100 years, researchers from various disciplines have been enthralled and occupied by the study of number words, that is, names of numbers in natural languages. As Andreea Calude writes:

"Phylogenetic modelling shows that low-limit number words are pre-served across thousands of years, a pattern consistently observed in several

language families. Cross-linguistic frequencies of use and experimental studies also point to widespread homogeneity in the use of number words. Yet linguistic typology and field documentation reports caution against positing a privileged linguistic category for number words, showing a wealth of variation in how number words are encoded across the world. In contrast with low-limit numbers, the higher numbers are characterized by a rapid and morphologically consistent pattern of expansion, and behave like grammatical phrasal units, following language-internal rules. Taken together, the evidence suggests that numbers are at the cross-roads of language history. For languages that do have productive and consistent number systems, numerals one to five are among the most reliable available linguistic fossils of deep history, defying change yet still bearing the marks of the past, while higher numbers emerge as innovative tools looking to the future, derived using language-internal patterns and created to meet the needs of modern speakers." (Calude, 2021)

Writing about history of arithmetic it is necessary to point out that mathematicians created several new mathematical fields intrinsically related to arithmetic. One of them is the *arithmetics of algebras* (Dickson, 1924). It studies the subalgebras of integer elements in algebras. Such elements naturally form arithmetics. Examples of the arithmetics of algebras are the arithmetic of complex integers, arithmetic of integer hypernumbers or arithmetic of integer quaternions.

Another field inherently related to arithmetic is *arithmetic geometry* (Cornell and Silverman, 1986; Lorenzini, 1996). Arithmetic geometry is an arithmetical counterpart of *algebraic geometry*. While algebraic geometry studies sets of solutions of multivariable polynomial equations in algebraically closed fields such as the field of complex numbers C, arithmetic geometry is interested instead in the solutions of such equations in the ring Z of integers and such fields as the field Q of rational numbers, the finite field F_p of p elements and their finite extensions.

An interesting phenomenon in the history of mathematics is that the proper algebra, i.e., symbolic algebra, was superseded by *geometric algebra*, in which algebraic by their essence problems were solved using geometrical tools. Much later, the dual mathematical field — *algebraic geometry* — emerged. Actually, analytic geometry of Descartes and Fermat can be called algebraic geometry because it studied geometrical objects using algebra and arithmetic. However, algebraic geometry acquired its contemporary meaning only in the 19th century when mathematicians started studies of zeroes of polynomials because the mains objects in this field are algebraic varieties, which are defined by zeroes of polynomials in multidimensional Euclidean spaces.

One more mathematical field naturally related to arithmetic is *arithmetic combinatorics*. It explores combinatorial properties of the basic arithmetical operations — addition, subtraction, and multiplication — estimating the sums, differences, and products of finite sets in different kinds of prearithmetics, or even infinite sets such as arithmetic progressions (Erdös and Szemerédi, 1983; Balog and Szemerédi, 1994; Elekes, 1997; Gowers, 1998; Łaba, 2008). By its methods, it is possible to characterize arithmetic combinatorics as a marriage of number theory, harmonic analysis, combinatorics, and ideas from ergodic theory, which aims to understand simple systems generated by the operations of addition and multiplication as well as their interaction (Green, 2009). The name of this field was coined by the Australian–American mathematician Terence Tao (Tao and Vu, 2006).

An important subfield of arithmetic combinatorics is *additive combinatorics*, i.e., combinatorics in which only the operations of addition and subtraction are considered (Tao and Vu, 2006; Bibak, 2013). *Arithmetic geometry* is a branch of algebraic geometry studying arithmetic schemes, i.e., schemes (usually of finite type) over the spectrum Spec(Z) of the arithmetic of integers (Poonen, 2009). It includes such mathematical structures as *arithmetic curves* and *arithmetic varieties*.

Moreover, algebraic geometry as geometry of spaces defined by polynomial equations depends not only on the type of the defining equations, but also on the choice of the numbers where we look for solutions. It means that algebraic geometry as a whole depends on the underlying arithmetic. For a long time, those were the arithmetic of real numbers and the arithmetic of complex numbers. Around 20 years ago, the new field called tropical geometry emerged. It is algebraic geometry based on the arithmetic of *tropical numbers* or *tropical arithmetic* (Mikhalkin, 2006; Mikhalkin and Rau, 2018).

There are two kinds of tropical arithmetics and two kinds of tropical numbers. Tropical numbers of the first type consist of all real numbers and the symbol ∞ (infinity). The corresponding arithmetic has two operations — addition, the role of which is performed by the operation of taking the maximum, for which ∞ is the largest element, and multiplication, the role of which is played by the addition of real numbers extended so that adding any real number to ∞ gives ∞.

Tropical numbers of the second type consist of all real numbers and the symbol $-\infty$ (negative infinity). The corresponding arithmetic has two operations — addition, the role of which is performed by the operation of taking the minimum, for which $-\infty$ is the least element, and multiplication, the role of which is played by the addition of real numbers extended so that adding any real number to $-\infty$ gives $-\infty$.

These two tropical arithmetics are isomorphic. They are important because both of them provide the base for idempotent analysis (Maslov, 1987; Maslov and Samborskii, 1992; Kolokoltsov and Maslov, 1997).

Many other arithmetics have been created. Mathematicians built the theory of *arithmetic differential equations* (Buium, 2005; 2010a) and *arithmetic non-commutative geometry* (Marcolli, 2005). They constructed *arithmetics on curves* (Mazur, 1986) and *arithmetics on fractals* (Aerts *et al.*, 2016a; 2018; Czachor, 2019). The arithmetic of dynamical systems or *arithmetic dynamics* is the study of number-theoretic properties of dynamical systems (Silverman, 2007; Benedetto *et al.*, 2019). It is related to the classical arithmetic geometry and classical complex dynamics. Arithmetic dynamics has two parts:

- Local arithmetic dynamics, also called p-adic nonarchimedian dynamics, is the dynamical system theory over p-adic fields, such as the field of real p-adic numbers or of complex p-adic numbers.
- Global arithmetic dynamics studies discrete dynamical systems.

We see that arithmetic is present in the majority of mathematical fields — geometry, topology, differential equations, combinatorics, the theory of dynamical systems, etc., which have an arithmetical subfield, e.g., arithmetic geometry, arithmetic topology, arithmetic differential equations, arithmetic combinatorics, arithmetic differential geometry, etc. It is possible to treat all these fields as parts of the mathematical area, which we can call *extended arithmetic*.

Note that a mathematical field can be a subfield of two or more other mathematical fields. For instance, arithmetic combinatorics is a subfield of combinatorics and extended arithmetic.

Moreover, arithmetic proliferated as the foundation of physics where *arithmetic quantum physics* was developed (Volovich, 1987; Altaisky and Sidharth, 1998; Varadarajan, 2002, 2004; Czachor, 2016). Non-Diophantine arithmetics provided novel efficient tools for the development of new directions in quantum physics and cosmology (Burgin and Czachor, 2020; Czachor, 2017; 2020a,b; Czachor and Nalikowski, 2021). In addition, the *arithmetic geometry of principal bundles* is closely related to the constructions of quantum field theory (Kim, 2017).

Chapter 4

Discovery of Non-Diophantine Arithmetics and their Applications

Arithmetic must be discovered in just the same sense in which Columbus discovered the West Indies, and we no more create numbers than he created the Indians.

Bertrand Russell

Before the creation of set theory, mathematicians regarded the arithmetic of natural numbers N as the intrinsic foundations of mathematics. The prominent German mathematician Leopold Kronecker wrote: "*God made the natural numbers, all the rest is the work of man.*" As a result, laws of this arithmetic, such as $2 + 2 = 4$, were treated as an obvious absolute self-evident truth, which did not need any testing or explanation. For instance, this was an opinion of the great mathematician and philosopher René Descartes although he theorized that this equality might, in fact, have no reality outside the mind (Descartes, 1641).

Laymen have been even more unrelenting in these issues. Almost all people had and even now have no doubts that $2 + 2 = 4$ is the most evident truth in the world. Philosophers used this equality as an example of unquestionable truth. For instance, analyzing the development of Popper's ideas, Frank Hutson Gregory wrote: "It is difficult to conceive how simple statements of arithmetic, such as "$2 + 2 = 4$", could ever be shown to be false" (Gregory, 1996). Mathematicians supported this attitude. For instance, according to the *American Mathematical Monthly* (April, 1999, p. 375), "Although other sciences and philosophical theories change their 'facts' frequently, $2 + 2$ remains 4."

In contrast to this, the arithmetical expression $2 + 2 = 5$ was used for a long time to indicate a high level of absurdity ($2 + 2 = 5$, Wikipedia).

For instance, Ephraim Chambers giving a definition of the word *absurd* in his *Cyclopædia, or an Universal Dictionary of Arts and Sciences*, uses this expression as paradigmatic example of something absurd:

"*Thus, a proposition would be absurd, that should affirm, that two and two make five; or that should deny 'em to make four.*" (Chambers, 1728)

The same opinion is strong even now as the title "2 + 2 never equals 5" of an article by James Lindsay shows (Lindsay, 2020).

In opposition to the widely held opinion, some people felt more positive towards this expression (cf., for example, Byron, 1974; Allais, 1895). For instance, the famous English poet George Gordon Byron, 6th Baron Byron (1788–1824) wrote:

"*I know that two and two make four — and should be glad to prove it too if I could — though I must say if by any sort of process I could convert 2 plus 2 into five it would give me much greater pleasure.*"

However, the negative opinion in mathematics and beyond prevailed and was still popular in the 20th century as even the best mathematicians did not see any alternative to the Diophantine arithmetic. For instance, in the dystopia "1984" by the English novelist George Orwell (1903–1950), the arithmetical equality $2 + 2 = 5$ is treated as an obviously false dogma connoting official deception of the totalitarian state, secret surveillance and brazenly misleading terminology.

In contrast to this, in another dystopia "We" by Evgeniy Zamyatin, the arithmetical equality $2 + 2 = 4$ is treated as a principal dogma of the totalitarian state, which cannot be questioned because even uncertainty in the truthfulness of this arithmetical statement is punished (Zamyatin, 1924).

However, time passed, and starting from the beginning of the 21st century, the expression $2 + 2 = 5$, along with another similar expression $1 + 1 = 3$, has become the symbol and metaphor of synergy — an important natural and social phenomenon when the union (sum) of the parts is larger than those parts simply taken together (Cambridge Business Dictionary; Grimsley, 2018).

Being simpler, the expression $1 + 1 = 3$ has been brought into play in various spheres including business and industry (cf., for example, Beechler, 2013; Gottlieb, 2013; Grant and Johnston, 2013; Hattangadi, 2017; Jude, 2014; Kress, 2015; Marks and Mirvis, 2010; Phillips, 2008; Murphy and Miller, 2010; Renner, 2013; Ritchie, 2014), economics and finance (cf., for example, Burgin and Meissner, 2017), psychology and sociology (cf., for example, Boksic, 2019; Brodsky *et al.*, 2004; Bussmann, 2013; Enge, 2017; Frame and Meredith, 2008; Jaffe, 2017; Klees, 2006; Mane, 1952; Trott, 2015)), library studies (cf., for example, Marie, 2007), biochemistry and bioinformatics (cf., for example, Kroiss *et al.*, 2009), computer science (cf., for example, Derboven, 2011; Glyn, 2017; Lea, 2016; Meiert, 2015), physics (cf., for example, Lang, 2014), chemistry (cf., for example, Murphy, 1999), medicine (cf., for example, Lawrence, 2011; Phillips, 2016; Trabacca *et al.*, 2012; Caesar and Cech, 2019), pharmacology (cf., for example, Cohen, 2011), and pedagogy (cf., for example, Nieuwmeijer, 2013). In this context, Vidya Hattangadi writes that in mathematical terms, synergy is also explained as $1 + 1 = 3$ (Hattangadi, 2017).

Note that it is possible to suggest the following non-mathematical interpretation of this expression where $1 + 1 = 3$ means that there are three symbols: 1, +, and 1. Another possible interpretation that comes to mind is: the union (addition) of one man and one woman gives birth to the third person — a child. As the result, the union (addition) of two people gives three people as the result. However, these and other similar interpretations do not agree with the rules of the conventional Diophantine arithmetic where $1 + 1$ is always equal to 2.

Some researchers also considered negative synergy, which was represented in mathematical terms as $2 + 2 = 3$ (Grimsley, 2018). This expression was also unacceptable from the point of view of conventional mathematics and its Diophantine arithmetic but became legitimate from the broader perspective of non-Diophantine arithmetics.

In the sarcastic history of $2 + 2 = 5$ written as parody on the history of mathematics, it is mockingly conveyed that "Riemann developed an arithmetic in which $2 + 2 = 5$, paralleling the Euclidean $2 + 2 = 4$ arithmetic" (Houston, 1990). However, this never happened and only much later than Riemann lived, many arithmetics, in which $2 + 2 = 5$, were constructed by Mark Burgin, who called them non-Diophantine arithmetics (Burgin, 1977; 1997). The family of non-Diophantine arithmetics is so immense that in some of them $1 + 1 = 3$ is true while in others $2 + 2 = 3$ is correct. This essentially contradicts what people know about arithmetic.

Indeed, we know that people's experience with numbers and, especially, with natural numbers is profound. From ancient times, much longer than they have worked with the Euclidean geometry, people have believed and continue to believe that only one arithmetic of natural numbers — the Diophantine arithmetic — has always existed and no other arithmetic can ever exist. In this arithmetic, it always $1 + 1 = 2$ and $2 + 2 = 4$.

However, do you think that we know everything about numbers and counting? The answer is negative because there were outstanding thinkers who doubted the absolute character of the conventional Diophantine arithmetic, giving examples when this arithmetic did not correctly describe certain systems and processes. It is possible to trace the roots of this fundamental problem of arithmetic relevance to people's everyday life and experience to ancient Greece. Long ago, there was a group of philosophers, who were called Sophists and lived from the second half of the fifth century B.C.E. to the first half of the fourth century B.C.E. Sophists asserted relativity of human knowledge and elaborated various brainteasers, explicating complexity and diversity of the real world. One of them, the famous Greek philosopher Zeno of Elea (ca. 490–430 B.C.E.), who was said to be a self-taught country boy, invented very impressive paradoxes, in which he challenged the popular knowledge and intuition related to such fundamental essences as time, space, and number (Yanovskaya, 1963). One of them is the *paradox of the heap* or the *Sorites paradox* (as σωρος [soros] means a heap in Greek), which is described in the following way.

(1) One million grains of sand, for example, make a heap.
(2) If we take away [or add] one grain, this will be actually the same heap.

Analyzing this situation, the famous American popular-science and popular-mathematics writer Martin Gardner (1914–2010) explained:

"Repeated applications of premise 2 (each time starting with one less number of grains), will eventually allow us to arrive at the conclusion that 1 grain of sand makes a heap. On the face of it, there are three ways to avoid that conclusion. Object to the first premise (deny that one million grains makes a heap, or more generally, deny that there are heaps), object to the second premise (it is not true for all collections of grains that removing one grain cannot make the difference between it being a heap or not), or accept the conclusion (1 grain of sand can make a heap). Few, if any, reply by accepting the conclusion. In addition to advocating a response, philosophers who work on this paradox also try to explain why it is that the premise one would have to deny seems so plausible, despite being false." (Gardner, 2005)

However, the main problem was not the reasoning but the contradiction with what the conventional arithmetic told people. Namely, the conventional arithmetic asserts that taking any number and adding or subtracting one, we get a new number. In the same way, the conventional arithmetic tells that taking any big number, say 10,000,000, and adding one, we get a new number. At the same time, the heap of sand that contains 10,000,000 grains remains actually the same if one grain is added to it.

The reader may ask why we are interested in puzzles that were suggested thousands years ago and look artificial to the modern reader. However, the paradox of a heap has a direct analogy in our times both in science and everyday life. For example, you are buying a car for $30,000. Then suddenly, when you have to pay, the price is changed and becomes one dollar larger. Do you think that the new price is different from the initial one or do you consider it practically the same price? It is natural to suppose that any sound person has the second opinion. Consequently, we come to the same paradox: if k is the price of the car in cents, then in the conventional Diophantine arithmetic $k + 1$ is not equal to k, while in reality they are equal as prices because k is much larger than 1.

Moreover, imagine that you are going to receive your salary in cents, the sum of which will be equal to the amount that you receive now but which will be given to you once a year. Do you think it will be the same salary or not if you get one cent less or one cent more? For many people, there is a difference how often they receive their salary, although the sum remains the same. This also does not agree with the rules the conventional Diophantine arithmetic.

Here is one more situation from real life. Imagine that you have $100,000,000 and somebody gives you $1. Will you say that now you have $100,000,001? No, you will probably assume that you have the same amount of money. This contradicts the rules of the Diophantine arithmetic where 100,000,000 and 100,000,001 are different numbers. These problems are of a psychological nature, but they have nontrivial implications for economic modeling.

The described examples show that in some cases we encounter inconsistency between the real life and the Diophantine arithmetic. There are two basic ways to deal with inconsistencies: one is to elaborate an inconsistent system and try to work with it, and another way is to create new mathematical structures, eliminating inconsistencies. The book "*How Mathematicians Think*" of the Canadian mathematician and philosopher William Byers has the subtitle "*Using Ambiguity, Contradiction, and Paradox to Create Mathematics*" because mathematical reasoning, according to Byers, is not completely algorithmic, computational or based on proof systems. It primarily uses creative ideas to shed new light on mathematical objects and structures, propelling in such a way mathematical progress (Byers, 2007). Ambiguities, contradictions, and paradoxes play the central role in the emergence of creative ideas. Being unsolvable when they appeared, many problems and paradoxes find their solutions on a higher level of cognition.

As we show further, non-Diophantine arithmetics solve the paradox of the heap and other paradoxes, which emerge when we compare some real-life situations with the rules of the conventional Diophantine arithmetic.

The same happened with many profound ideas of ancient philosophers. For instance, now all educated people know that material things are built of atoms. However, idea of atoms was introduced much earlier than atoms were really discovered in nature. Outstanding philosophers Democritus (ca. 460–370 B.C.E.) and Leucippus (fl. 5th century B.C.E.) from ancient Greece suggested the idea of atoms as the least physically indivisible particles of matter.

Leucippus was a student of the Sophist Zeno of Elea (ca. 490–430 B.C.E.), and although Aristotle wrote about Leucippus, there was a controversy related to the existence of Leucippus.

Democritus was a citizen of Abdera known as the "laughing philosopher" due to making emphasis on the value of "cheerfulness." He wrote on many fields, which included mathematics, physics, music, ethics, and cosmology. His works did not survive and we have information about them from works of other philosophers.

We know that for a long time, the atomistic doctrine was considered false due to the fact that scientists were not able to go sufficiently deep into the matter. Nevertheless, the development of scientific instruments and experimental methods made possible to discover such micro-particles that were and are called atoms, although they possess very few of those properties that were ascribed to them by ancient philosophers.

Another great idea of ancient Greece was the world of Ideas or Forms whose existence was postulated by Plato. In spite of the attractive character of this idea, the majority of scientists and philosophers believe that the world of ideas does not exist, because nobody had any positive evidence in support of it. The crucial argument of physicists is that the main methods of verification in modern science are observations and experiments, and nobody has been able to find this world by means of observations and experiments. Nevertheless, some modern thinkers, including such outstanding intellectuals as philosophers Karl Raimund Popper (1902–1993) and Alain Badiou, logicians Gottlob Frege (1848–1925), Georg Kreisel (1923–2015) and Kurt Gödel (1906–1978), mathematicians Alain Connes, David Mumford, and René Frédéric Thom (1923–2002), computer scientist

Gregory Chaitin, and physicists Werner Karl Heisenberg (1901–1976) and Roger Penrose, continued to believe in the world of Ideas giving different interpretations of this world but suggesting no ways for their experimental validation.

However, science is developing, and this development led to the discovery of the world of structures (Burgin, 2012a). On the level of ideas, this world may be associated with the Platonic world of Ideas, in the same way as atoms of the modern physics may be related to the atoms of Democritus. Existence of the world of structures is proved by means of observations and experiments (Burgin, 2017b). This world of ideal structures forms the structural level of the world as a whole. Each system, phenomenon or process, either in nature or in society, has some structure. These structures exist like tables, chairs, or buildings, and form the structural level of the world. When it is necessary to investigate or to create some system or process, it is possible to do this only by means of knowledge of the corresponding structure. Structures determine the essence of things in the same way as Aristotle ascribed to forms of things. Consequently, structures unite Ideas of Plato with Forms of Aristotle, eliminating contradictions that existed between their teachings about reality.

Still, while Greek sages and subsequent thinkers posed questions about the conventional Diophantine arithmetic, they suggested no answers that would allow improving the situation and eliminating contradictions. As a result, for more than two thousand years these problems were forgotten and everybody was satisfied with the conventional Diophantine arithmetic. The reason was that in spite of all the problems and paradoxes, this arithmetic has remained very and very useful in the practical activity and theoretical investigations of people.

The famous German scientist Herman Ludwig Ferdinand von Helmholtz (1821–1894) was may be the first thinker, who in modern times questioned absolute authority of the conventional arithmetic. In his work *"Counting and Measuring"*, Helmholtz considered an important problem of applicability of arithmetic to physical phenomena, although at that time people knew only one arithmetic (Helmholtz, 1887). This was a natural approach of a scientist, who even mathematical statements tested by the main criterion of science — observation and experiment.

Herman Helmholtz was born in Potsdam, Germany and studied at Medicinisch-chirurgisches Friedich-Wilhelm-Institute getting a medical doctorate. He worked at several universities in Germany — at the Pussian

University of Königsberg, University of Bonn, University of Heidelberg, and Humboldt University in Berlin.

Herman Helmholtz contributed to a variety of fields. In physics, he elaborated the theory of energy conservation making essential contribution to thermodynamics and electrodynamics. In psychology and physiology, he developed theories of vision and mathematics of the eye, suggested important ideas on color vision, perception of sound and space, and sensation of tone. Helmholtz also contributed to the philosophy of science.

Interestingly, the scientific approach to arithmetic of Herman Helmholtz essentially differed from the opinion of the great German mathematician Carl Friedrich Gauss (1777–1855), who assumed that arithmetic was purely aprioristic in contrast to geometry, which should be ranked with mechanics. It means that people create arithmetic as they want while geometry obeys the laws of nature.

Taking into account the role and nature of mathematics, it is possible to suggest that people can create or discover arithmetics without any connection to nature but when people want to apply arithmetic to a natural system, it is necessary to take into account the properties of this system.

The first observation of Helmholtz was that the commonly utilized arithmetic, that is, the Diophantine arithmetic of natural numbers, had to be applicable to practical experiences because the concept of number was derived from some practice. However, it is easy to find many situations when this assumption is not true. To mention but a few situations described by Helmholtz, we give the following examples.

One raindrop added to another raindrop does not make two raindrops but forms only one raindrop. Mathematically, this situation is described by the equality $1 + 1 = 1$.

It is interesting that this situation is exactly reflected in set theory where the union of one set with another set is only one set and not two sets (Abian, 1965). This means that when mathematicians combine sets, they also have the rule $1 + 1 = 1$ in terms of arithmetic. Moreover, another combination of sets called intersection also merges two sets into one once more demonstrating that it is possible that $1 + 1 = 1$. Even more, in algebra and arithmetic there are many systems, such as groups, semigroups, rings fields, Boolean algebras, and prearithmetics, with binary operations. A binary operation takes two elements and blends them into one element, and this once more brings us to the rule $1 + 1 = 1$.

One more example is when one mixes two equal volumes of water, one at 40° Fahrenheit and the other at 50° Fahrenheit, one does not get two volumes at 90° Fahrenheit. This also contradicts the conventional arithmetic where $40 + 50 = 90$.

In a similar way, the conventional arithmetic fails to describe correctly the result of combining gases or liquids by volumes. For example, one quart of alcohol and one quart of water yield about 1.8 quarts of vodka (Kline, 1980).

Later the famous French mathematician Henri Léon Lebesgue (1875–1941) facetiously indicated (cf., for example, Kline, 1980) that if one puts a lion and a rabbit in a cage, one will not find two animals in the cage later on. In terms of numbers, it will mean $1 + 1 = 1$.

Henri Lebesgue was born in Beauvais, France, and studied at the École Normale Supérieure and the Sorbonne where he specialized in mathematics. During his life, Lebesgue worked at the University of Rennes, University of Poitiers, Sorbonne and College de France. He was also a member of the Académie des Sciences. His main achievement was the theory of measure and integration, which essentially extended all other theories in this area of mathematics.

The famous Austrian–New Zealand–British philosopher Karl Raimund Popper (1902–1994) also pointed out that the equality $2 + 2 = 4$ is not always true as a physical fact. He wrote:

More important is the application in the second sense. In this sense, "$2 + 2 = 4$" *may be taken to mean that, if somebody has put two apples in a basket, and then again two, and has not taken any apples out of the basket, there will be four in it. In this interpretation* "$2 + 2 = 4$" *helps us to calculate, i.e., to describe certain physical facts, and the symbol "+" stands for a physical manipulation — for physically adding certain things to other things.... But in this interpretation* "$2 + 2 = 4$" *becomes a physical theory, rather than a logical one; and as a consequence, we cannot be sure whether it remains universally true. As a matter of fact, it does not.... It may hold for apples, but it hardly holds for rabbits. If you put $2 + 2$ rabbits in a basket you may soon find 7 or 8 in it.* (Ryle et al., 1946)

Karl Popper was born in Vienna and studied at the University of Vienna. He worked at the Canterbury University College of the University of New Zealand in Christchurch, and the London School of Economics, which was was a constituent of the University of London. The main achievement of Popper was the development of an important direction in the philosophy of science, which he called "critical rationalism." It was based on the necessity of observations in science and the *principle of falsifiability* of scientific statements and theories. It means that a statement T is falsifiable if there is a possibility to show that T is false.

Popper also wrote on open society and liberal democracy, origin and evolution of life, free will, and epistemology investigating the problems of truth and suggesting the construction of three worlds — the physical world, the world of mind and mental states, and the world of knowledge — as the global structure of the world as a whole (Popper, 1974, 1979). Popper world's structure was criticized and its limitation were eliminated by the introduction of the Existential Triad as the global structure of the world as a whole (Burgin, 1997b, 2012a).

Coming back to arithmetic, we find that similar to Popper's opinion on arithmetic was expressed by the great French mathematician Jules Henri Poincaré (1854–1912), who believed that arithmetics had to be tested by experiments pointing out that one did not "prove" $2 + 2 = 4$, one "checked" it (Gonthier, 2008). This is exactly what some researchers did. Performing definite mental experiments, Helmholtz, Lebesgue and Kline demonstrated that there are situations when two plus two is not equal to four.

Henri Poincaré was born in Nancy, Lorraine, France, and studied at the École Polytechnique and the École des Mines graduating as a mining engineer. He worked at the University of Caen, Sorbonne, and École Polytechnique where he was a chair. He liked teaching a course in a new area each year including lectures in optics, electricity, the equilibrium of fluid masses, the mathematics of electricity, astronomy, thermodynamics, light, and probability. At the same time, Poincaré worked at the Ministry of Public Services as an engineer as well as inspector general of the Corps de Mines.

In spite of this, Poincaré was very prolific in his research in mathematics, science, and philosophy being called a polymath and the last universalist. He made many important contributions to pure and applied mathematics, mathematical physics, and celestial mechanics. In pure mathematics, his works laid the foundation of topology, while he created the new field — the qualitative theory of differential equations, made important contributions to group theory, number theory (higher arithmetic), and the theory of automorphic functions.

In applied mathematics and mathematical physics, Poincaré contributed to quantum theory, relativity theory, physical cosmology, electricity, fluid mechanics, optics, telegraphy, thermodynamics, potential theory, and elasticity.

To be able to achieve his outstanding results, Poincaré kept very precise working hours and tended to develop his results from the first principles at first, as rule, solving the problem in his head and only then writing down his results.

In contrast to the opinions of Poincaré and Popper, the great German mathematician Carl Friedrich Gauss, whose mathematical achievements and life were considered in Chapter 2, had a different attitude to numbers assuming that they had nothing to do with the physical reality. In his 1830 letter to the well-known German astronomer, mathematician, physicists, and geodesist Friedrich Wilhelm Bessel (1784–1846), Gauss wrote:

"We must admit with humility that, while number is purely a product of our minds, space has a reality outside our minds, so that we cannot completely prescribe its properties a priori."

Because of this opinion was shared by the vast majority of mathematicians, very few (if any) researchers paid attention to the work of Helmholtz on arithmetic as well as to observations of other mathematicians pointing at only partial adequacy of the conventional arithmetic. As a result, because no alternative to the conventional arithmetic was still suggested, the problems with the conventional arithmetic were once more forgotten.

It took almost hundred years to revive doubts of Helmholtz that the Diophantine arithmetic of natural numbers provides absolute truth. The most extreme view on the arithmetic was expressed by the mathematical direction called ultraintuitionism where it was postulated that there was only a finite quantity of natural numbers (Yesenin-Volpin, 1960, 1970). In a similar way, the Dutch mathematician David van Dantzig (1900–1959) gives reasons for the assumptions that only some of natural numbers, such as $10^{10^{10}}$, may be considered finite (van Dantzig, 1956). Consequently, all other mathematical entities that are traditionally called natural numbers are only some expressions but not numbers. These arguments were later supported and extended in (Blehman *et al.*, 1983). As the American mathematician and computer scientist Samuel Buss indicated, a number of logicians and philosophers (cf., for example, Parikh, 1971; Sazanov, 1980; 1981; Nelson, 1986) have doubted the concrete existence of very large numbers, such as $67^{257^{729}}$, which were formed using exponential terms (Buss, 1996).

In addition, we know that people and even computers operate only with finite sets of numbers while any computer arithmetic is finite (Flynn and Oberman, 2001). Even more, mathematicians found that from the operational perspective, not all natural numbers are the same while in the conventional Diophantine arithmetic all natural numbers — big and small — behave in the same way.

As a matter of fact, much earlier than non-Diophantine arithmetics were discovered, the English mathematician John Edensor Littlewood (1885–1977) considered an example demonstrating how the rules of non-Diophantine arithmetics (in spite of that they were unknown at that time) can be imposed upon the real world (Littlewood, 1953). Several similar and even more lucid examples are given by the American mathematicians and philosophers Philip Davis and Reuben Hersh (Davis and Hersh, 1986) and by the American mathematics historian, philosopher and educator Morris Kline (Kline, 1967). For instance, when a cup of milk is added to a cup of popcorn then only one cup of mixture will result because the cup of popcorn will very nearly absorb a whole cup of milk without spillage. So, in this case we also have $1 + 1 = 1$. It is impossible in the conventional arithmetic but it is true in some non-Diophantine arithmetics.

To make the situation, when ordinary addition is inappropriate, more explicit, an absurd but not unrelated question is formulated: If the Mona Lisa painting is valued at \$10,000,000, what would be the value of two Mona Lisa paintings?

Let us consider more examples of situations when the Diophantine arithmetic does not work.

(1) A market sells a can of tuna fish for \$1.05 and two cans for \$2.00. So, we have $a + a \neq 2a$.

(2) In a similar way, coming to a supermarket, you can buy one gallon of milk for \$2.90 while two gallons of the same milk will cost you only \$4.40. Once more, we have $a + a \neq 2a$.

(3) Even more, coming to a supermarket, you can see an advertisement "Buy one, get one free." It actually means that you can buy two items for the price of one. Such advertisement may refer almost to any product: bread, milk, juice etc. For example, if one gallon of orange juice costs \$2, then we come to the equality $2 + 2 = 2$. It is impossible in the conventional arithmetic but it true for some non-Diophantine arithmetics.

Another property of the Diophantine arithmetic was also challenged. Some researchers although being moderate in their criticism of the conventional arithmetic, suggested that not all natural numbers are the same in contrast to the presupposition of the conventional arithmetic that the set of natural numbers is uniform (Mannoury, 1909; Poincaré, 1913;

Littlewood, 1953; Kolmogorov, 1961; Birkhoff and Barti, 1970; Rashevsky, 1973; Dummett, 1975; Knuth, 1976; Wittgenstein, 1983). On these lines, mathematicians separated all natural numbers into several classes without changing the conventional arithmetic. For example, Kolmogorov suggested that in solving practical problems, it is worth to separate *small, medium, large,* and *super-large* numbers (Kolmogorov, 1961, 1979).

We remind that a natural number k is called *small* if it is possible in practice to list and work with all combinations and systems such that are built from k elements each of which has two inlets and two outlets.

A natural number m is called *medium* if it is possible to count to and work directly with this number. At the same time, it is impossible to list and work with all combinations and systems that are built from m elements each of which has two or more inlets and two or more outlets.

A natural number n is called *large* if it is impossible to count a set with this number of elements. However, it is possible to elaborate a system of denotations for these elements.

If even this is impossible, then a number is called *super-large.*

According to this classification, 3, 4, and 5 are small numbers, 100, 120, and 200 are medium numbers, while an example of a large number is given by the quantity of all visible stars. Really, if we invite 4 people, we can consider all their possible positions at a dinner table. If you come to some place where there are 100 people, you can shake hands with everybody although it might take too much time. What concerns the visible stars, you cannot count them, although, a catalog of such stars exists. Using this catalog, it is possible to find information about any of these stars.

This classification of numbers is based on people's counting abilities. Consequently, borders between classes are vague and unstable. Higher counting abilities make borders between classes higher. For example, 10 is a medium number for an ordinary individual, but a small number for a computer. However, some numbers belong to a definite class of this typology in all known situations. For example, 300 is a medium number both for people and computers.

In a similar way to what has been done by Kolmogorov and on the akin grounds, the English mathematician John Edensor Littlewood (1885–1977) separated all natural numbers into an infinite hierarchy of classes (Littlewood, 1953).

Littlewood was born in Rochester, Kent, UK and studied in the University of Cambridge. He worked in number theory, analysis, and differential equations making significant contributions to the field of ballistics.

The described perception that small numbers are more concrete while large numbers are more abstract was mathematically formalized by the American mathematician, logician and philosopher Rohit Jivanlal Parikh who introduced the concept of feasible numbers (Parikh, 1971). To reflect operational properties of natural and integer numbers he built a mathematical theory, in which whole or integer numbers were divided into two classes — feasible and not feasible. In a similar way, operations of addition and multiplication were considered feasible, exponentiation was treated as not feasible because its application to feasible numbers resulted in numbers that were not feasible. In the context of the conventional Diophantine arithmetic, this theory is inconsistent while non-Diophantine arithmetics provide means for construction of a consistent theory of feasible numbers and operations.

In spite of problems with consistency, the approach of Parikh was further developed by different researchers. While Parikh formalized feasibility of arithmetical operations, Vladimir Sazonov elaborated a formal approach to *feasible numbers*, as well as to *middle* and *small* numbers based on ideas of Parikh (Sazonov, 1995). According to his approach, the inequality log log $n < 10$ is a formal condition defining feasible numbers. In a similar but logically consistent way, it is possible to delineate feasible numbers by choosing relevant non-Diophantine arithmetics.

Another direction motivated by ideas of Parikh was related to the bounded arithmetics, in which such operations as addition and multiplication were formally admissible but not exponentiation was not a permissible arithmetical operation (Buss, 1986; 1986a; Krajiček, 1995). The main efforts in this area have been directed at exploration of proof complexity (Carbone, 1999; Buss, 1991).

Non-Diophantine arithmetics suggest another solution to the problem of numerical feasibility. To solve this problem, it is not necessary to exclude almost all natural numbers from the arithmetic. It is sufficient to redefine operations with numbers. Namely, there are non-Diophantine arithmetics, in which only a finite quantity of numbers is accessible starting with the number 1 and applying the basic arithmetical operations (Burgin, 1997). It is possible to name these numbers *accessible* or *feasible* and to call other natural numbers *trans-feasible*. This approach gives an arithmetic compatible with the Peano axioms and does not entail contradictions as in the case of the approach of Parikh and some other mathematicians (Parikh, 1971; Meyer and Mortensen, 1984; Priest, 1994, 1994a, 1996, 1997, 2000; Mortensen, 1995; Priest, 2003a; Rosinger, 2008).

It is interesting that feasibility of addition and multiplication in contrast to exponentiation is intrinsically related to the concept of tractability
in computer science. According to the contemporary classification (Balcazar
et al., 1998; Hopcroft *et al.*, 2007; Burgin, 2010c; Kozen, 1997), problems
that can be solved by algorithms with polynomial time complexity are
deemed tractable while those that can be solved only by algorithms with
exponential time complexity are treated as intractable. Note that polynomials are built using addition and multiplication of numbers and variables.

A worthy of note development of this understanding was that psychologists found that people usually represent and process small and large numbers in a different way (Trick and Pylyshyn, 1994; Spelke and Barth, 2003;
Cantlon and Brannon, 2006; Goldman, 2010). For instance, reaction time
and performance in solving problems with large numbers are determined
by their ratio while for small numbers, their size plays the main role.

However, these classifications of natural numbers did not go beyond the
conventional Diophantine arithmetic while examples when this arithmetic
was not adequate in representing various real-life situations challenged
beliefs about uniqueness of the Diophantine arithmetic. Such examples
are described in the books of the noted American mathematics historian,
philosopher and educator Morris Kline (1908–1992). Let us consider one of
these examples from (Kline, 1967).

If a farmer has two herds consisting of 10 and 25 heads of cows,
respectively, he knows by adding 10 and 25 that the total number of cows
is 35. That is, he need not count his cows. Suppose, however, he brings the
two herds of cows to market where they are selling for $100 apiece. Will
a herd of 10 cows which might bring $1000 and a herd of 25 cows which
might bring $2500 together bring in $3500? Every businessman knows that
when supply exceeds demand, the price may drop, and hence 35 cows may
bring in only $3000. In some idealized world, the value of the cows may
continue to be $3500, but in actual situations this need not be true.

Consequently, continues Kline, mathematicians are, of course, free to
introduce the symbols 1, 2, 3, ..., where 2 means $1 + 1, 3$ means $2 + 1$,
and so on (Kline, 1967). We can even deduce from this that $2 + 2 = 4$.
But the question is not whether the mathematician can set up definitions
and axioms and deduce conclusions. It is necessary to know whether this
system necessarily expresses truths about the physical world.

The American mathematician and philosopher Reuben Hersh (1927–
2020) argues that even laws of arithmetic are uncertain by considering
a hotel that is missing a thirteenth floor. Take an elevator up eight floors,

then go five floors more, and you reach floor fourteen. Hersh apparently thinks this violates the equation $8 + 5 = 13$. What he has done, of course, is a jump from pure arithmetic to applied arithmetic, where applications are often uncertain.

Hersh studied mathematics at New York University and worked at the University of New Mexico writing papers on partial differential equations, probability, random evolution, and linear operator equations as well as popular articles and books on mathematics.

Two beans plus two beans make four beans only if you assign to beans what the famous German–Austrian–American philosopher Rudolf Carnap (1891–1970) called a correspondence rule. In this case, the rule is that each bean corresponds to 1. In the case of Hersh's elevator, if you assume that every floor corresponds to 1, then 8 floors plus 5 floors is sure to make 13 floors. Without correspondence rules, applications of mathematical truths are indeed uncertain.

Rudolf Carnap studied at the University of Jena and the University of Berlin. Later he worked at the University of Vienna and the University of Prage. In 1935, he emigrated to the United States where he worked at the University of Chicago and the University of California at Los Angeles (UCLA) writing on scientific knowledge, cognition, thermodynamics, foundations of probability, and inductive logic.

Ideas related to arithmetic were naturally compared with the situation in geometry. According to Kline, discovery of non-Euclidean geometries had taught mathematicians that geometry does not offer ultimate truths (Kline, 1967). That was the reason why many turned to the ordinary number system and the developments built upon it and maintained that this part of mathematics still offers unquestionable truths. The same thought is often expressed today by people who, wishing to give an example of an absolute truth, quote $2 + 2 = 4$. However, examination of the relationship between our ordinary number system and the real-life situations to which it is applied vividly demonstrates that it does not always offer truths. For instance, there are biological processes that cannot be correctly described using the conventional Diophantine arithmetic (Cleveland, 2008).

Various examples when $2 + 2$ does not equal 4 are considered by David Gershaw, who writes, "Believe it or not, sometimes $2 + 2$ does not equal 4. It depends on what type of measurement scale you are using" (Gershaw, 2015)

Following the advice of Poincaré who explained that one did not "prove" $2 + 2 = 4$, but one "checked" it (cf., Gonthier, 2008), we describe several experiments with geometrical shapes (rectangles) and pieces of land.

These experiments provide new examples from mathematics and real life demonstrating how the result of addition of similar objects can be equal to different numbers of the corresponding objects depending on the conditions of how addition is performed in practice.

The first example describes mathematical experiments with addition of rectangles where □ is the symbol of a rectangle.

Let us consider a rectangle □ ABCD (cf., Figure 4.1(a)). Adding the rectangle □ EFGH to the rectangle □ ABCD without other changes, e.g., without moving, we have two rectangles: □ ABCD and □ EFGH (cf., Figure 4.1(b)). In this case, we obtain the equality $1 + 1 = 2$.

However, adding the rectangle □ EFGH to the rectangle □ ABCD by moving □ EFGH to □ ABCD and identifying segments CD and FE, we have three rectangles: □ ABCD, □ EFGH and □ ABGH (cf., Figure 4.1(c)). In this case, we obtain the equality $1 + 1 = 3$.

At the same time, adding the rectangle □ EFGH to the rectangle □ ABCD by moving □ EFGH to □ ABCD and eliminating segments CD and FE, we have one rectangle □ ABGH (cf., Figure 4.1(d)). In this case, we obtain the equality $1 + 1 = 1$.

However, we can get even more rectangles. Adding the rectangle □ EFGH to the rectangle □ ABCD by moving □ EFGH to □ ABCD so that the segment CD comes into the rectangle □ EFGH while the segment FE comes into the rectangle □ ABCD, we have six rectangles □ ABCD, □ ABEF, □ EFGH, □ EFCD, □ CDGH, and □ ABGH (cf., Figure 4.1(e)). In this case, we obtain the equality $1 + 1 = 6$.

These cases different possible results of addition of two rectangles, i.e., the sum $1 + 1$ can be equal only to 1, 2, 3 and 6. It would be interesting to know whether it is possible to get other numbers of rectangles by adding two rectangles. In general, we can consider an arithmetic of rectangles and study related combinatorial problems in a general form. For instance, what numbers of rectangles it is possible to get by adding n rectangles to m rectangles.

The next similar example comes from real life and describes addition of pieces of land.

Alice has a piece of land A in California and buys a piece of land B in Florida (cf., Figure 4.2(a)). In this case, she has 1 piece $+ 1$ piece $= 2$ pieces.

Bob has a piece of land A and buys another piece of land B, which has a common border with the first one (cf., Figure 4.2(b)). In this case, he has 1 piece $+ 1$ piece $= 1$ piece D (cf., Figure 4.2(c)).

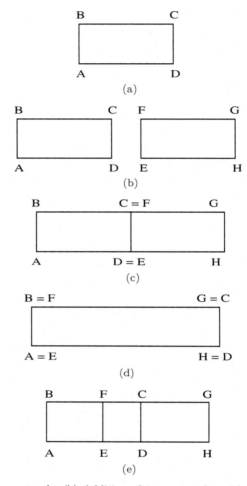

Fig. 4.1. (a) One rectangle. (b) Addition of two rectangles without changes. In this case, we have $1 + 1 = 2$. (c) Addition of two rectangles with identification. In this case, we have $1 + 1 = 3$. (d) Addition of two rectangles with elimination. In this case, we have $1 + 1 = 1$. (e) Addition of two rectangles with overlapping. In this case, we have $1 + 1 = 6$.

In a similar way, it is possible to add three or more pieces of land and get only one piece of land as a result.

It is interesting that while the first situation is reflected in the conventional Diophantine arithmetic, the second situation is represented by operations with sets in set theory where the union of one set with another

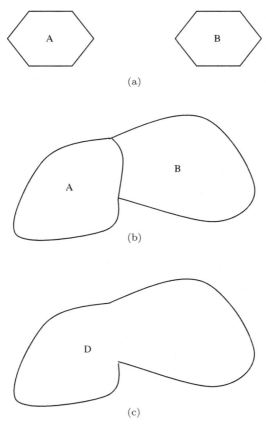

Fig. 4.2. (a) Addition of two disjoint pieces of land. In this case, we have $1 + 1 = 2$. (b) Addition of two pieces of land with a common border. (c) In the second case, we have $1 + 1 = 1$.

set is only one set and not two sets (Abian, 1965). The difference between these two situations hints that there are two arithmetics: in one of them (the Diophantine arithmetic) $1 + 1 = 2$, while in the other (a non-Diophantine arithmetic), we have $1 + 1 = 1$.

Figure 4.3 shows how it is possible to obtain 1, 3, 4, 5, 6, 7, 8, and 9 rectangles by adding 2 rectangles to 2 rectangles. Figure 4.1(e) hints that adding 2 rectangles to 2 rectangles, it is possible to get even more than nine rectangles. An interesting problem is finding all numbers, which can be obtained in such a way. For instance, is it possible to get 11 rectangles, by adding 2 rectangles to 2 rectangles?

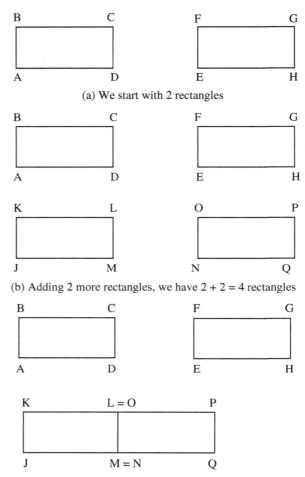

(a) We start with 2 rectangles

(b) Adding 2 more rectangles, we have 2 + 2 = 4 rectangles

(c) Adding to the initial ones (from part (a)), 2 more rectangles, we have 2 + 2 = 5 rectangles. Find all 5 rectangles

Fig. 4.3. (*Continued*)

In general, we see that the described examples bring us to the additive arithmetic of rectangles. An interesting problem in this arithmetic is given two natural numbers m and n, find what number of rectangles it is possible to obtain by adding m rectangles to n rectangles.

We see that the Diophantine arithmetic models only one of these cases, namely, Figure 4.3(b). Consequently, we can ask a question whether it is

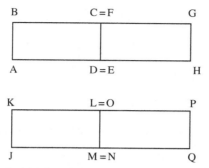

(d) Adding to the initial ones (from part (a)), 2 more rectangles, we have 2 + 2 = 6 rectangles. Find all 6 rectangles

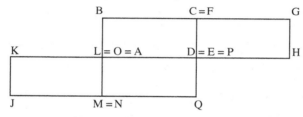

(e) Adding to the initial ones (from part (a)), 2 more rectangles, we have 2 + 2 = 7 rectangles. Find all 7 rectangles

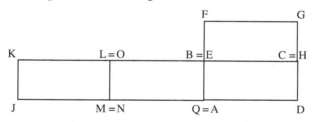

(f) Adding to the initial ones (from part (a)), 2 more rectangles, we have 2 + 2 = 8 rectangles. Find all 8 rectangles

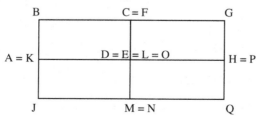

(g) Adding to the initial ones (from Part (a)), 2 more rectangles, we have 2 + 2 = 9 rectangles. Find all 9 rectangles

Fig. 4.3.   (*Continued*)

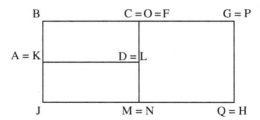

(h) Adding to the initial ones (from part (a)), 2 more rectangles, we have
2 + 2 = 5 rectangles. Find all 5 rectangles

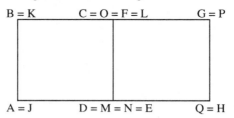

(i) Adding to the initial ones (from part (a)), 2 more rectangles, we have
2 + 2 = 3 rectangles. Find all 3 rectangles.

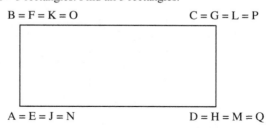

(j) Adding to the initial ones (from part (a)), 2 more rectangles, we have
2 + 2 = 1 rectangle

Fig. 4.3.   (a) Two rectangles are given. (b) Addition of two rectangles to two rectangles
without changes gives us four rectangles. In this case, we have $2 + 2 = 4$. (c) Addition of
two rectangles with equating (gluing together) gives us five rectangles. In this case, we
have $2 + 2 = 5$. (d) Addition of two rectangles with equating (gluing together) gives us six
rectangles. In this case, we have $2 + 2 = 6$. (e) Addition of two rectangles with equating
(gluing together) gives us seven rectangles. In this case, we have $2 + 2 = 7$. (f) Addition
of two rectangles with equating (gluing together) gives us eight rectangles. In this case,
we have $2 + 2 = 8$. (g) Addition of two rectangles with equating (gluing together) gives
us nine rectangles. In this case, we have $2 + 2 = 9$. (h) Addition of two rectangles to two
rectangles with partial equating (gluing together) and partial elimination gives us five
rectangles. In this case, we have $2 + 2 = 5$. (i) Addition of two rectangles to two rectangles
with partial equating (gluing together) and partial elimination gives us three rectangles.
In this case, we have $2 + 2 = 3$. (j) Addition of two rectangles to two rectangles with
elimination gives us one rectangle. In this case, we have $2 + 2 = 1$.

always viable to reflect addition of mathematical objects, such as rectangles, with the rules of the Diophantine arithmetic.

Exploring the history of mathematics, it is possible to divide all mathematicians who were wise enough to distrust the complete adequacy of the conventional arithmetic into three groups. Representatives of the first group, such as Helmholtz, Kolmogorov or Littlewood, only explained that in practice natural numbers and operations with them were different from those which were known from mathematics (Helmholtz, 1887; Littlewood, 1953; van Dantzig, 1956; Kolmogorov, 1961; Birkhoff and Barti, 1970; Knuth, 1976). Representatives of the second group, such as Gasking, Kline or Rashevsky, in addition conjectured that different arithmetics existed but people did not know what they were and how to build them (Gasking, 1940; Kline, 1967; Rashevsky, 1973). Finally representatives of the third group constructed or discovered non-Diophantine arithmetics (Burgin, 1977).

It is necessary to note that even before Kline and Rashevsky, the Australian philosopher Aidan Trist Douglass Gasking (1911–1994) not only suggested a possibility of existence of other arithmetics (of natural numbers) but even described how operations in such arithmetics can correctly describe some situations in people's practice (Gasking, 1940). Analyzing relations between mathematics and the physical world, he demonstrated how "queer" operations, as he called them, such as $4 \times 6 = 12$ or $3 \times 4 = 24$, could give useful results when applied to practical tasks in the material world. Writing about one of his "queer" arithmetics, Gasking also gave a description of the nonstandard multiplication $\otimes$ of even numbers defined by the formula

$$m \otimes n = \tfrac{1}{2}(m + 2)(n + 2).$$

However, Gasking did not give the entire description of any of the arithmetics he discussed.

The famous Guatemala-American philosopher Hector-Neri Castaneda (1924–1991) further analyzed argumentation of Gasking towards the necessity of new arithmetics (Castaneda, 1959). Castaneda explains that Gasking uses essentially two arguments:

(1) *"We could use any mathematics and compensate for deviations from the present one by means of an adequate technique of counting or measuring.*

(2) *We could use any mathematics different from the present one and modify our physics."* (Castaneda, 1959)

At the same time, according to Castaneda, Gasking *"does not define a queer arithmetic, but it is clear from his discussion that he does not mean one or more changes of labels. In other words, to use the formula '6×4 = 12' is not merely to use '6' to mean the number 3, and perhaps '3' or 'w' to mean the number 6 .... By a 'queer arithmetic' Gasking means something more exciting. Presumably, we are to think of queer arithmetics on the analogy with non-Euclidean geometries. Provisionally, just to fix the sense of the argument, we may say that a queer arithmetic is a system of propositions about natural numbers in which we have a different multiplication table."* (Castaneda, 1959)

We see that Castaneda and later Rashevsky suggested that unusual arithmetics would be analogous to non-Euclidean geometries although Castaneda tried to argue that Gasking's claim that "queer" arithmetics describe reality was ungrounded.

An important observation of Castaneda is that *"counting is, in fact, the simplest form of measuring, and every measuring includes it or presupposes it."* (Castaneda, 1959)

Hector Castaneda was born in San Vicente, Guatemala, and later immigrated to the United States receiving PhD from the University of Minnesota and working at Duke University and Wayne University. He developed the guise theory of intentionality, language and perception. According to this theory called abstract ontology, the entire realm of thought includes both real and imagined objects, which are called guises. In turn, this implies that all objects of ordinary perception are such guises or systems of guises and thoughts about guises are thoughts about the real world.

In the same venue related to numbers and counting, the well-known American mathematician and philosopher Philip Davis (1923–2018) posed a challenging question: Is One and One Really Two? (Davis, 1972). In his paper, Davis writes that the *"ordinary arithmetic is one of the most elementary of the mathematical disciplines"* while *"among the theorems of arithmetic are various sums"* meaning such equalities as $1 + 1 = 2$, $33 + 67 = 100$ *or* $11111 + 22222 = 33333$. *With this in mind, he arrives to the conclusion that "the arithmetic of* [meaning *operations with*, MB] *excessively large numbers can be carried out only with diminishing fidelity."* In essence, it means that while operations with small and medium numbers are performed in the same way as it is done in the conventional Diophantine arithmetic, correct representation of operations with "excessively large" numbers demands new arithmetic.

Philip Davis was born in Lawrence, Massachusetts, and earned his degrees in mathematics from Harvard University. He worked in numerical analysis and approximation theory writing many popular papers and books on the history and philosophy of mathematics.

Similar to Davis, Boran Berčić also asked, what made it true that $2 + 2 = 4$? (Berčić, 2005). As if to answer this question, Poincaré suggested that arithmetics had to be tested by experiments pointing out that one did not "prove" $2 + 2 = 4$, but one "checked" it (cf., Gonthier, 2008).

The Russian-American physicist Felix Lev expresses similar considerations:

"Let us pose a problem of whether $10 + 20$ *equals 30. Then we should describe an experiment which should solve this problem. Any computer can operate only with a finite number of bits and can perform calculations only modulo some number p. Say $p = 40$, then the experiment will confirm that $10 + 20 = 30$ while if $p = 25$ then we will get that $10 + 20 = 5$. So the statements that $10 + 20 = 30$ and even that $2 + 2 = 4$ are ambiguous because they do not contain information on how they should be verified."* (Lev, 2017)

Here is one more example from real life demonstrating the situation where the Diophantine arithmetic does not work (cf. Ramsey, 2011).

The Belgian Draft Horse is specific breed recognized for its size and strength. In general, a Belgian Draft horse is weighing around 2000 pounds, standing nearly 6 feet tall and being capable of pulling 8000 pounds in competition.

At the same time, if a pair of these horses is yoked together, then they are capable of pulling 24,000 pounds. Arithmetically, it means $8000 + 8000 = 24,000$. This is incorrect in the Diophantine arithmetic where $8000 + 8000 = 16,000$.

Even more, if a pair of these horses is raised and trained together, they can pull 32,000 pounds. Arithmetically, it means $8000 + 8000 = 32,000$. This is also incorrect in the Diophantine arithmetic.

Thus, we can see that there are many cases when the conventional Diophantine arithmetic does not correctly describe situations that emerge in science, mathematics and exist in everyday life. As a result, we come to the problem of finding or constructing arithmetics, which correctly represent these situations. This problem implicitly existed almost for two and a half millennia. In the 20th century, it was explicitly formulated by some researchers (cf., for example, Gasking, 1940; Kline, 1967; Rashevsky, 1973;

Rotman, 1997). However, the belief in the uniqueness of the Diophantine arithmetic was so strong that it took a lot of time at first to understand the problem and then to solve it. This happened near the end of the 20th century. Discovery of non-Diophantine arithmetics took more time than discovery of non-Euclidean geometries due to the following reasons. First, the conventional Diophantine arithmetic was so solidly ingrained in the mentality of people that it was extremely hard to imagine other possibilities. Second, people in general and mathematicians in particular even did not understand that there is such a problem. Third, people usually try to minimize their actions, to stay where they are and even attack those who introduce innovations.

Finally, it is necessary to remark that although a quantity of various arithmetics (arithmetic of real numbers, arithmetic of complex numbers, residual arithmetic, arithmetic of algebraic numbers, arithmetic of cardinal numbers, arithmetic hypernumbers, nonstandard arithmetic, computer arithmetic, arithmetic of surreal numbers, arithmetic of computable numbers, etc.). have been constructed. All previously introduced arithmetics and algebras did not change the arithmetic of natural numbers and thus, did not solve problems related to the Diophantine arithmetic. As Wojciech Krysztofiak wrote: *"up to the present there has not been constructed any alternative arithmetic of natural numbers"* (Krysztofiak, 2012). Only non-Diophantine arithmetics of natural numbers were able to provide tools for solving the discussed and similar problems in a rigorous mathematical way.

In other words, although ancient Greeks found examples of apparent disparities between the standard Diophantine arithmetic of natural numbers and some real-life situations involving (natural) numbers, it took almost 2.5 millennia to discover (construct) the first classes of non-Diophantine arithmetics of natural numbers.

Writing about the discovery of non-Diophantine arithmetics, we allow the reader to choose how to understand innovations in mathematics. Those who believe in the Platonic realm of mathematical structures (cf., for example, Bernays, 1935; Mazur, 2008a; Burgin, 2017c) can assume that non-Diophantine arithmetics were discovered in the world of structures. Those who deem mathematics as a remarkable creation of the human mind can assume that non-Diophantine arithmetics were invented and constructed. Those who advocate a dualistic approach to mathematics, suggesting that some parts of it exist and are discovered while other parts are created by mathematicians, can apply their approach to non-Diophantine arithmetics, reasoning that non-Diophantine arithmetics were constructed in the physical world, in which people live, but properties of these arithmetics were discovered.

Note that now the existence of the Platonic realm is scientifically clarified, explained and validated as the world of abstract structures (Burgin, 2012, 2017b). Mathematical structures, as well as other structures, form a part of this world as it is demonstrated in (Burgin, 1994, 2017c, 2018). As a result, mathematics exists as a scientific field, such as physics or biology, having theoretical, experimental and applied components. However, while the domains of physics and biology represent fragments of the material world, the domain of mathematics lies in the world of structures encompassing all mathematical structures.

Exploring the world of structures in 1973, the renowned Russian mathematician Pyotr Konstatinovich Rashevsky (1907–1983) published a paper, in which he described the paradox of a heap and explicitly formulated the problem of the necessity of building arithmetics of natural numbers different from the conventional one (Rashevsky, 1973).

This paper had an interesting history as the author of the book learned much later. Rashevsky was a member of the Editorial Board of one of the main mathematical journals in the Soviet Union. Naturally, after writing his paper, he brought it to this journal called *"Uspehi Matematicheskih Nauk,"* which in English means *Achievements of Mathematics.* However, the other members of the Editorial Board did not want to publish such a radical text. Rashevsky insisted and at last a compromise was achieved. The paper was published outside the main body of the journal's content, in a much smaller font in comparison with other papers in that issue, with a notice that the Editorial Board is not responsible for its content and the paper was published only for discussion.

Mark Burgin read that paper and started thinking how it might be possible to build arithmetics of natural numbers, which could be different from the conventional Diophantine one. His main contention was that if it had been possible to discover/construct non-Euclidean geometries, then it might be feasible to find/elaborate non-Diophantine arithmetics. And indeed, the first class of non-Diophantine arithmetics of natural numbers called perspective or more exactly, direct perspective arithmetics was discovered (constructed) by Burgin in 1975. However, it was not easy to publish such a revolutionary result. From the history of mathematics we know that in the case of non-Euclidean geometries, the top mathematician of his time Carl Friedrich Gauss did not want to publish his results on this topic understanding that bigots would attack him.

In his letter to the well-known German astronomer, mathematician, physicists, and geodesist Wilhelm Bessel in 1829, Gauss wrote:

"In leisure hours now and then I have again been reflecting on a subject which with me is now nearly forty years old; I mean the first principles of geometry; I do not know if I have ever told you my views on that matter. Here too I have carried many things to farther consolidation, and my conviction that we cannot lay the foundation of geometry completely a priori become if possible firmer than before. Meantime it will be long before I bring myself to work out my very extensive researches on this subject for publication, perhaps I shall never do so during my lifetime; for I fear the outcry of the Boeotians, were I to speak out my views on the question."

Note that Boeotians were notorious as the most stupid people in ancient Greece.

In contrast to this, the great Russian mathematician Nikolay Ivanovich Lobachevsky (1792–1856) was able to publish his results in the Surveys of the Kazan University only because he was the Chancellor of this university, and the great Hungarian mathematician János Bolyai (1802–1860) was able to publish his work only as a very short attachment to his father's textbook.

Nikolay Ivanovich Lobachevsky was born in Nizny Novgorod and graduated from Kazan University, in which he taught finally becoming its Chancellor. One of his teachers at the university was the German mathematician Johan Christian Martin Bartels (1769–1836), who was teaching Gauss and became his friend when he lived in Germany before coming to Russia. It is interesting that before Lobachevsky became the Chancellor, professors from his university forbade publication of his groundbreaking work on non-Euclidean geometries. In addition, when Lobachevsky was already the Chancellor and his works on non-Euclidean geometries, he applied for the membership to the Russian Academy of sciences. His application was rejected because of the negative opinion of other Russian mathematicians and he was even mocked in general public newspapers.

János Bolyai was born in Hungarian town Kolozsvár, Ausro-Hungary (now Kluz, Romania). His father was the well-known Hungarian mathematician Farkas Bolyai, who studied and befriended Gauss. Farkas Bolyai taught János Bolyai mathematics. Although János was a gifted mathematician and had great interest in mathematics, he was not able to pursue a mathematical career due to the family circumstances. As a result, János Bolyai graduated from the Imperial and Royal Military Academy in Vienna and after this served in the army until retirement. Although he was very talented and made the landmark discovery of non-Euclidean geometries, János Bolyai never worked as a mathematician and never published any other work in mathematics.

Similar to non-Euclidean geometries, it was not easy to publish the work on non-Diophantine arithmetics. That is why it was natural that the first paper by Burgin on non-Diophantine arithmetics was published only in 1977 in a very compressed form (Burgin, 1977). In that paper, he suggested calling these arithmetics non-Diophantine. It was reasonable to call the conventional arithmetic of natural numbers by the name the *Diophantine arithmetic* due to the foremost contribution of the ancient Greek mathematician Diophantus specifically to arithmetic of natural numbers, making first steps in transformation arithmetic into algebra as it is described in Chapter 1.

It is important to understand that non-Diophantine arithmetics do not change numbers but employ new operations with these numbers. For instance, any non-Diophantine arithmetic of natural numbers contains the same natural numbers as the Diophantine arithmetic but addition and/or multiplication of them can be very different. Moreover, non-Diophantine arithmetics of natural numbers satisfy all five Peano axioms for natural numbers (Burgin and Czachor, 2020).

To better understand that numbers in the non-Diophantine arithmetic of natural numbers are the same as in the Diophantine arithmetic of natural numbers, we can deem that a natural number simply has diverse roles in different arithmetics and thus, performs differently in distinctive arithmetics. This is similar to the situation with people where the same woman can be an accountant at her job, a mother and daughter at home and a guest at somebody else's house.

Thus, it is possible to define a non-Diophantine arithmetic of natural numbers as the set of all natural numbers ordered in the standard way but in which one or both operations of addition and multiplication are different from addition and multiplication in the conventional Diophantine arithmetic.

As there are infinitely many non-Diophantine arithmetics, this gives supportive evidence to the opinion of the famous British philosopher, logician, and mathematician Bertrand Russell, who understood more than hundred years ago that Peano axioms *"are capable of an infinite number of different interpretations, all of which will satisfy the five primitive propositions"* (Russell, 1919).

Note that although a multitude of various arithmetics, such as arithmetic of real numbers, arithmetic of complex numbers, residual arithmetic, arithmetic of algebraic numbers, arithmetic of cardinal numbers, arithmetic of ordinal numbers, computer arithmetic and arithmetic of computable numbers, as well as the deluge of diverse algebras have appeared as the further development of the arithmetic of natural numbers, neither of those mathematical structures solved the problems with the Diophantine arithmetic, which were discussed by different researches, such as Helmholtz, Lebesgue, Kline, and Rashevsky. Some of these problems were considered above. Only non-Diophantine arithmetics provide tools for solving these problems in a rigorous mathematical way (Burgin, 2001, 2001a).

The new step in the development in this area was made when second class of non-Diophantine arithmetics of natural numbers, called dual or, more exactly, dual perspective arithmetics, was discovered (constructed) by Mark Burgin in 1979 and published in 1980, in an exceedingly compressed form (Burgin, 1980). Only much later, these results were included in the book *"Non-Diophantine Arithmetics Or What Number is* $2 + 2$*"* in a more detailed form (Burgin, 1997).

It is possible to find the most expanded exposition of the theory of non-Diophantine arithmetics and their applications in the book "Non-Diophantine arithmetics in mathematics, physics and psychology" by Mark Burgin and Marek Czachor (Burgin and Czachor, 2020).

Contradicting the knowledge accumulated by mathematicians through millennia of research as well as to the mundane experience of all people, non-Diophantine arithmetics drastically change people's understanding not only of mathematics but also of the whole world because numbers reflect reality. In spite of this (or, maybe, because of this), the discovery of non-Diophantine arithmetics was not noticed by the majority of mathematicians and philosophers of mathematics (not speaking about general public) since it contradicted to such a big extent the conventional knowledge. Usually, instead changing what they learned from the very young age, people prefer not to see new groundbreaking discoveries.

It is interesting that the discovered classes of non-Diophantine arithmetics validated some predictions of those who wrote about necessity of new types of arithmetic. For instance, being a geometer, Rashevsky predicted that new arithmetics would form parametric families as it had been with non-Euclidean geometries. Namely, he wrote:

"*It must not be expected that* [such a, MB] *hypothetical theory* [of new arithmetics, MB], *if it would be ever destined to see the light of day, will be unique; on the contrary, it will have to depend on certain 'parameters' (with a role distantly reminiscent of the radius of Lobachevsky space when we repudiate Euclidean geometry in favor of non-Euclidean). It may be expected that in the limiting case the hypothetical theory should coincide with the existing one.*" (Rashevsky, 1973)

Pyotr Konstatinovich Rashevsky was born in Moscow and graduated from Moscow State University. He taught mathematics at Moscow Pedagogical Institute, Institute of Railway Engineers, and Moscow State University. His main mathematical works were in differential geometry.

The first prediction of Rashevsky about new arithmetics came out true. Indeed, the first and second discovered classes of non-Diophantine arithmetics form parametric families (classes) although the parameter is not numerical as in the case of non-Euclidean geometries (cf., Gray, 1979; Trudeau, 1987) but functional. It means that in any such arithmetic, properties and laws of its operations depend on a definite function $f(x)$ (Burgin, 1977, 1980, 1997). The conventional, Diophantine arithmetic is a member of both parametric families with the parameter equal to the identity function $f(x) = x$. This distinction of arithmetic from geometry shows higher intrinsic complexity and fundamentality of arithmetic in comparison with geometry. Geometry is reducible to arithmetic as it was proved by René Descartes.

The second prediction of Rashevsky was also validated because in the limiting case, i.e., when the functional parameter was the identity function $f(x) = x$, the corresponding arithmetic from the parametric family coincided with the conventional Diophantine arithmetic.

It is interesting to know that in contemporary mathematical terms, the technique for building the first families (classes) of non-Diophantine arithmetics is based on mathematical structures called fibred spaces and bundles, which appeared only in the 20th century (Husemöller, 1994). These structures are utilized as a tool for operating with numbers in non-Diophantine arithmetics. A similar technique is used for defining operations with hypernumbers when it is impossible to define these operations straightforwardly in the standard fashion (Burgin, 2010a, 2011, 2012c, 2015). This technique is also used by the prominent physicist Paul Benioff for scaling in gauge theory and geometry (Benioff, 2011, 2012, 2012a, 2014,

2015, 2016). The scaling is defined for addition and multiplication by two functions: $h(x) = (1/n)x$ and the identity function $g(x) = x$ from R into R or from C into C. This construction of scaling is inherently based on some types of non-Diophantine arithmetics and prearithmetics described in this book. In essence, Benioff's method implicitly utilizes non-Diophantine arithmetics with nonstandard multiplication in sets of real and complex numbers without changing addition of these numbers.

The history of non-Diophantine arithmetics clearly resembles what happened with non-Euclidean geometries. Indeed, for quite a while, the discovery of non-Euclidean geometries was not known to the prevalent part of the mathematical community. The same is going on with non-Diophantine arithmetics. For instance, more than twenty years after the discovery of non-Diophantine arithmetics, the American mathematician Brian Rotman put forward the problem to elaborate arithmetics essentially different from the conventional one (Rotman, 1997). He based his suggestion on a series of examples demonstrating that many laws of the conventional arithmetic are not true in different real-world situations. Rotman called those hypothetical structures non-Euclidean arithmetics, although he did not describe them. However, it is more natural to call the conventional arithmetic by the name the *Diophantine arithmetic* than by the name the *Euclidean arithmetic* because Diophantus contributed much more to the development of the arithmetic of natural numbers than Euclid. Consequently, new arithmetics of natural numbers acquired the name *non-Diophantine arithmetics*.

Similar to the Euclidean geometry, the Diophantine arithmetic was unique and nonchallengeable for a very long time when people did not known other arithmetics. Its position in human society has been and is now even more stable and firm than the position of the Euclidean geometry before the discovery of the non-Euclidean geometries. Really, all people use the Diophantine arithmetic for counting. Utilization of numbers and arithmetical operations make all people some kind of consumers of mathematics. At the same time, Euclidean geometry is only studied at school and in real life rather few specialists use it. It is arithmetic, and not geometry, which is considered as a base for the whole mathematics in the intuitionistic approach. As a result, the discovery of non-Diophantine arithmetics changes our understanding of the world similar to the transformation in minds of people caused by the discovery of non-Euclidean geometries. For millennia, people in general and mathematicians in particular believed there existed only one geometry, which was absolute and supreme. The discovery of non-Euclidean geometries in the 19th century disproved this misconception demonstrating that the world is much more opulent than it had seemed before. In a similar way, for millennia, people in general and mathematicians in particular believed there was only one arithmetic, which was absolute and ultimate. The discovery of non-Diophantine arithmetics in the 20th century disproved this misconception demonstrating that the world was much more affluent and plentiful than it had seemed before.

At the same time, it is necessary to correctly understand ontological implications of these two discoveries. One can have an impression that these discoveries undermine the existence of the independent mathematical structures, which dwell in the eternal immutable World of Structures or Plato Ideas. For instance, mathematical Platonists argue that "2 plus 2 equals 4" is an eternal truth that would be true even if the Big Bang had never occurred and the universe did not exist. In contrast to this, there are non-Diophantine arithmetics where "2 plus 2 equals 5" and they belong to the World of Structures.

However, the eternal truth of "2 plus 2 equals 4" still remains but not as an absolute equality. It becomes relative depending on the necessary modification. Namely, what is true and unchangeable is

"There is a mathematical structure called the Diophantine arithmetic, in which 2 plus 2 equals 4"

or

"2 plus 2 equals 4 in the Diophantine arithmetic"

It is interesting that when non-Euclidean geometries were discovered, their discoverers, Carl Friedrich Gauss and Nikolay Ivanovich Lobachevsky, performed physical experiments trying to find whether such geometries exist in reality (Livanova, 1969). In contrast to this, people in general and mathematicians in particular have already used and are using some elements from non-Diophantine arithmetics. Moreover, there are many examples demonstrating that various practical calculations are performed according to the rules of non-Diophantine arithmetics and people unconsciously use them. In spite of this evidence, acceptance of non-Diophantine arithmetics is extremely slow due to the powerful bias that "2 plus 2 equals 4" is an eternal truth, which does not depend on any context.

Power of people's stereotypes is vividly demonstrated by the book (Blehman *et al.*, 1983). At first (in Section 1.2.4), the authors of that book explain with many examples and references that our intuition of natural numbers and arithmetic can be very misleading in various situations. After this (in Section 1.2.5), they announce that it is completely impossible that two times two is not equal to four. The authors are even trying to prove this utilizing a probabilistic reasoning. Here are their arguments (Blehman *et al.*, 1983, p. 50).

Really, the statement that two times two is equal to four may be taken as an example of the most evident truth. Although, nobody doubts that this is a true equality, it is possible to evaluate formally probability that in reality two times two is equal to five, while the standard statement that two times two is equal to four is a result of a constantly repeated arithmetical mistake. Let us suppose that any individual performing multiplication with numbers that are less than ten can decrease the result by one with the probability 10^{-6}. This corresponds to several such mistakes during his or her life. If we assume that through the whole history of mankind, 10^{10} people performed the multiplication "two times two" 10^6 times during the life of each of them, then the probability that they repeated this mistake of decreasing the result is less than $10^{-10^{17}}$. Thus, the authors conclude, *the probability is so small that the event is absolutely impossible and we see that two times two is equal to four.*

This is an explicit example of incorrect probabilistic reasoning.

The most famous example of appealing that nobody saw something and concluding that it is the absolute truth, is attributed to Aristotle. He asserted

All swans are white.

All people known to Aristotle saw only white swans. He saw only white swans. So, Aristotle gave this as an example of absolute truth. Based on their experience, Europeans had believed in this until they came to Australia where they found black swans and disproved this statement.

The same change happened with the statement

Two plus two is equal to four.

Indeed, all people have known this arithmetical equality from their early childhood. So, everybody believed that this is an absolute truth. However, when non-Diophantine arithmetics were discovered, it was found that in some of them two plus two is not equal to four (Burgin, 1977, 1997; Burgin and Czachor, 2020). Transparent examples of violation of this "law" are presented in Figure 4.3.

When we consider non-Diophantine arithmetics, it is possible to think that they are absolutely formal constructions like many other mathematical objects, which are very far from the real world. However, let us recollect that the discovery of non-Euclidean geometries met similar skepticism and mistrust. Carl Friedrich Gauss understood very well this situation with extreme innovations and in spite of being acknowledged as the greatest mathematician of his time, he did not dare to publish his results concerning non-Euclidean geometries. Another reason for this was that he was not able to find anything that was similar to them in nature. Lobachevsky called his geometry imaginable. Nevertheless, it was demonstrated later that the real physical space fitted non-Euclidean geometries, and that the Euclidean geometry did not have such essential applications as the non-Euclidean ones.

In this respect, the situation with non-Diophantine arithmetics is different. In spite of the relatively short time, which has passed after their discovery, it has been demonstrated that many real phenomena and processes exist that match the non-Diophantine arithmetics (cf., for example, Burgin, 1992, 2001, 2001a, 2007; Czachor, 2015, 2017, 2017a; Burgin and Meissner, 2017; Tolpygo, 1997). Moreover, mathematicians and scientists implicitly used some elements of non-Diophantine arithmetics without understanding their essence.

Discussing Diophantine arithmetics, we see that the most proper application of the term Diophantine describes the conventional arithmetic of natural numbers because Diophantus was an ancient Greek mathematician while ancient Greeks new only natural numbers. That is why now we will discuss only non-Diophantine arithmetics of natural numbers elaborating a working definition of them.

Definition 4.1. An arithmetic is called *non-Diophantine* if it contains all natural numbers, addition and/or multiplication in which do not coincide with addition and/or multiplication in the Diophantine arithmetic.

The development of the concept of a *non-Diophantine arithmetic* shows that there are three classes of non-Diophantine arithmetics.

- *Parametric non-Diophantine arithmetics* are constructed based on a parameter, e.g., a function, which determines transformation of addition and/or multiplication in the uniform way (Burgin, 1977, 1997). Each value of the parameter gives specific non-Diophantine arithmetic.

- In *regular non-Diophantine arithmetics*, (non-Diophantine) addition and multiplication are defined by a distinctive rule.
- In *singular non-Diophantine arithmetics*, non-Diophantine addition and/or non-Diophantine multiplication are defined only for a finite collection of numbers while operations with other number are the same as in the Diophantine arithmetic.

There is a more exact specification of ultra-singular non-Diophantine arithmetics.

Definition 4.2. If non-Diophantine addition and/or non-Diophantine multiplication are defined not more than for n pairs of numbers, then this arithmetic is called *n-singular non-Diophantine arithmetic*.

Let us consider some examples.

Example 4.1. Let us take the set N of all natural numbers and define the following operations:

$$x \oplus y = x + y \text{ when } x, y > 10,$$

$$x \oplus y = \max(x, y) \text{ otherwise}$$

$$x \otimes y = x \times y \text{ for all natural numbers } x \text{ and } y.$$

This is a 10-singular non-Diophantine arithmetic.

Example 4.2. Let us take the set N of all natural numbers and define the following operations:

$$x \oplus y = x + y \text{ when } x, y > 1,$$

$$x \oplus 1 = 1 \oplus x = x,$$

$$x \otimes y = x \times y \text{ when } x, y > 1,$$

$$x \otimes 1 = 1 \otimes x = 1.$$

This is a 1-singular non-Diophantine arithmetic.

Example 4.3. Let $\mathbf{N}_{\min}$ be the set N with the operations $\oplus = \min$ and $\otimes = +$, which is the usual addition in $\mathbf{N}$. This is a regular non-Diophantine arithmetic of natural numbers, which is a subarithmetic of the prearithmetic $\mathbf{R}_{\min}$ used in idempotent analysis (Maslov, 1987; Maslov and Samborskii, 1992; Kolokoltsov and Maslov, 1997).

Example 4.4. Let $\boldsymbol{N}_{\max}$ be the set N with the operations $\oplus = \max$ and $\otimes = +$, which is the usual addition in $\boldsymbol{N}$. This is a regular non-Diophantine arithmetic of natural numbers, which is a subarithmetic of the prearithmetic $\boldsymbol{R}_{\max}$ used in idempotent analysis (Maslov, 1987; Maslov and Samborskii, 1992; Kolokoltsov and Maslov, 1997).

Definition 4.2 implies the following result.

Lemma 4.2. *(a) Any n-singular non-Diophantine arithmetic is singular. (b) The class of singular non-Diophantine arithmetics is the union of the classes of n-singular non-Diophantine arithmetics where n goes from 1 to infinity.*

Non-Diophantine arithmetics in general and parametric non-Diophantine arithmetics can possess many unusual properties. For instance, recently the expressions $1 + 1 = 3$ and $2 + 2 = 5$ have become a popular metaphor for synergy in a variety of areas: in business and industry (Beechler, 2013; Brown, 2015; Grant and Johnston, 2013; Jude, 2014; Kress, 2015; Ritchie, 2014; Marks and Mirvis, 2010; Phillips, 2008; Ritchie, 2014), in economics and finance (Burgin and Meissner, 2017), in anthropology, psychology and sociology (Boksic, 2017; Brodsky, 2004; Bussmann, 2013; Enge, 2017; Frame and Meredith, 2008; Jaffe, 2017; Mane, 1952), library studies and informatics (Marie, 2007), in studies of creativity (Trott, 2015), in biochemistry and bioinformatics (Kroiss *et al.*, 2009), in theory and practice of organizations (Klees, 2006), in technology (Gottlieb, 2013), in computer science (Derboven, 2011; Glyn, 2017; Lea, 2016), in networking (Meiert, 2015), in physics (Lang, 2014), in biology and medicine (Lawrence, 2011; Trabacca *et al.*, 2012; Archibald, 2014; Phillips, 2016; Jaffe, 2017), in agriculture (Riedell *et al.*, 2002), in pedagogy (Nieuwmeijer, 2013) and in politics (Van de Voorde, 2017).

Other unusual expressions have also appeared in scientific publications. Examples are $1 + 1 = 1$ (Carroll and Mui, 2009; Morris, 2017), $1 + 1 = 4$ (Flegenheimer, 2012), *one plus one makes more than two* (Pascoe, 2017), *one plus one equals three-fourths* (Ries, 2014), *one plus one equals one and a half* (Covey, 2004), *negative one plus negative one equals negative three* (Katsenelson, 2015) and $2 + 2 = 5$ (Cambridge Dictionary).

At the same time, those who use the expressions $1 + 1 = 3$ and $2 + 2 = 5$ think that it is incorrect mathematics because in the Diophantine arithmetic, $1 + 1 = 2$ and $2 + 2 = 4$. They believe, for example, that because synergetic relations can be described by the formula $1 + 1 = 3$, these relations defy the laws of mathematics. Some people even call them anti-mathematical formulas. However, the discovery of non-Diophantine arithmetics demonstrated that synergetic relations defy only the laws of the Diophantine arithmetic but not of mathematics because in mathematics there are non-Diophantine arithmetics, in which $1 + 1 = 3$ or $2 + 2 = 5$.

In addition to this, non-Diophantine arithmetics solve some problems that remained unsolved from the time of ancient Greece, just to mention the "paradox of a heap" we have already encountered in the previous section. Indeed, the heap is not changing if we add one grain. Consequently, if we take the number k of the grains in the heap, then adding 1 to k does not change k. This contradicts the main law of the Diophantine arithmetic stating that for an arbitrary number k, the number $k + 1$ is not equal to k, and gives birth to a paradox if we have only one arithmetic. Non-Diophantine arithmetics solve the paradox.

In a similar way, paradoxes from ancient Greece are sometimes revived in modern physics. For example, theory of chaos encounters many difficult problems caused by insufficiency of modern discrete mathematics to correctly represent chaotic dynamics. Arbitrary small changes in external parameters or/and initial conditions cause essential changes in the behavior of a system, making the classical difference calculus inefficient for simulating a chaotic motion (Gontar and Ilin, 1991). The problem leads us, as emphasizes the Russian-Israeli mathematician Vladimir Gontar, to the paradoxes of ancient Greeks: is it mathematically and logically possible to formulate a contradiction-free description of the process of approaching an object when the distance to this object contains an infinite number of segments, involving an infinite number of steps necessary to reach this object? (Gontar, 1993)

Non-Diophantine arithmetics suggest a new understanding of this problem. It is possible that we can approach the object only to a definite distance, a kind of minimal length. It is also reasonable to do this in a finite number of steps. All consequent steps cannot make the distance to the object smaller. For instance, it might be impossible to go above 1000 by adding 1. In terms, of non-Diophantine arithmetics, this means that 1000 is much bigger than 1 (formally $1000 \gg 1$), that is, by adding 1 to 1 to 1 and so on, it is impossible to get exactly to 1000. This situation emerges in many non-Diophantine arithmetics and performing simulation, it is only necessary to choose one of them that better fits other conditions implied by the simulated system.

In essence, some of non-Diophantine arithmetics possess similar properties to those of transfinite numbers arithmetics built by the great German mathematician Georg Cantor (1845–1918). For example, a non-Diophantine arithmetic may have a sequence of numbers $a_1, a_2, \ldots, a_n, \ldots$ such that for any number b that is less than some a_n the equality $a_n + b = a_n$ is valid. This is an important property of infinity, which is formalized by transfinite (cardinal) numbers. The equality $a^2 = a$ is another interesting property of some transfinite numbers. This equality may be also true in some non-Diophantine arithmetics. Thus, non-Diophantine arithmetics provide mathematical models in which finite objects — natural numbers — acquire features of infinite objects — transfinite numbers. In such a way, it is possible to model and to describe behavior of infinite entities in finite domains.

At the same time, non-Diophantine arithmetics allow rigorous representation of finite arithmetics, such as computer arithmetics, which have the largest number. In the context of non-Diophantine arithmetics, the largest number of finite arithmetic is modeled by an inaccessible number of an appropriate non-Diophantine arithmetic. So, although other numbers exist but they cannot be reached by conventional computations.

Besides, working with numbers, different automata change rules of the Diophantine arithmetic and as a result, actually use some non-Diophantine arithmetics. For example, computer arithmetic is a special case of non-Diophantine arithmetics (Parhami, 2010). This is a result of round-off procedures and existence of the largest number in this arithmetic. Consequently, if we want to build better models for numerical computations than we have now, it is necessary to utilize relevant non-Diophantine arithmetics in these models.

Examples also demonstrate that non-Diophantine arithmetics are important for business and economics. Some economical problems and inconsistencies caused by the conventional arithmetic are considered in (Tolpygo, 1997). As some studies of economy show, sometimes finite quantities possess properties of infinite numbers with respect to people's practice (cf., for example, Birkhoff and Barti, 1970). Consequently, when one applies mathematics to solve such problems, the results are often mathematically correct but practically misleading. Utilization of non-Diophantine arithmetics eliminates those problems and inconsistencies.

There are other features of non-Diophantine arithmetics, which are different from the properties of the Diophantine arithmetic. For instance, we know from school that the main laws of the Diophantine arithmetic are as follows:

(1) Addition is commutative, i.e., $a + b = b + a$.
(2) Multiplication is commutative, i.e., $a \cdot b = b \cdot a$.
(3) Addition is associative, i.e., $(a + b) + c = a + (b + c)$.

(4) Multiplication is associative, i.e., $(a \cdot b) \cdot c = a \cdot (b \cdot c)$.
(5) Multiplication is distributive with respect to addition, i.e., $a \cdot (b + c) = a \cdot b + a \cdot c$.
(6) Zero is a neutral element with respect to addition, i.e., $a + 0 = 0 + a = a$.
(7) One is a neutral element with respect to multiplication i.e., $a \cdot 1 = 1 \cdot a = a$.

Naturally, we can ask whether these laws are valid for non-Diophantine arithmetics. Exploring them we find that addition and multiplication are always commutative. However, zero is not always a neutral element with respect to addition and one is not always a neutral element with respect to multiplication in all non-Diophantine arithmetics. At the same time, the laws of associativity and distributivity fail in the majority of non-Diophantine arithmetics. Only special conditions on the functional parameter of the non-Diophantine arithmetic in question provide validity of these laws (Burgin, 1997).

Besides, the Diophantine arithmetic possesses the so-called Archimedean property, which is important for proofs of many results in arithmetic and number theory. It states that if we take any two natural numbers m and n, in spite that n may be enormously larger than m, it is always possible to add m enough times to itself, i.e., to take the sum $m + m + \cdots + m$, so that the result will be larger than n. This property is also invalid in the majority of non-Diophantine arithmetics. The Archimedean property is important for proving that sets of all natural and prime numbers are infinite. Thus, having in general no Archimedean property in non-Diophantine arithmetics, we encounter such arithmetics that have only a finite number of accessible elements, or such infinite arithmetics that have only a finite set of prime numbers (Burgin, 1997).

One more unusual property of non-Diophantine arithmetics is related to physics. Physicists often use the relation $a \ll b$, which means that a is much smaller than b. However, this relation does not have an exact mathematical meaning and is used informally. In contrast to this, non-Diophantine arithmetics provide rigorous interpretation and formalization for such relations. Namely, $a \ll b$ if and only if $b + a = b$.

Note that this is impossible in the conventional mathematics because for any number $a > 0$, the sum $b + b$ is larger than b. At the same time, there are non-Diophantine arithmetics, in which $b + a = b$ is true for different numbers a and b when $0 < a < b$, i.e., $a \ll b$.

Interestingly, this property reflects some basic features of nature. Physicists (cf., for example, Penrose, 1972; Zeldovich *et al.*, 1990; Czachor, 2017) emphasize that fundamental problems of modern physics are dependent on our ways of counting and calculation. Mathematicians also call attention to connections between the worlds of quantum physics and number theory expressing even more radical view that number theory can be treated as the ultimate physical theory (cf., for example, Varadarajan, 2002, 2004; Volovich, 2010).

This idea correlates with problems of modern physical theories in which physical systems are described by chaotic processes. Taking into account the fact that chaotic solutions are obtained by computations, physicists ask (Cartwrite and Piro, 1992; Gontar, 1997) whether chaotic solutions of the differential equations, which model different physical systems, reflect the dynamic laws of nature represented by these equations or whether they are solely the result of an extreme sensitivity of these solutions to numerical procedures and computational errors.

It is even clearer that properties of non-Diophantine arithmetics, which reflect the way people count, influence functioning of economy and are important for economical and financial models (cf., for example, Tolpygo, 1997). Thus, it would be useful to build models of economical systems and processes employing an appropriate non-Diophantine arithmetic.

In the first discovered class of non-Diophantine arithmetics called projective arithmetics, there is an interesting property. Specifically, projective arithmetics allow us to formalize and make rigorous concepts such as *much smaller* (denoted by $\ll$) and *much larger* (denoted by $\gg$). These relations are defined as follows:

A number m is *much smaller* than a number n ($m \ll n$) if $n + m = n$.

In this case, the number n is *much larger* than the number m ($n \gg m$).

There are many non-Diophantine arithmetics, which are essentially different from the Diophantine arithmetic N. For instance, we know that there are infinitely many prime numbers in N. At the same time, it is proved that for any $n > 0$, there is a non-Diophantine arithmetic that has exactly n prime numbers and there are infinitely many non-Diophantine arithmetics with only one prime number. We know that in N, half of numbers are even and the other half consists of odd numbers. At the same time, it is proved that there are infinitely many non-Diophantine arithmetics with only one odd number.

According to Fermat's Last Theorem, also called Fermat's conjecture, which was proved by the famous English–American mathematician Andrew Wiles with the help of his graduate student Richard Taylor (Taylor and Wiles, 1995; Wiles, 1995), the equation $x^n = y^n + z^n$ cannot have positive integer solutions for any natural number n greater than 2. At the same time, it is proved that there are infinitely many non-Diophantine arithmetics of natural numbers, in which for any natural number n, the equation $x^n = y^n + z^n$ has infinitely many solutions.

Andrew Wiles was born in 1953 in Cambridge, England, and studied at Merton College, Oxford, and Clare College, Cambridge. Coming to the United States, he worked at the Institute of Advanced Study in Princeton and Princeton University, as well as at the Institut des Haute Ètudes Scientifiques and at École Normale Supérieure in Paris and the University of Oxford in England. Wiles became famous after he proved Fermat's Last Theorem.

Non-Diophantine arithmetics also allow elimination of several inconsistencies and misconceptions related to arithmetic. For instance, the Romanian-South African mathematician Elemer Elad Rosinger (1937–2019) explains:

"...we have been doing inconsistent mathematics for more than half a century by now, and in fact, have quite heavily and essentially depended on it in our everyday life. Indeed, electronic digital computers, when considered operating on integers, which is but a part of their operations, act according to the system of axioms given by

- (PA): the usual Peano Axioms for N,

plus the ad-hock axiom, according to which

- (MI): there exists M in N, $M \gg 1$, such that $M + 1 = M$.

Such a number M, called *"machine infinity"*, is usually larger than 10^{100}, however, it is inevitably inherent in every electronic digital computer, due to obvious unavoidable physical limitations. And clearly, the above mix of (PA) + (MI) axioms is inconsistent. Yet we do not mind flying on planes designed and built with the use of such electronic digital computers." (Rosinger, 2008)

In a similar way, the American logician Robert Meyer (1932–2009) and Australian philosopher Chris Mortensen built various inconsistent models of arithmetic (Meyer and Mortensen, 1984), while the Australian-American philosopher Graham Priest developed axiomatic systems for inconsistent arithmetics (Priest, 1997, 2000).

Even before Priest, the Belgian mathematician and philosopher of science, Jean Paul Van Bendegem developed an inconsistent axiomatic arithmetic by changing the Peano axioms so that a number that is the successor of itself exists (Van Bendegem, 1994). The fourth Peano axiom states that if $x + 1 = y + 1$, then x and y are the same number. In the system of Van Bendegem, starting from some number n, all its successors will be equal to n. Then the statement $n = n + 1$ is considered as both true and false at the same time. This makes the new arithmetic inconsistent. It is possible to rigorously eliminate these inconsistencies using non-Diophantine arithmetics.

In general, there are two basic ways to deal with inconsistencies: one is to elaborate an inconsistent system and try to work with it and another way is create new mathematical structures, eliminating inconsistencies. Ambiguities, contradictions, and paradoxes play the central role in the emergence of creative ideas. One of the main kinds of contradictions is existence of two seemingly contradictory perspectives in a mathematical problem (Byers, 2007). For instance, Peano axioms imply infiniteness of the arithmetic, while the existence of the largest number implies it finiteness. However, this paradox vanishes with the discovery of non-Diophantine arithmetics and weak arithmetics. Actually all these inconsistencies and contradictions exist only in the absence of non-Diophantine arithmetics. For instance, the machine arithmetic analyzed by the Israeli–South African mathematician Elemer Elad Rosinger (1938–2019) does not satisfy Peano axioms (Rosinger, 2008) not because it is inconsistent, but since it is non-Diophantine. In essence, the machine arithmetic satisfies the axioms of the corresponding non-Diophantine arithmetic, and operating on integers, electronic digital computers perform according to this system of axioms. In the setting of non-Diophantine arithmetics, the existence of the largest number means that all larger numbers are not accessible in the machine arithmetic by means of the basic arithmetical operations — addition and multiplication.

The rule $n = n + 1$ of Van Bendegem, which is considered above, is natural for many non-Diophantine arithmetics and causes no inconsistencies and contradictions there. It simply means $1 \ll n$. In other words, starting with the number 1 and repeating the operation of adding 1, the machine will never get a number that is larger than n.

It is possible to compare this situation with artificially derived inconsistencies and contradictions with numbers when people knew only natural numbers and positive fractions. Getting information about negative numbers, mathematicians who lived at that time would be able to build an inconsistent formal system by taking two "axioms":

- Only positive numbers exist;

and

- There are negative numbers.

Naturally these "axioms" give a contradiction. Now we know that the first "axiom" is valid only for natural numbers and positive fractions if we consider numbers known at that time. However, integer numbers combine both positive and negative numbers without any inconsistency.

Mathematicians who lived in the 19th century and earlier were also able to build an inconsistent formal system in geometry, combining together two sets of axioms:

- All postulates of the Euclidean geometry.

and the postulate which is true for the geometry on a sphere, which is considered a geometrical model of the Earth:

- Any two straight lines intersect with one another.

However, now we know that spherical geometry is non-Euclidean and does not have any contradictions in it.

The great German philosopher Immanuel Kant (1724–1804) asserted that the world described by science was a world of sense impressions organized and controlled by the mind in accordance with innate categories of space and time (Kant, 1786). Consequently, continued Kant, there never would be a world's description other than Euclidean geometry and Newtonian mechanics. As we now know, non-Euclidean geometry better describes space and time in comparison with Euclidean geometry, while quantum mechanics substituted Newtonian mechanics in the description of the microworld.

Immanuel Kant was born in Königsberg, Prussia (now Kaliningrad, Russia) and studied at the University of Königsberg where he later taught. Kant created an influential philosophical system based on his doctrine of transcendental idealism and encompassing almost all areas of philosophy including epistemology, or theory of knowledge.

One may say that reasoning about geometry, Kant was a philosopher and not an expert in mathematics. However, the well-known Irish mathematician William Rowan Hamilton (1805–1865), certainly one of the outstanding mathematicians of the 19th century, expressed similar consideration in 1837 after the works of Lobachevsky (1829) and Bolyai (1932) had been already published but were not known to the majority of mathematicians. Namely, Hamilton wrote:

"No candid and intelligent person can doubt the truth of the chief properties of *Parallel Lines*, as set forth by Euclid in his *Elements*, two thousand years ago; though he may well desire to see them treated in a clearer and better method. The doctrine involves neither obscurity nor confusion of thought, and leaves in the mind no reasonable ground for doubt, although ingenuity may usefully be exercised in improving the plan of the argument."

Even in 1883, more than 50 years after the discovery of non-Euclidean geometries, another famous English mathematician Arthur Cayley (1821–1895) in his presidential address to the British Association for the Advancement of Science affirmed:

"*My own view is that Euclid's twelfth axiom* [usually called the fifth or parallel axiom or postulate] *in Pfayfair's form of it does not need demonstration, but is part of our notion of space, of the physical space of our experience...*"

Similar situation exists now with non-Diophantine arithmetics. Mathematical community at large and other people do not know about non-Diophantine arithmetics although properties of non-Diophantine arithmetics were discussed by the Argentinian mathematician Sergio Amat Plata (Amat Plata, 2005) while non-Diophantine arithmetics of real and complex numbers were explicitly applied to physics (Czachor, 2016, 2017, 2017b, 2020, 2020a; Czachor and Posiewnik, 2016; Chung and Hassanabadi, 2019, 2020; Chung and Hounkonnou, 2020; Burgin and Czachor, 2020), psychology (Czachor, 2017a; Burgin and Czachor, 2020), cryptography (Czachor, 2020b), fractal theory (Aerts *et al.*, 2016, 2016a, 2018; Czachor, 2019), economics and finance (Burgin and Meissner, 2017).

Many financial processes have better description in terms of non-Diophantine arithmetics in comparison with operations in the Diophantine arithmetics. Let us take making deposits in the bank as simple examples of such operations.

Example 4.5. Ann does not have money in the Alpha bank and deposits $1000 for one year with the simple interest rate 2%. This operation is modeled by the non-Diophantine addition \$0 ⊕ \$1000 where this addition is defined by the formula

$$x \oplus y = 0.02x + 0.02y.$$

Example 4.6. Britney has an account of \$100 with the compound interest rate 3% compounded monthly in the Alpha bank and deposits \$1000 to the same account. This operation is modeled by the non-Diophantine addition \$100 ⊕ \$1000 where this addition depends on the period of time and after one year, it is defined by the formula

$$x \oplus y = (1 + 0.0025)^{12}x + (1 + 0.0025)^{12}y.$$

Example 4.7. Claude has an account of \$200 with the compound interest rate 3% compounded semiannually in the Alpha bank and deposits \$2000 to another account with the compound interest rate 6% compounded monthly. This operation is modeled by the non-Diophantine addition \$200 ⊕ \$2000 where this addition depends on the period of time and after one year, it is defined by the formula

$$x \oplus y = (1 + 0.005)^{2}x + (1 + 0.005)^{12}y.$$

These examples are extremely simple and use well-known formulas. So, the question can be asked why we need non-Diophantine arithmetics if we can use only formulas for the unconventional operations in the Diophantine arithmetic.

The answer can be given if we take lessons from the human history. One historical situation is about negative numbers and the arithmetic of integers. Historians found that negative numbers emerged when they were interpreted as debts. However, it was possible to work with debts without negative numbers by using subtraction instead of addition. However, now we can see that negative numbers essentially increased efficiency of operation. Moreover, the usefulness and effectiveness of negative numbers greatly increased after the arithmetic of integers had been elaborated.

Another lesson comes from astronomy. For a long time, the Ptolemaic model was the best for the Solar system. However, to make it more exact, it was necessary to introduce epicycles as additional formulas. In contrast to this, the Copernican model became much simpler and more exact after the elliptical shape of the planets' orbits was discovered.

These examples suggest the following answer to the question about non-Diophantine arithmetics. Having the structure of an arithmetic, we have a uniform representation for all unconventional operations.

This representation allows more flexible and efficient application of arithmetical structures and operation. In addition, when all such operations are organized as the system of arithmetic, it is possible to better learn their properties and relations.

Alongside direct applications of non-Diophantine arithmetics, there were many implicit applications, in which many particular families of non-Diophantine arithmetics appeared under different guises such as applications of non-Diophantine arithmetics of real and complex numbers to physics (Benioff, 2011–2013, 2015, 2016, 2016a), psychology (Bachem, 1952; Baird, 1975, 1997; Baird and Noma, 1978; Falmagne, 1985), decision-making (McEneaney, 1999), and economics (Tolpygo, 1997). Writing about the Pythagorean Law, Lucio Berrone also employs operations from non-Diophantine arithmetics (Berrone, 2010). Different kinds of non-Diophantine arithmetics of real numbers are used in idempotent analysis (Maslov, 1987; Maslov and Samborskii, 1992; Maslov and Kolokoltsov, 1997; Kolokoltsov and Maslov, 1997), in tropical geometry, cryptography and analysis (Litvinov, 2007; Itenberg *et al.*, 2009; Speyer and Sturmfels, 2009; Grigoriev and Shpilrain, 2014; Maclagan and Sturmfels, 2015; Zhang *et al.*, 2018). In turn, all these areas have a variety of practical and theoretical applications.

In spite of all these and previous publications even in 2012, after many years passed when works on non-Diophantine arithmetics were already published, researchers thought that up to the present any alternative arithmetic of natural numbers has not been constructed (Krysztofiak, 2012).

This situation is entirely mirrored in the contemporary mathematics. Mathematicians assume that arithmetic, which comprises mostly representation and operation with numbers, is the field, which was completed long ago. Now it is trivial and aimed at the beginners, while number theory, which comprises investigation of properties of natural numbers, is a respected field of theoretical mathematics with its highly complicated, "deep" and abstract problems. However, knowing only one arithmetic N of natural numbers, mathematicians forget that this arithmetic (in the modern sense) is the base for number theory. If you change arithmetic, you will need new theory of numbers as numbers will change their properties bringing us to the non-Diophantine number theory (Burgin, 2018c). For example, the definition of a prime number is completely based on the definition of the operation of multiplication and for any $n = 2, 3, 4, 5, \ldots$, there are non-Diophantine arithmetics, in which there are exactly n prime numbers, while in the Diophantine arithmetic, there are infinitely many prime numbers.

With respect to the usability of non-Diophantine arithmetics, it is possible to mention attempts of some mathematicians to build new physical theories based exclusively on numbers suggesting to study physical reality at the Planck scale using number fields of p-adic numbers (Volovich, 1987) or modular arithmetics (Lev, 1989, 1993) as the fundamental physical objects. However, there are arguments for suggesting that non-Diophantine arithmetics, which have functional parameters providing adequate adjustment of the theory to different physical scales, are more relevant for representing quantum objects at the Planck scale than p-adic numbers.

The most basic class of numbers — natural numbers — came from practical activity of people, which involved counting. Counting, in turn, brought people to arithmetical operations of addition and multiplication, which resulted in creation (discovery) of the Diophantine arithmetic, or more exactly, the Diophantine arithmetic of natural numbers. The subsequent development of arithmetic went by adding new types of numbers. That is why all these arithmetics naturally bear the common name the Diophantine arithmetic. When 0 was added, the Diophantine arithmetic of whole numbers was created. When that arithmetic was extended by negative numbers, the Diophantine arithmetic of integer numbers emerged. Introduction of rational numbers gave birth to the Diophantine arithmetic of rational numbers. Construction of real numbers from rational numbers brought mathematicians to the Diophantine arithmetic of real numbers while the discovery of complex numbers resulted in the Diophantine arithmetic of complex numbers.

Naturally, the same situation is reflected in the system of non-Diophantine arithmetics. Namely, there are Diophantine and non-Diophantine arithmetics not only of natural numbers but also of whole, integer, rational, real and complex numbers. In this big family of arithmetics, the Diophantine arithmetic of natural numbers is the most fundamental because all other arithmetics are built using natural numbers as their foundation.

The development of non-Diophantine arithmetics was based on the very general abstract algebraic structure called prearithmetic. In essence, any arithmetic is some kind of prearithmetics. Many other mathematical structures studied in algebra, such as rings, semirings, fields, lattices, linear algebras, Ω-groups, Ω-ring, Ω-algebras, topological rings, topological fields, normed rings, normed algebras, normed fields, tropical semirings, and subtropical algebras, also are prearithmetics.

Chapter 5

Non-Diophantine Operations with Numbers and Related Constructions in the History of Mathematics and Science

One of the endlessly alluring aspects of mathematics is that its thorniest paradoxes have a way of blooming into beautiful theories.

Philip J. Davis

Elements of non-Diophantine operations with numbers appeared in ancient times but it took millennia to explicitly discover (construct) non-Diophantine arithmetics. In a similar way, people have been living on the surface of the Earth, the geometry of which is non-Euclidean, but for a long time they believed that there was only one unique geometry — Euclidean geometry.

The first appearance of non-Diophantine addition, which was different from the addition in the conventional (Diophantine) arithmetic of natural numbers, can be tracked to the Pythagorean Theorem, which has the name of the famous Greek mathematician Pythagoras (6th century B.C.E.). However, this theorem was known in ancient Babylon and Egypt between 2000 and 1700 B.C.E. (Neugebauer, 1969), that is, more than thousand years before Pythagoras was born. It means that the non-Diophantine addition appeared around four millennia ago.

The Pythagorean Theorem states that in a right triangle, the square of the hypotenuse is equal to the sum of the squares of the other two sides, that is,

$$c^2 = a^2 + b^2. \tag{5.1}$$

This equality implies that the hypotenuse is equal to the square root of the sum of the squares of the other two sides, i.e.,

$$c = (a^2 + b^2)^{1/2}. \tag{5.2}$$

Formulas (5.1) and (5.2) are equivalent. It means that the first formula implies the second one while the second formula implies the first one. So, it is possible to treat the second formula as another form of the Pythagorean Theorem.

At the same time, in the non-Diophantine arithmetic **A** that is exactly projective (cf., for example, Burgin, 2019) with respect to the Diophantine arithmetic $\mathbf{R}^+$ of non-negative real numbers with the generator function $f(x) = x^2$ (cf., for example, Burgin and Czachor, 2020), addition is defined by the following formula:

$$a \oplus b = \sqrt{a^2 + b^2} = (a^2 + b^2)^{1/2}.$$

This is just the expression equivalent to the Pythagorean Theorem. It means that, in the context of non-Diophantine arithmetics, we have the following form of the Pythagorean Theorem:

The length of the hypotenuse is equal
to the non-Diophantine sum of the two other sides.

This shows that implicitly elements of non-Diophantine arithmetics have been used for millennia by mathematicians and engineers.

Later a similar formula based on the non-Diophantine sum of two numbers was used for defining Euclidean distance in a plane. This explicitly demonstrates that non-Diophantine arithmetics form the foundation of the Euclidean geometry in two dimensions and as it is possible to show, even in more dimensions. Namely, the formula for distance between points $x = (x_1, x_2, x_3, \ldots, x_n)$ and $y = (y_1, y_2, y_3, \ldots, y_n)$ in the n-dimensional Euclidean space has the form

$$\mathrm{d}(x,y) = [(y_1 - x_1)^2 + (y_2 - x_2)^2 + (y_3 - x_3)^2 + \cdots + (y_n - x_n)^2]^{1/2}.$$

This is also an operation from a definite non-Diophantine arithmetic (Burgin, 1997; Burgin and Czachor, 2020). Namely, this is a non-Diophantine addition of the differences between the coordinates of these points.

Interestingly, there is also geometry on a plane, which uses the Manhattan, or taxicab, distance (cf., for example, Krause, 1987) where the distance between two points in a plane is the conventional sum of the differences between the coordinates of these points. It is expressed by the formula

$$d((a, b), (c, d)) = |c - a| + |d - b|.$$

Consequently, this geometry is based on the Diophantine arithmetic while the Euclidean geometry is based on the non-Diophantine arithmetic with the generator function (parameter) $f(x) = x^2$. The taxicab distance determines a non-Euclidean geometry. Thus, we come to an interesting situation: the Euclidean geometry is based on a non-Diophantine arithmetic while a non-Euclidean geometry is based on the Diophantine arithmetic.

However, utilization of non-Diophantine operations did not change the situation in mathematics. Indeed, the history of mathematics reveals that until the discovery of non-Diophantine arithmetics at the end of the 20th century, appearance of other mathematical systems called *arithmetic* did not change the predominant position of the Diophantine arithmetic of natural numbers. In particular, two plus two always remained four because nobody explicitly built non-Diophantine arithmetics of natural numbers.

One more constructive feature of non-Diophantine arithmetics also appeared in mathematics long before non-Diophantine arithmetics were discovered or constructed. Namely, operations in non-Diophantine arithmetics are usually constructed by functional reduction to the conventional operations in the Diophantine arithmetic (Burgin, 1977, 1997; Burgin and Czachor, 2020). Although non-Diophantine arithmetics were discovered (constructed) only at the end of the 20th century, the history of mathematics showed that an important case of such reduction appeared several centuries before this discovery with the invention of logarithms by John Napier (1550–1617) in the 17th century. Logarithms allowed reduction of conventional multiplication to conventional addition and then returning back by exponentiation. Performing addition is much simpler than performing multiplication. Thus, this reduction simplified calculations by means of slide rules and logarithm tables, which were efficiently used by scientists and engineers for several centuries before proliferation of calculators and computers.

Note that Napier did not build another arithmetic but only used the construction similar to those used in non-Diophantine arithmetics.

Next steps in the direction of non-Diophantine arithmetics were made only in the 20th century. According to contemporary knowledge, in a general form, non-Diophantine arithmetical operations were studied by the great Russian mathematician Andrey Nikolayevich Kolmogorov (1903–1986), who introduced and studied the *non-Diophantine average*, which is usually called the *quasi-arithmetic mean, generalized f-mean, Kolmogorov–Nagumo average*, and *Kolmogorov mean* (Kolmogorov, 1930). The noted Japanese mathematician Mitio Nagumo (1905–1995) and the well-known Italian statistician and actuary Bruno de Finetti (1906–1985) also found and investigated this arithmetical concept (Nagumo, 1930; de Finetti, 1931), and it was used in the book (Hardy *et al.*, 1934).

Mitio Nagumo was born in Yamagata Prefecture, Japan, studied at the Imperial University of Tokyo and after graduation taught at that university as well as at Sophia University in Japan. His main research was in the field of differential equations.

Bruno de Finetti was born in Innsbruck, Austria, and studied mathematics at Polytecnico di Milano. After this he worked as an actuary and statistician at National Institute of Statistics in Rome, at insurance company in Trieste, at the University of Trieste and Sapienza University of Rome acquiring international reputation in the field of probability.

To understand the construction of Kolmogorov, Nagumo, and de Finetti, we remind that the conventional *arithmetical average*, also called the *arithmetical mean*, of n numbers $x_1, x_2, \ldots, x_n$ is defined as

$$(x_1 + x_2 + \cdots + x_n)/n.$$

However, different applications demanded other kinds of average and researchers elaborated a variety of types of average. For instance, the *geometrical average*, also called the *geometrical mean*, of n numbers $x_1, x_2, \ldots, x_n$ is defined as

$$(x_1 \cdot x_2 \cdot \ldots \cdot x_n)^{1/n}.$$

Mathematicians and scientists strive to reduce multiplicity of structures to one unifying structure. To achieve this goal, Kolmogorov demonstrated that all known types of averages have the form

$$h((g(x_1) + g(x_2) + \cdots + g(x_n))/n),$$

where g and h are some strictly monotone functions and h is the inverse of g, i.e., and $h = g^{-1}$ (Kolmogorov, 1930).

This is exactly the form of average in non-Diophantine arithmetics of real numbers with the generator g because between the Kolmogorov–Nagumo average and conventional average exists exact projectivity (cf. Burgin, 1997). Thus, it would be natural to call them *non-Diophantine averages*, or *means*. These averages also coincide with α-*average* defined by Grossman and Katz in their non-Newtonian calculus (Grossman and Katz, 1972). Now such operations are often called *quasi-arithmetic means* or *Kolmogorov–Nagumo averages*. They are studied and used by many authors.

Kolmogorov–Nagumo average unifies many particular cases. For instance, geometrical mean is obtained from Kolmogorov–Nagumo average if we take logarithm as $f(x) = \ln x$ and exponent as $g(x) \exp x = e^x$. Indeed,

$$(\ln x_1 + \ln x_2 + \ldots + \ln x_n)/n$$
$$= (\ln x_1 + \ln x_2 + \ldots + \ln x_n) \cdot (1/n)$$
$$= \ln(x_1 \cdot x_2 \cdot \ldots \cdot x_n) \cdot (1/n)$$

and

$$\exp[\ln(x_1 \cdot x_2 \cdot \ldots \cdot x_n) \cdot (1/n)] = (x_1 \cdot x_2 \cdot \ldots \cdot x_n)^{1/n}.$$

Kolmogorov–Nagumo average had many applications. For instance, they were efficiently used in decision making and utility theory (Marichal, 2009). In addition, researchers demonstrated that Kolmogorov–Nagumo averages appeared very useful in thermostatistics, which is a scientific field from the intersection of thermodynamics and statistics (Naudts and Czachor, 2001; Czachor and Naudts, 2002; Naudts, 2002, 2011; Borges, 2004; Dukkipati *et al.*, 2005; Amblard and Vignat, 2006; Masi, 2007; Dukkipati, 2010). Later it was demonstrated that Kolmogorov–Nagumo averages are formed in the specific class of non-Diophantine arithmetics called the *generalized arithmetics of Kolmogorov and Nagumo* (Jizba and Korbel, 2020). These arithmetics are used in the studies of entropy, which is one of the most important concepts in physics, information theory and statistics. In particular, they allow re-establishing the entropic parallelism between information theory and statistical inference (Jizba and Korbel, 2020).

This was the beginning of the process in which under different guises and names, elements of non-Diophantine arithmetics at first implicitly and then explicitly proliferated in a variety of the applications of mathematics.

The process of proliferation of some particular elements and bits of non-Diophantine arithmetics started with elaboration and utilization of the unconventional operations of addition and multiplication of real numbers. At first, non-Diophantine addition appeared. It was called *quasiaddition*, denoted by $\oplus$ and introduced by the prominent Austrian-American mathematician and philosopher Karl Menger (1902–1985) in the form of *triangular norms* (*t-norms* for short) in probabilistic metric spaces (Menger, 1942).

By definition, a *t-norm* is a function T: $[0,1] \times [0,1] \to [0,1]$, which in essence, is a binary operation that satisfies the following properties:

- *Commutativity*: $T(a,b) = T(b,a)$.
- *Monotonicity*: $T(a,b) \leq T(c,d)$ if $a \leq c$ and $b \leq d$.
- *Associativity*: $T(a,T(b,c)) = T(T(a,b),c)$.
- The number 1 is the *identity element*, that is, $T(a,1) = a$.

In the algebraic context, a *t*-norm is a binary operation on the closed unit interval $[0,1]$ defining an abelian, totally ordered semigroup with the neutral identity element 1.

The American mathematicians Berthold Schweizer (1929–2010) and Abe Sklar provided the axioms for *t*-norms, as they are used today (Schweizer and Sklar, 1960). Many results concerning *t*-norms were obtained in probabilistic metric spaces.

The main ways of construction of *t*-norms include using generators, defining parametric classes of *t*-norms, rotations, or ordinal sums of *t*-norms. Namely, with the additive generator $f(x)$, a *t*-norm $T(x,y)$ is defined as

$$T(x,y) = f^{(-1)}(f(x) + f(y)).$$

This is exactly the form of addition in real number prearithmetics and non-Diophantine arithmetics of real numbers with the generator f.

With the multiplicative generator $h(x)$, a *t*-norm $T(x,y)$ is defined as

$$T(x,y) = h^{(-1)}(h(x) \cdot h(y)).$$

This is exactly the form of multiplication in real number prearithmetics and non-Diophantine arithmetics of real numbers with the generator h.

Mathematicians found different conditions when it is possible to represent general *t*-norms as the non-Diophantine addition (*quasiaddition* $\oplus$) or non-Diophantine multiplication (*quasimultiplication* $\otimes$) in the interval

[0, 1] (Aczel, 1966; Ling, 1995). Later these operations were extended to arbitrary finite real intervals $[a, b]$ (Maslov and Samborskij, 1992; Pap, 1995; Kolokoltsov and Maslov, 1997) and to the infinite interval $\mathbf{R}^+ = [0, \infty)$ (Marinova, 1986; Kolesarova, 1993, 1996). Note that mathematical systems with quasiaddition, which is usually denoted by $\oplus$, and conventional multiplication, or with quasiaddition and quasimultiplication, which is denoted by $\otimes$ or by $\odot$, are special cases of prearithmetics (Burgin, 1977; 1997).

Researchers use t-norms as an important tool for the interpretation of the conjunction in fuzzy logics allowing evaluation of the truth degrees of compound formulas, and are useful for semantics of the intersection of fuzzy sets (Alsina *et al.*, 1983). Besides, t-norms are also utilized in fuzzy control to formulate assumptions of rules as conjunctions (fuzzy intersections) of fuzzy sets called antecedents or premises, in decision making, in statistics as well as in the theory of measure and game theory (Klement *et al.*, 2000).

The Hungarian physicist Lajos Jánossy (1912–1078) used non-Diophantine operations of quasiaddition and quasimultiplication for exploring foundations of probability theory (Jánossy, 1955) and the Hungarian-Canadian mathematician János Derzõ Aczél (1924–2020) studied their properties. Janusz Matkowski and Tadeusz Świątkowski explored when quasiaddition was defined by a convex homeomorphism (Matkowski and Świątkowski, 1993).

It is necessary to understand that although quasiaddition was often denoted by the symbol $\oplus$, the operation itself could be different. For instance in some cases, it was defined as in the theory of t-norms and non-Newtonian calculus, i.e., $x \oplus y = f^{(-1)}(f(x) + f(y))$, while in other cases, it could be as in tropical analysis and idempotent analysis, i.e., $x \oplus y = \max(x, y)$ or $x \oplus y = \min(x, y)$.

The non-Diophantine addition (quasiaddition $\oplus$) was employed for defining new classes of measures and integrals, i.e., $\oplus$-measures and $\oplus$-integrals (Marinova, 1986; Kolesarova, 1993, 1996; Pap, 2002). The special case of $\oplus$-integrals when $\oplus$ is *max* was studied by Shilkret (1971) and later in such full-sized mathematical theory as tropical analysis (Litvinov, 2007; Speyer and Sturmfels, 2009).

The non-Diophantine addition $\oplus$ has been intensively used in psychophysics (cf., for example, Czachor, 2017a). Falmagne and Doble analyzed utilization of the non-Diophantine addition and other unconventional, i.e., non-Diophantine, arithmetical operations in scientific theories (Falmagne and Doble, 2015).

Traces of non-Diophantine arithmetics can be also found in the works of Kaniadakis on generalized statistics and statistical mechanics (Kaniadakis, 2001; 2002; 2005; 2006).

One of the basic relations in the theory of prearithmetics and non-Diophantine arithmetics is projectivity (Burgin, 1977, 1997, 2019). It has three basic forms and allows building new prearithmetics and non-Diophantine arithmetics from the existing ones (Burgin, 1992, 2010). Let us consider exact definitions.

We take two arithmetics $\mathbf{A} = (A; \oplus, \otimes)$ with addition $\oplus$ and multiplication $\otimes$, $\mathbf{B} = (B; +, \circ)$ with addition $+$ and multiplication $\circ$ and two mappings $g\colon A \to B$ and $h\colon B \to A$ where A and B are sets of numbers in arithmetics $\mathbf{A}$ and $\mathbf{B}$.

Then, the arithmetic $\mathbf{A}$ is *weakly projective* with respect to the arithmetic $\mathbf{B}$ if there are following relations between operations in $\mathbf{A}$ and in $\mathbf{B}$:

$$a \oplus b = h(g(a) + g(b)),$$
$$a \otimes b = h(g(a) \circ g(b)).$$

The mapping g is called the *projector* and the mapping h is called the *coprojector* for the pair $(\mathbf{A}, \mathbf{B})$. When the projector g is inverse function of the coprojector h and $\mathbf{B}$ is the Diophantine arithmetic, the projector g is called the *generator* (*generator function*) of the arithmetic $\mathbf{A}$.

The arithmetic $\mathbf{A}$ is *projective* with respect to the arithmetic $\mathbf{B}$ if it is weakly projective with respect to the prearithmetic $\mathbf{B}$ and $hg = 1_A$, where 1_A is the identity mapping of A.

The arithmetic $\mathbf{A}$ is *exactly projective* with respect to the arithmetic $\mathbf{B}$ if it is projective with respect to the prearithmetic $\mathbf{B}$ and $gh = 1_B$, where 1_B is the identity mapping of B.

It is possible that arithmetics have other operations. For example, the arithmetic $\mathbf{Z}$ of integer numbers has also subtraction, while the arithmetic $\mathbf{R}$ of real numbers has additionally subtraction and division. In this case, all types of projectivity imply the same relations between additional operations (Burgin, 1997, 2019).

Various particular cases of projectivity relation implicitly appeared in many publications of different authors. For instance, in the theory of Extensive Measurement (Roberts and Luce, 1968; Krantz *et al.*, 1971), a measurement scale is defined on an ordered semigroup G a strictly increasing function $h\colon G \to \mathbf{R}^+$, which makes multiplication in G projective with respect to addition of real numbers, i.e., it is determined by the formula

$$x \times y = h^{-1}(h(x) + h(y)).$$

Here $\times$ is the semigroup operation and $+$ is the usual addition in $\mathbf{R}^+$.

The considered above non-Diophantine addition (often called *quasiaddition* by different authors) $\oplus$ and non-Diophantine multiplication (often called *quasimultiplication* by different authors) $\otimes$ were also successfully utilized for defining *pseudo-integral* (Weber, 1984; Maslov, 1987; Sugeno and Murofushi, 1987; Kolokoltsov and Maslov, 1997; Benvenuti *et al.*, 2002; Pap, 1993, 2002) and even more general pan-integral (Yang, 1985; Yang and Song, 1985; Pap, 2002; Wang and Klir, 2009; Pap and Štrboja, 2010, 2013).

Michael Carroll introduced and studied an infinite sequence of binary operations on real and complex numbers, each of which can be treated as non-Diophantine addition or multiplication (Carroll, 2001). Let us look at his definitions for real numbers.

Taking two real numbers $x, y \in \mathbf{R}$ and an integer number $n \in \mathbf{Z}$, the system of binary operation $\oplus_n$ is recursively defined as follows.

For $n = 0$, it is defined as

$$x \oplus_0 y = x + y.$$

For each $n \le 0$, it is defined as

$$x \oplus_{n-1} y = \ln(e^x \oplus_n e^y) = \ln[\exp(x) \oplus_n \exp(y)].$$

For each $n \ge 0$, $x \ge 0$ and $y \ge 0$, it is defined as

$$x \oplus_{n+1} y = \exp[\ln(x) \oplus_n \ln(y)].$$

Each pair of these operations for n and m less than 0 defines a non-Diophantine arithmetic of real numbers while each pair of these operations for n and m larger than 0 defines a non-Diophantine arithmetic of positive real numbers.

Note that ordinary multiplication coincides with $\oplus_1$, that is

$$x \oplus_1 y = \exp[\ln(x) \oplus_0 \ln(y)] = \exp[\ln(x) + \ln(y)] = \exp[\ln(x \cdot y)] = x \cdot y.$$

Carroll also finds different properties of these operations. For instance, it is proved (Theorem 2.1) that for all $n \in \mathbf{Z}$,

(i) operation $\oplus_n$ is associative;
(ii) operation $\oplus_n$ is commutative;
(iii) operation $\oplus_n$ is distributive over operation $\oplus_{n-1}$.

Using these operations, Carroll also defines non-Newtonian derivatives of functions and explores their properties.

Yamano (2002) and Umarov *et al.* (2006) defined $\otimes_q$-multiplication as $x \otimes_q y = \exp_q(\log_q(x) + \log_q(y))$ and used it with the corresponding $\otimes_{2q}$-factorization to formulate a Central Limit Theorem for non-extensive statistical mechanics and the q-Hammersley–Clifford Theorem.

Non-Diophantine arithmetics of real numbers are used for obtaining different kinds of entropy and in particular, Tsallis entropy (Tsallis, 1988; Dukkipati, *et al.*, 2005; Sunehag, 2007). The connection between these arithmetics and the Diophantine arithmetic of real numbers is established the projectivity relation used in the theory of Non-Diophantine arithmetics (Burgin, 1997; Burgin and Czachor, 2020). This relation corresponds to a measurement scale used the entropy reflects uncertainty in measuring the information content of the even (Sunehag, 2007).

An important step in applications of non-Diophantine arithmetics of real numbers was done with the development of non-Newtonian calculi, which were discovered (constructed) in at least three distinct settings by different researchers. The first were American mathematicians Michael Grossman and Robert Katz, who aimed at the further development of the classical Newtonian calculus. In 1970, they found that it was possible to build a variety of non-Newtonian calculi by changing the Diophantine arithmetic of real numbers to non-Diophantine arithmetics of real numbers (Grossman and Katz, 1972). These arithmetics are built using the construction of the exact projectivity defined above (cf., also Burgin, 1997; Burgin and Czachor, 2020). Namely, a one-to-one mapping $f : \mathbf{R} \to \mathbf{R}$ of real numbers, which can be called the generative function, is chosen and the new operations with real numbers are defined in the following way:

$$\textit{Addition: } a \oplus b = f^{-1}(f(a) + f(b)),$$
$$\textit{Multiplication: } a \otimes b = f^{-1}(f(a) \cdot f(b)),$$
$$\textit{Subtraction: } a \ominus b = f^{-1}(f(a) - f(b)),$$
$$\textit{Division: } a \oslash b = f^{-1}(f(a) \div f(b)).$$

These operations allow defining new kinds of derivatives and integrals of real functions which gives the new kind of calculus, which is called non-Newtonian (Grossman and Katz, 1972; Grossman, 1979a, 1983). Although the new arithmetic, which is also a field, of real numbers is isomorphic to the conventional Diophantine arithmetic (field) of real numbers, the integrals and derivatives in these structures can be essentially different.

Non-Newtonian calculi have roots in the classical mathematics such as the product integral (Volterra, 1887, 1887a) and logarithmic differentiation (cf., for example, Bali, 2005; Bird, 1993; Krantz, 2003). The non-Newtonian calculi of Grossman and Katz have many applications in different areas including decision making, dynamical systems, differential equations, chaos theory, economics, marketing, finance, fractal geometry, image analysis and electrical engineering. Because all non-Newtonian calculi are based on non-Diophantine arithmetics of real numbers, these arithmetics have applications in all mentioned areas.

Later the next discovery/elaboration of more general non-Newtonian calculus called *g-calculus* (generalized calculus) was independently made by the Serbian mathematician Endre Pap (Pap, 1993). As any calculus, it is based on arithmetical operations of pseudo-addition (non-Diophantine addition) and pseudo-multiplication (non-Diophantine multiplication). They are determined by a generative function as in the non-Newtonian calculus but this function can be defined on and/or take values not necessarily the whole real line but in some interval such as [0, 1].

The Slovak mathematicians Radko Mesiar and Ján Rybařík modified the initial construction and studied *g*-calculus, the foundation of which was formed by the non-Diophantine arithmetics of real numbers from the interval [0, 1] with four non-Diophantine operations: addition $\oplus$, subtraction $\ominus$, division $\oslash$ and multiplication $\otimes$ (Mesiar and Rybařík, 1993).

g-calculus has been applied to nonlinear ordinary and partial differential equations, difference equations, generalized functions (distributions), idempotent analysis, measure theory and Laplace transforms. Thus, non-Diophantine arithmetics of real numbers have applications in all these areas.

In addition, non-Diophantine arithmetics of real numbers constitute the base of the new mathematical area called *pseudo-analysis* (Pap and Vivona, 2000; Štajner-Papuga *et al.*, 2006). However, it is analysis only in a different from the classical domain. Namely, it is analysis in the domain of prearithmetics and non-Diophantine arithmetics. Pseudo-analysis allows treating three important problems — nonlinearity, uncertainty and optimization — in a unified way solving various nonlinear partial and ordinary differential equations, difference equations, improving decision making, and estimating insurance risks.

The arithmetical base of pseudo-analysis is a semiring, which is an algebraic structure with two operations also called pseudo-addition $\oplus$ and pseudo-multiplication $\otimes$ while elements of which constitute a closed interval $[a, b]$ in the extended real line $[-\infty, \infty]$, i.e., the interval $[a, b]$ can contain not only real numbers but also symbols $-\infty$ and ∞. Pseudo-addition (non-Diophantine addition) $\oplus$ is commutative, non-decreasing, associative and the zero (neutral) element denoted by **0**. Pseudo-multiplication (non-Diophantine multiplication) $\otimes$ is commutative, positively non-decreasing, associative, distributive over $\oplus$, has the unit element **1** and the identity $\mathbf{0} \otimes x = \mathbf{0}$. Note that any semiring is a special case of prearithmetics (Burgin, 2019; Burgin and Czachor, 2020).

There are three canonical forms of such semirings (Pap, 2008).

Form 1. Operations $\otimes$ and $\oplus$ are defined in the following way:

$$a \oplus b = \sup(a, b),$$
$$a \otimes b = g^{-1}(g(a) \cdot g(b)),$$

where the generator $g : [a, b] \to [0, \infty]$ is a continuous and strictly increasing bijection.

Form 2. Operations $\otimes$ and $\oplus$ are defined in the following way:

$$a \oplus b = g^{-1}(g(a) + g(b)),$$
$$a \otimes b = g^{-1}(g(a) \cdot g(b)),$$

where the generator $g : [a, b] \to [0, \infty]$ is continuous and strictly monotone.

(a) If g is a strictly increasing generator, then $0 = a$.
(b) If g is a strictly decreasing generator, then $0 = b$.

Form 3. Operations $\otimes$ and $\oplus$ are defined in the following way:

$$a \oplus b = \sup(a, b),$$
$$a \otimes b = \inf(a, b).$$

These non-Diophantine operations provide means for defining not only generalizations of conventional Riemann integral but also of the fuzzy integral called Choquet integral.

It is possible to suggest that a better name for pseudo-analysis can be *extended analysis* with its two parts: *regular analysis* with Leibniz-Newton's calculus as its part and *irregular analysis*, which would include nonstandard analysis, non-Newtonian calculi, idempotent analysis (Maslov and Samborskij, 1992; Kolokoltsov and Maslov, 1997), tropical analysis (Litvinov, 2007; Speyer and Sturmfels, 2009), non-commutative and non-associative pseudo-analysis pseudo-analysis (Pap and Vivona, 2000; Pap and Štajner-Papuga, 2001), non-Newtonian functional analysis and idempotent functional analysis (Litvinov *et al.*, 2001). Idempotent analysis has various applications, which include optimization, optimal design of computer systems and media, optimal organization of data processing, dynamic programming, computer science, discrete mathematics, and mathematical logic. Pseudo-analysis has applications in utility theory, nonlinear equations, the arithmetic of fuzzy numbers, information theory, system theory, option pricing, large deviation principle, cumulative prospect theory, artificial intelligence, and physics. Thus, non-Diophantine arithmetics of real numbers have applications in all these areas.

Besides, some elements of non-Diophantine arithmetics also implicitly appeared in signal analysis (Oppenheim *et al.*, 1968; Oppenheim and Schafer, 1975; Childers *et al.*, 1977). Namely, in the area of nonlinear filtering and cepstral analysis, products are replaced by sums using logarithms for defining logarithmic integral. This approach is called homomorphic filtering (Oppenheim *et al.*, 1968).

Almost 20 years later after Pap, coming from problems in physics, the Polish physicist Marek Czachor independently built and studied new kinds of non-Newtonian calculi, which were also based on non-Diophantine arithmetics of real numbers, finding diverse applications in physics, psychology, and cryptography (Czachor, 2016, 2017, 2019, 2020b; Aerts *et al.*, 2018).

If we compare the approaches of Grossman and Katz, Pap, and Czachor, we see that the construction of g-calculus is more general than the construction of Grossman and Katz because g-calculus includes arithmetics and calculi on all real numbers as well as on finite intervals of real numbers, while the construction of Grossman and Katz includes only arithmetics and calculi on all real numbers. The construction of Czachor is even more general because it includes arithmetics and calculi on complex numbers as well as on arbitrary sets. The construction of the pseudo-analysis is more general then the construction of g-calculus because it includes not only arithmetics as the base for the calculi but also specific semirings, which are special cases of prearithmetics (Burgin, 2019).

Another difference between these approaches is the dissimilarity of their roots. The approach of Grossman and Katz, as we saw, is rooted in classical mathematics, namely, it stems from the product integral and logarithmic differentiation. In contrast to this, the approach of Pap originated from the theory of probabilistic metric spaces and fuzzy set theory being based at the beginning on t-norms. The distinction of the approach of Czachor is that it is explicitly based on non-Diophantine arithmetics. These differences show that these three approaches came forward independently.

As the classical calculus in mathematics uses real and complex numbers, the development of non-Newtonian calculus and idempotent calculus brought forth non-Diophantine arithmetics of real and complex numbers. Emergence of these arithmetics was caused by the needs to extend the calculus and to expand its applications.

In contrast to this, the discovery of non-Diophantine arithmetics of natural numbers stems from the goal to solve fundamental problems related to counting, which were first formulated in ancient Greece (e.g., the paradox of the heap or the Sorites paradox) and then put forward from time to time by the most insightful researchers, in particular, by Herman Helmholtz, Douglass Gasking, Rudolf Carnap, Pyotr Rashevsky or Morris Kline. Counting makes use of natural numbers, which are the most basic in mathematics according to the opinion of numerous mathematicians. As a result, non-Diophantine arithmetics of natural numbers were discovered (constructed) providing new ways of counting. The construction used for this purpose also allowed building non-Diophantine arithmetics of whole numbers, of integer numbers, of rational numbers, of real numbers and of complex numbers. Besides, the concept of prearithmetic was introduced forming the foundations for non-Diophantine arithmetics and comprising all known kinds of non-Diophantine arithmetics. In particular, specializing prearithmetics to real numbers, it is possible to obtain non-Diophantine arithmetics of real numbers such that go beyond those arithmetics used in non-Newtonian calculus. In turn, this would allow building and utilizing new kinds of non-Newtonian calculus.

Another big area in mathematics *idempotent analysis* based on different types of non-Diophantine arithmetics was originated by the prominent Russian mathematician Victor Pavlovich Maslov, who introduced several types of non-Diophatine additions $\oplus$ and multiplications $\odot$ in his studies of pseudodifferential equations (Maslov, 1987). Let us consider some of these systems.

Example 5.1. Let $\mathbf{R}_{\max}$ be the set A be the set of all real numbers $\mathbf{R}$ with the additional element $-\infty$ and the operations $\oplus = \max$ and $\odot = +$, which is the usual addition in $\mathbf{R}$ and define $\mathbf{0} = -\infty$ and $\mathbf{1} = 0$. This structure is very useful in idempotent analysis (Maslov, 1987; Maslov and Samborskii, 1992; Kolokoltsov and Maslov, 1997). By construction, $\mathbf{R}_{\max}$ is a commutative idempotent semi-ring and thus, a prearithmetic.

Example 5.2. Let $\mathbf{R}_{\min}$ be the set $A = \mathbf{R} \cup \{+\infty\}$ with the operations $\oplus = \min$ and $\odot = +$, which is the usual addition in $\mathbf{R}$ and defining $\mathbf{0} = +\infty$ and $\mathbf{1} = 0$. This structure is also very useful in idempotent analysis (Maslov, 1987; Maslov and Samborskii, 1992; Kolokoltsov and Maslov, 1997). By construction, $\mathbf{R}_{\min}$ is a commutative idempotent semi-ring and thus, a prearithmetic.

As a mathematical field, idempotent analysis includes idempotent integration theory, idempotent linear algebra, idempotent spectral theory, and idempotent functional analysis, is a branch of analysis based on replacing the conventional Diophantine arithmetic by a prearithmetic on an interval of real numbers or a non-Diophantine arithmetic on all real numbers with such operations as maximum or minimum instead of conventional addition (Maslov, 1987; Maslov and Samborskii, 1992; Maslov and Kolokoltsov, 1997). Idempotent analysis studies and applies functions taking values in idempotent semirings, which are special cases of prearithmetics (Burgin, 1977, 1997, 2019), while the corresponding function spaces are semimodules over semirings, which are non-Grassmannian linear spaces over prearithmetics (Burgin and Czachor, 2020). In turn, semimodules are special cases of quasilinear spaces, which are constructed based on non-Diophantine arithmetics.

Idempotent semirings, i.e., semirings, with idempotent addition when $a \oplus a = a$ for all a, were introduced and applied to problems in computer science, optimization, control, and discrete mathematics (cf., for example, Kleene, 1956; Pandit, 1961; Vorobjev, 1963; Zimmermann, 1981; Gondran and Minoux, 1979; Cuninghame-Green, 1995; Gaubert, 1997; Cohen *et al.*, 1999).

An impetus to the development of idempotent analysis was given by the observation that some problems that are nonlinear in the traditional sense, i.e., over the Diophantine arithmetic of real or complex numbers, turn out to be linear over a suitable prearithmetic or non-Diophantine arithmetic (Maslov, 1987; Maslov and Kolokoltsov, 1997). This helps solving many equations because this linearity considerably simplifies the explicit construction of solutions. As a result, idempotent analysis has various applications, which include optimization, optimal design of computer systems and media, optimal organization of data processing, dynamic programming, computer science, discrete mathematics, and mathematical logic. Consequently, non-Diophantine arithmetics of real numbers have applications in all these areas.

Thus, we can see how different kinds and types of non-Diophantine arithmetics and prearithmetics, at first, were used by mathematicians without understanding their essence while in the 20th century these arithmetics came to the forefront of mathematics being consciously discovered and utilized for the further development of mathematics as well as for solving a variety of problems in diverse areas of theory and practice.

Appendix

List of Arithmetics Used in Mathematics and Beyond

Affine arithmetic
Algebraic arithmetic
Algorithmic arithmetics
Applied (popular) arithmetic
Approximate arithmetic
Arithmetic of 1-motives
Arithmetics of algebras
Arithmetic of algebraic curves
Arithmetic of algebraic tori
Arithmetic of bicomplex numbers
Arithmetic of calculus
Arithmetic of cardinal multinumbers
Arithmetic of cardinal numbers
Arithmetic of chemistry
Arithmetic of complex hypernumbers
Arithmetic of computable numbers
Arithmetic of Coxeter groups
Arithmetic of del Pezzo surfaces
Arithmetic of differentiation
Arithmetic of dynamical systems
Arithmetic of elliptic curves
Arithmetic of feasible numbers
Arithmetic of functions
Arithmetic of functionals

History of Numbers and Arithmetic

Arithmetic of fuzzy numbers
Arithmetic of graphs
Arithmetic of higher-dimensional algebraic varieties
Arithmetic of hyperbolic 3-manifolds
Arithmetic of hyperreal numbers
Arithmetic of hypercomplex numbers
Arithmetic of intuitionistic fuzzy numbers
Arithmetic of infinitesimals
Arithmetic of infinities
Arithmetic of L-functions
Arithmetic of matrices
Arithmetic of multicardinal numbers
Arithmetic of multinumbers
Arithmetic of named numbers
Arithmetic of nonnegative real numbers
Arithmetic of number representations in computers
Arithmetic of number rings
Arithmetic of octonions
Arithmetic of operators
Arithmetic of ordinal numbers
Arithmetic of orthogonal groups
Arithmetic of pebbles
Arithmetic of pharmacy
Arithmetic of polynomials
Arithmetics of p–adic numbers
Arithmetic of quaternions
Arithmetic of real hypernumbers
Arithmetic of sedenions
Arithmetic of surfaces
Arithmetic of surreal numbers
Arithmetic of tessarines
Arithmetic of trees
Arithmetic of vectors
Arithmetic of Weil curves
Arithmetic on curves
Arithmetics on fractals (fractal arithmetic)
Arithmetic with big integers
Arithmetic with unsigned binary integers
Arithmetic of Z-numbers

Basic arithmetic
Biological arithmetic
Binary arithmetic
Binary fixed-point arithmetic
Bounded arithmetic
Business arithmetic
Calpanic arithmetic
Character arithmetic
Chemical arithmetic
Cognitive arithmetic
Color imaging arithmetic
Commercial arithmetic
Compound unit arithmetic
Computer arithmetics
Constructive arithmetic
Decimal floating-point arithmetic
Diophantine arithmetic of complex numbers
Diophantine arithmetic of natural numbers
Diophantine arithmetic of rational numbers
Diophantine arithmetic of real numbers
Diophantine arithmetic of whole numbers
Economic arithmetic
Educational arithmetic
Elementary arithmetic
Engineers arithmetic
Exact arithmetic
Experimental arithmetic
Exponential arithmetic
Exact real arithmetic
Financial arithmetic
First-order arithmetic
Floating-point arithmetic
Fuzzy arithmetic
Geometric arithmetic
Generic arithmetics
Generated arithmetics
Generalized arithmetics
Genetic arithmetic
Harmonic arithmetic

Higher arithmetic (Number theory)
Higher-order arithmetic
High-speed arithmetic
Indian arithmetic
Interval arithmetic (also range arithmetic or digital range arithmetic)
Large number arithmetic
Location arithmetic
Mechanics' arithmetic
Medical arithmetic
Mental arithmetic
Merchant arithmetic
Mercantile arithmetic
Modular (or residue or clock) arithmetics
Nonclassical arithmetic
Nonstandard arithmetics
Non-Diophantine arithmetics of complex numbers
Non-Diophantine arithmetics of natural numbers
Non-Diophantine arithmetics of rational numbers
Non-Diophantine arithmetics of real numbers
Non-Diophantine arithmetics of whole numbers
Non-symbolic arithmetic
Numeric arithmetics
Number arithmetics
Peano (or Peano-Dedekind) arithmetic
Phonon arithmetic
Physical arithmetic
Political arithmetic
Pointer arithmetic
Practical arithmetic
Predicative arithmetic
Presburger arithmetic
Primary arithmetic
Primitive recursive arithmetic
Quadratic arithmetic
Rational arithmetic
Rational arithmetic with big integers
Recursive arithmetic
Restricted arithmetic
Robinson arithmetic

Scaled arithmetic
Second-order arithmetic
Skolem arithmetic
Signed arithmetic on integers
Signed arithmetic with fixed-point numbers
Symbolic arithmetic
Synergy arithmetic
Theoretical arithmetic
Third-order arithmetic
Transfinite arithmetics
Transreal arithmetic
Tropical arithmetic
Universal arithmetic
Unrestricted arithmetic

Bibliography

$1 + 1 = 3$, Urban Dictionary (https://www.urbandictionary.com/define.php? term=1%2B1%3D3).

$2 + 2 = 5$, Cambridge Business Dictionary (http://dictionary.cambridge.org/us/ dictionary/english/2-2-5).

$2 + 2 = 5$, Wikipedia (https://en.wikipedia.org/wiki/2_%2B_2_%3D_5).

Abbasbandy, S. and Hajjari, T. (2009) A new approach for ranking of trapezoidal fuzzy numbers, *Comput. Math. Appl.*, V. 57, pp. 413–419.

Aberth, O. (1968) Analysis in the computable number field, *J. Assoc. Comput. Machinery* (JACM), V. 15, No. 2, pp. 276–299.

Abian, A. (1965) *The Theory of Sets and Transfinite Arithmetic*, W. B. Saunders Company, Philadelphia/London.

Abiteboul, S. and Dowek, G. (2020) *The Age of Algorithms*, Cambridge University Press, Cambridge.

Abiyev, R. H., Uyar, K., Ilhan, U., Imanov, E. and Abiyeva, E. (2018) Estimation of food security risk level using Z-number-based fuzzy system, *J. Food Quality, Special Issue Natural Strategies to Improve Quality in Food Protection*, V. 2018, Article ID 7480910.

Abraham, U. and Bonnet, R. (1999) Hausdorff's theorem for posets that satisfy the finite antichain property, *Fund. Math.*, V. 159, pp. 51–69.

Abubakr, M. (2011) *On Logical Extension of Algebraic Division*, preprint in Logic in Computer Science, cs.LO (arXiv:1101.2798).

Ackermann, W. (1924) Begréndung des "tertium non datur" mittels der Hilbertschen Theorie der Widerspruchsfreiheit, *Math. Ann.*, V. 93, pp. 1–36.

Ackermann, W. (1940) Zur Widerspruchsfreiheit der Zahlentheorie, *Mathematische Annalen*, V. 117, 162–194.

Aczel, A. D. (2011) *A Strange Wilderness: The Lives of the Great Mathematicians*, Sterling Publishing Co., New York.

Aczel, A. D. (2014) Uncover the Origins of Numbers, *Smithsonian Magazine*, December 2014.

Aczel, A. D. (2015) *Finding Zero: A Mathematician's Odyssey to Uncover the Origins of Numbers*, Palgrave Macmillan, London/New York/Shanghai.

Aczél, J. (1955) A solution of some problems of K. Borsuk and L. Jánossy, *Acta Physica*, V. 4, 351.

Aczél, J. (1966) *Lectures on Functional Equations and Their Applications*, Mathematics in Science and Engineering, V. 19, Academic Press, New York.

Ada, 95 Reference Manual (electronic edition: http://www.grammatech.com/rm95html-1.0/rm9x-toc.html).

Adams, J. F. (1960) On the non-existence of elements of Hopf invariant one, *Annals of Mathematics*, V. 72, No. 1, pp. 20–104.

Adaval, R. (2012) Numerosity and consumer behavior, *J. Consumer Res.*, V. 39, No. 5, pp. xi–xiv.

Aerts, D., Czachor, M. and Kuna, M. (2016) Crystallization of space: Space-time fractals from fractal arithmetic, *Chaos, Solitons Fractals*, V. 83, pp. 201–211.

Aerts, D., Czachor, M. and Kuna, M. (2016) Fourier transforms on Cantor sets: A study in non-Diophantine arithmetic and calculus, *Chaos, Solitons Fractals*, V. 91, pp. 461–468.

Aerts, D., Czachor, M. and Kuna, M. (2018) Simple fractal calculus from fractal arithmetic, *Rep. Math. Phys.*, V. 81, pp. 357–370.

Agahi, H., Ouyang, Y., Mesiar, R., Pap, E. and Štrboja, M. (2011) Hölder and Minkowski type inequalities for pseudo-integral, *Appl. Math. Comput.*, V. 217, pp. 8630–8639.

Agrawal, A. and Jaffe, J. (2000) The post merger performance puzzle, *Adv. Mergers Acquisitions*, V. 1, pp. 119–156.

Agrillo, C. (2014) *Numerical and Arithmetic Abilities in Non-primate Species*, Oxford University Press, Oxford, UK.

Agrillo, C. and Bisazza, A. (2018) Understanding the origin of number sense: a review of fish studies, *Philos. Trans. R. Soc. Lond.*, B Biol. Sci., V. 373, 20160511.

Agrillo, C. and Bisazza, A. (2014) Spontaneous versus trained numerical abilities. A comparison between the two main tools to study numerical competence in non-human animals, *J. Neurosci. Methods*, V. 234, pp. 82–91.

Agrillo, C., Dadda, M. and Bisazza, A. (2006) Quantity discrimination in female mosquitofish, *Animal Cognition*, V. 10, No. 1, pp. 63–70.

Agrillo, C., Petrazzini, M. E. M. and Bisazza, A. (2015) At the root of math: Numerical abilities in fish, in *Evolutionary Origins and Early Development 37 of Number Processing*, Academic Press, London/SanDiego/Oxford, pp. 3–33.

A'Hearn, B., Baten, J. and Crayen, D. (2009) Quantifying quantitative literacy: Age heaping and the history of human capital, *J. Economic History*, V. 69, No. 3, pp. 783–808.

Ahmavaara, Y. (1965) The structure of space and the formalism of relativistic quantum theory, *J. Math. Phys.*, V. 6, pp. 87–93.

Aidagulov, R. R. and Shamolin, M. V. (2015) Polynumbers, norms, metrics, and polyingles, *J. Math. Sci.*, V. 204, No. 6.

Aigner, M. (1979) *Combinatorial Theory*, Springer-Verlag, New York.

Alberti, L. B. (1485) *De re ædificatoria...*, Florence, Niccolò di Lorenzo.

d'Alembert, J. Le Rond (1784) Diophante, in *Encyclopédie méthodique. Mathématiques*, tome premier, Paris (Planckoucke)/Li'ege (Plomteux), pp. 533–534.

Alexandroff, P. S. (1977) *Introduction to Set Theory and General Topology*, Nauka, Moscow (in Russian).

Aliev R. A., Alizadeh, A., Aliyev, R. R. and Huseynov, O. H. (2015) *The Arithmetic of Z-numbers, Theory and Applications*, World Scientific, Singapore.

Aliev, R. A., Huseynov, O. H. and Zulfugarova, R. X. (2016) Z-distance based IF-THEN rules, *Scientific World J.*, 1673537.

Alkhazaleh, S. and Salleh, A. R. (2012) Fuzzy soft multiset theory, *Abstr. Appl. Anal.*, Article ID 350600, 20 pp.

Alkhazaleh, S., Salleh, A. R. and Hassan, N. (2011) Soft multisets theory, *Appl. Math. Sci.*, V. 5, No. 72, pp. 3561–3573.

Allais, A. (1895) *Deux et deux font cinq* ($2 + 2 = 5$), Ollendorff, Paris.

Allenby, R. B. J. T. (1991) *Rings, Fields and Groups*, Edward Arnold, London.

Alling, N. L. (1962) On the existence of real-closed fields that are η_α-sets of power $\aleph_\alpha$, *Trans. Amer. Math. Soc.*, V. 103, pp. 341–352.

Alling, N. L. (1985) Conway's field of surreal numbers, *Trans. Amer. Math. Soc.*, V. 287, No. 1, pp. 365–386.

Alling, N. L. (1987) *Foundations of Analysis Over Surreal Number Fields*, North-Holland.

Alp, K. O. (2010) A comparison of sign and symbol (their contents and boundaries), *Semiotica*, V. 182, No.1/4, pp. 1–13.

Alsina, C., Trillas, E. and Valverde, L. (1983) On some logical connectives for fuzzy set theory, *J. Math. Anal. Appl.*, V. 93, pp. 15–26.

Alt, R. (2008) Stochastic arithmetic as a model of granular computing, in *Handbook of Granular Computing*, pp. 33–54.

Altaisky, M. V. and Sidharth, B. G. (1998) *p-Adic physics below and above Planck scales*, preprint in General Relativity and Quantum Cosmology (gr-qc) arXiv:gr-qc/9802034.

Amalric, M. and Dehaene, S. (2016) Origins of the brain networks for advanced mathematics in expert mathematicians, *Proc. Natl. Acad. Sci. USA*, V. 113, No. 18, pp. 4909–4917.

Amat Plata, S. (2005) ¿Mentimos a nuuestros hijos cuuando les decimos quue $1 + 1$ son 2? *Rev. Eureka Sobre Enseñanza Y Divulgación De Las Ciencia*, V. 2, No. 1, pp. 33–37.

Amblard, P.-O. and Vignat, C. (2006) A note on bounded entropies, *Physica A*: V. 365, No. 1, pp. 50–56.

American Chemical Society, (2008) *CAS Registry and CASRNs*. http://www.cas.org/expertise/cascontent/registry/regsys.html.

Anderson, N. H. (1964) Test of a model for number-averaging behavior. *Psychonomic Sci.*, V. 1, 191–192.

Anderson, R. L. (2004) It adds up after all: Kant's philosophy of arithmetic in light of the traditional logic, *Philos. Phenomenological Res.*, V. 69, No. 3, pp. 501–540.

Anderson, P., Black, A., Machin, D. and Watson, N. (2014) *The Business Book*: *Big Ideas Simply Explained*, Books, London, UK.

Andres, M., Michaux, N. and Pesenti, M. (2012) Common substrate for mental arithmetic and finger representation in the parietal cortex, *Neuroimage*, V. 62, pp. 1520–1528.

Andriambololona, R., Ranaivoson, R. T. and Rakotoson, H. (2016) Numeral System Change in Arithmetic and Matricial Formalism, *Pure Appl. Math. J.*, V. 5, No. 3, pp. 87–92.

Anghileri, J. (1995) Language, arithmetic and the negotiation of meaning, *Learning Math.*, V. 21, No. 3, pp. 10–14.

Anglin W. S. and Lambek J. (1995) Plato and Aristotle on Mathematics, in *The Heritage of Thales*, Undergraduate Texts in Mathematics (Readings in Mathematics), Springer, New York, NY, pp. 67–69.

Aniszewska, D. (2007) Multiplicative Runge–Kutta methods, *Nonlinear Dyn.*, V. 50, No. 1–2, pp. 265–272.

Annas, J. (1975) Aristotle, number and time, *Philos. Quart.*, V. 25, No. 99, pp. 97–113.

Ansari, D. (2016) The neural roots of mathematical expertise, *Proc. Natl. Acad. Sci. USA*, V. 113, No. 18, pp. 4887–4889.

Antell, S. and Keating, D. P. (1983) Perception of numerical invariance in neonates, *Child Development*, V. 54, pp. 695–701.

Anobile, G., Cicchini, G. M., and Burr, D. C. (2016). Number as a primary perceptual attribute: a review. *Perception*, V. 45, pp. 5–31.

Apostol, T. M. (1967) *Calculus*, V. 1: *One-Variable Calculus with an Introduction to Linear Algebra*, John Wiley & Sons, New York.

Aragona, J. and Juriaans, S. O. (2001) Some structural properties of the topological ring of Colombeau's generalized numbers, *Comm. Alg.*, V. 29, No. 5, pp. 2201–2230.

Archibald, J. (2014) *One Plus One Equals One: Symbiosis and the Evolution of Complex Life*, Oxford University Press, New York.

Argand, J.-R. (1806) *Essai sur une manière de représenter des quantités imaginaires dans les constructions géométriques*, chez Mme Vve Blanc, Paris.

Archimedes, (1880) *Archimedis Opera Omnia cum Commentariis Eutocii*, J. L. Heiberg, Ed., B. G. Teubner, Leipzig.

Archimedes, (2002) *The Works of Archimedes*, T. L. Heath, Ed., Dover Publications Inc, New York.

Aristotle, (1984) *The Complete Works of Aristotle*, Princeton University Press, Princeton.

Arithmetic, New World Encyclopedia, 2016.

Arsalidou, M. and Taylor, M. J. (2011) Is $2+2=4$? Meta-analyses of brain areas needed for numbers and calculations, *Neuroimage*, V. 54, pp. 2382–2393.

Arthur, R. (1999) Infinite number and the world soul; in defense of Carlin and Leibniz, *Leibniz Rev.*, V. 9, 105–116.

Arthur, R. (2001) Leibniz on infinite number, infinite wholes, and the whole world: A reply to Gregory Brown, *Leibniz Rev.*, V. 11, pp. 103–116.

Artin, M. (1991) *Algebra*, Prentice-Hall, Englewood Cliffs, NJ.

Arzt, P. (2014) *Measure theoretic trigonometric functions*, preprint in Mathematics, math.SP (arXiv:1405.4693).

Asady, B., Akbari, M. and Keramati, M. A. (2011) Ranking of fuzzy numbers by fuzzy mapping, *International Journal of Computer Mathematics*, V. 88, No. 8, pp. 1603-1618.

Ash, A. and Gross, R. (2008) *Fearless Symmetry: Exposing the Hidden Pattern of Numbers*, Princeton University Press, Princeton.

Ash, A. and Gross, R. (2012) *Elliptic Tales: Curves, Counting, and Number Theory*, Princeton University Press, Princeton.

Ash, A. and Gross, R. (2018) *Summing It Up: From One Plus One to Modern Number Theory*, Princeton University Press, Princeton.

Ashcraft, M. H. (1992) Cognitive arithmetic: A review of data and theory, *Cognition*, V. 44, pp. 75–106.

Ashcraft, M. H. (1995) Cognitive psychology and simple arithmetic: A review and summary of new directions, *Math. Cogn.*, V. 1, pp. 3–34.

Asher, M. (2002) *Mathematics Elsewhere: An Exploration of Ideas Across Cultures*, Princeton University Press, Princeton/Oxfordshire.

Ashkenazi, S. (2008) Basic numerical processing in left Intraparietal Sulcus (IPS) Acalculia, *Cortex*, V. 44, No. 4, pp. 439–448.

Asper, O. D. and Tsaprounis, K. (2018) Long reals, *J. Logic Anal.*, V. 10, No. 1, pp. 1–36.

Assad, J. A. and Maunsell, J. H. R. (1995) Neuronal correlates of inferred motion in primate posterior parietal cortex, *Nature*, v373, pp. 518–521.

Atanassov, K. T. (1983) Intuitionistic fuzzy sets, *VII ITKR's Session*, Sofia (deposited in Central Sci.-Technical Library of Bulg. Acad. of Sci., 1697/84) (in Bulgarian).

Atanasov, K. (1986) Intuitionistic fuzzy sets, *Fuzzy Sets Systems*, V. 20, No. 1, pp. 87–96.

Atkinson, K. (1989) *An Introduction to Numerical Analysis*, Wiley, New York.

Atkinson, R. L., Atkinson, R. C., Smith E. E. and Bem, D. J. (1990) *Introduction to Psychology*, Harcourt Brace Jovanovich, Inc., San Diego/New York/Chicago.

Augustine, (2018) *The Works of St. Augustine: A Translation for the 21st Century*, New City Press.

Ausejo, E. (2015) New Perspectives on Commercial Arithmetic in Renaissance Spain, in *A Delicate Balance: Global Perspectives on Innovation and Tradition in the History of Mathematics*, Trends in the History of Science, Birkhäuser, Cham.

Avigad, J. (2015) *Mathematics and language*, preprint in Mathematics History and Overview (math.HO), (arXiv:1505.07238).

Aydin, K., Ucar, A., Oguz, K. K., Okur, O. O., Agayev, A., Unal, Z., Yilmaz, S. and Ozturk, C. (2007) Increased gray matter density in the parietal cortex of mathematicians: A voxel-based morphometry study, *Am. J. Neuroradiol.* (*AJNR*), V. 28, No. 10, pp. 1859–1864.

Babanli, M. and Huseynov, V. (2016) Z-number-based alloy selection problem, *Procedia Comput. Sci.*, V. 102:C, pp. 183–189.

Baccheli, B. (1986) Representation of continuous associative functions, *Stochastica: Rev. Mat Aplicada*, V. 10, No. 1, pp. 13–28.

Bachem, A. (1952) Weber's law in physics and arithmetic, *Am. J. Psychol.*, V. 65, No. 1, pp. 106–107.

Bascelli, T., Błaszczyk, P., Kanovei, V., Katz, K. U., Katz, M. G., Kutateladze, S. S., Nowik, T., Schaps, D. M. and Sherry, D. (2016) Gregory's sixth operation, *Found. Sci.*, DOI 10.1007/s10699-016-9512-9.

Badaev S. and Goncharov S. (2008) Computability and numberings, in *New Computational Paradigms: Changing Conceptions of What is Computable*, Elsevier, Amsterdam, pp. 19–34.

Badets, A. and Pesenti, M. (2010) Creating number semantics through finger movement perception, *Cognition*, V. 115, pp. 46–53.

Badiou, A. (2008) *Number and Numbers*, (R. Mackay, Trans.), Polity, Cambridge, UK/Oxford, UK/Boston, USA.

Badiou, A. (2014) *Mathematics of the Transcendental*, Bloomsbury Academic, London/New York.

Bailey, D. H., Borwein, J. M., Calkin, N. J., Girgensohn, R., Luke, D. R. and Moll, V. H. (2007) *Experimental Mathematics in Action*, A K Peters, Wellesley, MA.

Bair, J., Błaszczyk, P., Ely, R., Henry, V., Kanovei, V., Katz, K. U., Katz, M. G., Kutateladze, S. S., McGaffey, T., Reeder, P., Schaps, D. M., Sherry, D. and Shnider, S. (2017) Interpreting the Infinitesimal Mathematics of Leibniz and Euler, *J. Gen. Philos. Sci.*, V. 48, No. 2, pp. 195–238.

Baird, J. C. (1975) Psychophysical study of numbers (IV): Generalized preferred state theory, *Psychol Res.*, V. 38, pp. 175–187.

Baird, J. C. (1975a) Psychophysical study of numbers (V): Preferred state theory of matching functions, *Psycho. Res.*, V. 38, pp. 188–207.

Baird, J. C. (1997) *Sensation and Judgment: Complementarity Theory of Psychophysics*, Lawrence Erlbaum Associates, Mahwah.

Baird, J. C. and Noma, E. (1975) Psychophysical study of numbers (I): Generation of numerical responses, *Psychol. Res.*, V. 37, pp. 281–297.

Baird, J. C. and Noma, E. (1978) *Fundamentals of Scaling and Psychophysics*, Wiley, New York.

Baker, A. (1990) *Transcendental Number Theory*, Cambridge University Press, Cambridge.

Balaguer, M. (2011) Fictionalism in the philosophy of mathematics, in *The Stanford Encyclopedia of Philosophy* (http://plato.stanford.edu/archives/fall2011/entries/fictionalism-mathematics/).

Balaguer, M. (2017) Philosophy of mathematics, *Encyclopædia Britannica*.

Balcazar, J. L., Diaz, J. and Gabarro, J. (1998) *Structural Complexity*, Springer-Verlag.

Bali, N. P. (2005) *Golden Differential Calculus*, Laxmi Publication, New Delhi, India.

Ballet, S., Perret, M. and Zaytsev, A. (Eds.) Algorithmic arithmetic, geometry, and coding theory, 14th *International Conference Arithmetic, Geometry, Cryptography and Coding Theory*, Contemporary Mathematics, V. 637, AMS, Providence, Rhode Island, 2013.

Balliett, L. D. (2018) *The Philosophy of Numbers: Their Tone and Colors* (Classic Reprint) Forgotten Books.

Balog, A. and Szemerédi, E. (1994) A statistical theorem of set addition, *Combinatorica*, V. 14, pp. 263–268.

Banerjee, R. and Pal, S. (2017) A computational model for the endogenous arousal of thoughts through Z*-numbers, *Inform. Sci.*, V. 405:C, pp. 227–258.

Banks, V. P. and Coleman, M. J. (1981) Two subjective scales of number, *Perception & Psychophysics*, V. 29, No. 2, pp. 95–105.

Banks, W. P. and Hill, D. K. (1974) The apparent magnitude of number scaled by random production, *J. Exp. Psychol. (Monograph Series)*, V. 102, pp. 353–376.

Bardeys, E. (1907) *Aufgabensammlung für Arithmetik, Algebra und Analysis*: Für Gymnasien. Reformausgabe, Ernst Bardey. B. G. Teubner, Leipzig.

Barker, S. (1967) Number, in *Encyclopedia of Philosophy* (P. Edwards, ed.), Macmillan Publishing Company, New York, V. 5, pp. 526–530.

Barner, D. and Bachrach, A. (2010) Inference and exact numerical representation in early language development, *Cogn. Psychol.*, V. 60, pp. 40–62.

Barner, D., Thalwitz, D., Wood, J. and Carey S. (2007) On the relation between the acquisition of singular-plural morpho-syntax and the conceptual distinction between one and more than one, *Dev. Sci.*, V. 10, pp. 365–373.

Barner, D., Wood, J., Hauser, M. and Carey, S. (2008) Evidence for a non-linguistic distinction between singular and plural sets in rhesus monkeys. *Cognition*, V. 107, No. 2, pp. 603–622.

Barras, C. (2001) How did Neanderthals and other ancient humans learn to count? *Nature*, V. 594, pp. 22–25.

Barrouillet, P. and Fayol, M. (1998) From algorithmic computing to direct retrieval: evidence from number and alphabetic arithmetic in children and adults, *Mem Cognit.*, V. 26, No. 2, pp. 355–368.

Barrouillet, P. and Lépine, R. (2005) Working memory and children's use of retrieval to solve addition problems, *J. Exp. Child. Psychol.*, V. 91, No. 3, pp. 183–204.

Bar-Shai, N., Keasar, T. and Shmida, A. (2011) The use of numerical information by bees in foraging tasks, *Behav. Ecol.*, V. 22, pp. 317–325.

Barth, H., Kanwisher, N. and Spelke, E. (2003) The construction of large number representations in adults, *Cognition*, V. 86, No. 3, pp. 201–221.

Barth, H., La Mont, K., Lipton, J., Dehaene, S., Kanwisher, N. and Spelke, E. (2006) Non-symbolic arithmetic in adults and young children, *Cognition*, V. 98, pp. 199–222.

Barton, B. (2008) *The Language of Mathematics: Telling Mathematical Tales*, Springer, New York.

Barton, N. (2020) Absence perception and the philosophy of zero, *Synthese*, V. 197, No. 9, pp. 3823–3850 (https://doi.org/10.1007/s11229-019-02220-x).

Bashirov, A. and Riza, M. (2011) On complex multiplicative differentiation, *TWMS J. Appl. Engi. Math.*, V. 1, No. 1, pp. 75–85.

Bashirov, A. E., Kurpınar, E. M. and Özyapıcı, A. (2008) Multiplicative calculus and its applications, *J. Math. Anal. Appl.*, V. 337, No. 1, pp. 36–48.

Bashirov, A. E., Mısırlı, E., Tandoğdu, Y. and Özyapıcı, A. (2011) On modeling with multiplicative differential equations, *Appl. Math.*, V. 26, No. 4, pp. 425–438.

Bashmakova, I. G. (1974) *Diophant und Diophantische Gleichungen*, VEB Deutscher Verlag der Wissenschaften, Berlin.

Bashmakova, I. G. (1981) Arithmetic of algebraic curves from diophantus to Poincaré, *Historia Mathematica*, V. 8, pp. 393–416.

Bateman, P. T., Erdős, P., Pomerance, C. and Straus, E. G. (1981) The arithmetic mean of the divisors of an integer, in *Proc. Conf. Analytic Number Theory*, Lecture Notes in Mathematics, V. 899, Springer-Verlag, pp. 197–220.

Bawden-Davis, J. (2014) *The 7 Financial Numbers Every Business Owner Should Know*, (https://www.americanexpress.com/en-us/business/trends-and-insights/articles/the-7-financial-).

Beck, C. and Schlögl, F. (1993) *Thermodynamics of Chaotic Systems*: *An Introduction*, Cambridge University Press, Cambridge.

Beck, A., Bleicher, M. N. and Crowe, D. W. (2000) *Excursions into Mathematics*, A K Peters, Natick, MA.

Beckenbach, E. F. (1956) *Modern Mathematics for the Engineer*, MacGraw-Hill, New York/Toronto/London.

Bector, C. R. and Chandra, S. (2005) *Fuzzy Mathematical Programming and Fuzzy Matrix Games*, Part of the Studies in Fuzziness and Soft Computing book series (STUDFUZZ, V. 169) Springer, Berlin/Heidelberg.

Beechler, D. (2013) *How to Create "1 + 1 = 3" Marketing Campaigns*, http://www.marketingcloud.com/blog/how-to-create-1-1-3-marketing-campaigns.

Bell, E. T. (1927) Arithmetic of logic, *Transactions of the American Mathematical Society*, V. 29, No. 3, pp. 597–611.

Bell, E. T. (1937) *Men of Mathematics*, Simon & Schuster, New York.

Bell, E. T. (2011) *The Magic of Numbers*, Dover Books on Mathematics, Dover Publications, New York.

Bell, J. L. (1981) Category theory and the foundations of mathematics, *British J. Philos. Sci.*, V. 32, pp. 349–358.

Bell, J. L. (2006) Abstract and variable sets in category theory, in *What is Category Theory?* Polimetrica International Scientific Publisher, Monza, Italy, pp. 9–16.

Bell, J. L. (2008) *A Primer of Infinitesimal Analysis*, Cambridge University Press, Cambridge.

Bell, J. L. (2019) *The Continuous, the Discrete and the Infinitesimal in Philosophy and Mathematics*, The Western Ontario Series in Philosophy of Science, Springer Nature, Switzerland.

Bellavitis, G. (1835) Saggio di applicazioni di un nuovo metodo di geometria analitica (Calcolo delle equipollenze), *Ann. Ddelle Sci. Regno Lombaro-Veneto. Padova*, V. 5, pp. 244–259.

Bellissima, F. (2014) Arithmetic, geometric and harmonic means in music theory, *Boll. Storia Delle Sci. Mat.*, V. 34, pp. 201–244.

Bellman, R. E., Kalaba, R. E. and Zadeh, L. A. (1964) *Abstraction and Pattern Classification*, Memorandum RM-4307-PR, The RAND Corporation, Santa Monica, California.

Belna, J.-P. (1996) *La Notion de nombre chez Dedekind, Cantor, Frege: Théories, conceptions, et philosophie*, Vrin, Paris.

Benacerraf, P. (1965) What numbers could not be? *Philos. Rev.*, V. 74, pp. 47–73.

Benacerraf, P. (1973) Mathematical truth, *J. Philos.*, V. 70, No. 19, pp. 661–679.

Benacerraf, P. (1996) What mathematical truth could not be — I, in *Benacerraf and His Critics*, Basil Blackwell & Bloomfield, Oxford, pp. 9–59.

Benacerraf, P. and Putnam, H. (Eds.) (1983) *Philosophy of Mathematics: Selected Readings*, Cambridge University Press, Cambridge.

Benci, V. (1995) I numeri e gli insiemi etichettati, *Conferenze del seminario di matematica dell'Universita' di Bari*, V. 261. Bari, Italy: Laterza, pp. 29.

Benci, V. and Di Nasso, M. (2003) Numerosities of labelled sets: a new way of counting, *Adv. Math.*, V. 173, pp. 50–67.

Benci, V. and Di Nasso, M. (2005) A purely algebraic characterization of the hyperreal numbers, *Proc. Amer. Math. Soc.*, V. 133, pp. 2501–2505.

Benci, V., Di Nasso, M. and Forti, M. (2006) An Aristotelean notion of size, *Annals of Pure and Applied Logic*, V. 143, pp. 43–53.

Benci, M., Di Nasso, M. and Forti, M. (2007) A Euclidean measure of size for mathematical universes, *Logique et Anal.*, V. 197, pp. 43–52.

Benedetto, R., DeMarco, L., Ingram, Jones, R., Manes, M., Silverman, J. H. and Tucker, T. J. (2019) Current trends and open problems in arithmetic dynamics, *Bull. Amer. Math. Soc.*, V. 56, No. 4, pp. 611–685.

Benioff, P. (2001) The Representation of natural numbers in quantum mechanics, *Phys. Rev.*, V. A63, Article No. 032305.

Benioff, P. (2001b) The Representation of numbers in quantum mechanics, *Algorithmica*, V. 34, pp. 529–559.

Benioff, P. (2002) Towards a coherent theory of physics and mathematics, *Found. Phys.*, V. 32, pp. 989–1029.

Benioff, P. (2005) Towards a coherent theory of physics and mathematics: The theory-experiment connection, *Found. Phys.*, V. 35, pp. 1825–1856.

Benioff, P. (2006) Complex rational numbers in quantum mechanics, *Int. J. Mod. Phys. B*, V. 20, pp. 1730–1741.

Benioff, P. (2007) A representation of real and complex numbers in quantum theory, *Int. J. Pure Appl. Math.*, V. 39, pp. 297–339.

Benioff, P. (2009) A possible approach to inclusion of space and time in frame fields of quantum representations of real and complex numbers, *Adv. Math. Phy.*, V. 1, ID 452738.

Benioff, P. (2011) New gauge field from extension of spacetime parallel transport of vector spaces to the underlying number systems, *Int. J. Theor. Phys.*, V. 50, 1887.

Benioff, P. (2012) Local availability of mathematics and number scaling: effects on quantum physics, in *Quantum Information and Computation* X (Donkor, E.; Pirich, A.; Brandt, H., Eds.) Proceedings of SPIE, V. 8400, SPIE, Bellingham, WA, 84000T (also arXiv:1205.0200).

Benioff, P. (2012a) Effects on quantum physics of the local availability of mathematics and space time dependent scaling factors for number systems, in *Advances in Quantum Theory* (I. I. Cotaescu, Ed.), InTech (also arXiv:1110.1388).

Benioff, P. (2013) Representations of each number Type that differ by scale factors, *Adv. Pure Math.*, V. 3, pp. 394–404.

Benioff, P. (2015) Fiber bundle description of number scaling in gauge theory and geometry, *Quan. Stud. Math. Found.*, V. 2, pp. 289–313.

Benioff, P. (2016) Space and time dependent scaling of numbers in mathematical structures: Effects on physical and geometric quantities, *Quantum Inform. Process.*, V. 15, No. 3, pp. 1081–1102.

Benioff, P. (2016a) Effects of a scalar scaling field on quantum mechanics, *Quantum Inform. Process.*, V. 15, No. 7, pp. 3005–3034.

Benioff, P. (2020) The "No information at a distance" prtinciple and local mathematics: some effects on physics and geometry, in *Theoretical Information Studies: Information in the World*, pp. 115–137.

Benioff, P. (2020a) Space Local mathematics and no information at a distance; some effects on physics and geometry, Proceedings, V. 47, No. 1, 3; doi:10.3390/proceedings47010003.

Bennett, A. A. (1921) Some arithmetic operations with transfinite ordinals, *The Amer. Math. Monthly*, V. 28, No. 11–12, pp. 427–430.

Benoit, P. (1988). The commercial arithmetic of Nicolas Chuquet, in C. Hay, ed., *Mathematics from Manuscript to Print*, 1300-1600, Clarendon, Oxford, pp. 96–116.

Bentley, P. (2008) *The Book of Numbers: The Secret of Numbers and How They Changed the World*, Firefly Books.

Benvenuti, P. and Mesiar, R. (2004) Pseudo arithmetical operations as a basis for the general measure and integration theory, *Inform. Sci.*, V. 160, No. 1–4, pp. 1–11.

Benvenuti, P., Mesiar, R. and Vivona, D. (2002) Monotone set functions-based integrals, in *Handbook of Measure Theory*, vol. II, Elsevier, North-Holland, pp. 1329–1379.

Berch, D. B. (2005) Making sense of number sense: implications for children with mathematical disabilities, *J. Learning Disabil.*, V. 38, No. 4, pp. 333–339.

Berčić, B. (2005) Zašto 2 + 2 = 4? *Filozofska Istraživanja*, V. 25, No. 4, pp. 945–961.

Berkeley, G. (1707) *Arithmetica absque algebra aut Euclide demonstrata and Miscellanea Mathematica*, A. & J. Churchill, London; J. Pepyat, Dublin.

Berkeley, G. (1734) *The Analyst, A Discourse Addressed to An Infidel Mathematician*, S Fuller, Dublin.

Berkeley, G. (1956) Letter to Molyneux [1709] in *The Works of George Berkeley, Bishop of Cloyne*, V. 8, Nelson, London.

Berkeley, G. (1989) *Philosophical Commentaries*, Garland, London.

Berlinghoff, W. P. and Gouvêa, F. Q. (2004) *Math Through the Ages: A Gentle History for Teachers and Others*, MAA, Washington, DC.

Berman, K. and Knight, J. (2008) *Financial Intelligence for Entrepreneurs: What You Really Need to Know About the Numbers*, Harvard Business Press, Brighton, MA.

Bernays, P. (1935) Sur le platonisme dans les mathematiques, *L'Enseignement math.*, V. 34, pp. 52–69.

Bernays, P. (1951) Über das Inductionsschema in der recursiven Zahlentheorie, in *Kontrolliertes Denken, Festschrift für Wilhelm Britzelmayr*, Freiburg/München, pp. 10–17.

Bernays, P. (1967) David Hilbert, in *The Encyclopedia of Philosophy*, V. 3, Macmillan publishing company and The Free Press, New York, pp. 496–504.

Berrone, L. (2010) The associativity of the Pythagorean law, *Amer. Math. Monthly*, V. 116, pp. 936–939.

Berteletti, I. and Booth, J. R. (2015) Perceiving fingers in single-digit arithmetic problems, *Front. Psychol.*, V. 6, pp. 226–241.

Bertrand, J. (1849) *Traité d'Arithmétique*, Hachette, Paris.

Bettazzi, R. (1887) Sul concetto di numero, *Periodico Mat. L'insegnamento Secondario*, V. 2, pp. 97–113, 129–143.

Bettazzi, R. (1890) *Teoria Delle Grandezze*, Spoerri, Pisa.

Bettazzi, R. (1891) Osservazioni sopra l'articolo del Dr. G. Vivanti Sull'infinitesimo attuale, *Rivista di Matematica*, V. 1, pp. 174–182.

Bettazzi, R. (1892) Sull'infinitesimo attuale, *Rivista Mat.*, V. 2, pp. 38–41.

Beutelspacher, A. (2016) *Numbers: Histories, Mysteries, Theories*, Dover Publications, New York.

Bhanu, M. S. and Velamma, G. (2015) Operations on Zadeh's Z-numbers, *IOSR J. Math.*, V. 11, No. 3, pp. 88–94.

Biagioli, F. (2016) *Space, Number, and Geometry from Helmholtz to Cassirer*, Springer, New York.

Bibak, K. (2013) Additive combinatorics with a view towards computer science and cryptography, in *Number Theory and Related Fields*: *in Memory of Alf van der Poorten*, Springer Proceedings in Mathematics & Statistics, New York, pp. 99–128.

Bidwell, J. (1993) Archimedes and pi-revisited, *School Sci. Math.*, V. 94, No. 3, pp. 127–129.

Bigelow, J. (1988) *The Reality of Numbers: A Physicalist's Philosophy of Mathematics*, Oxford University Press, New York.

Binbaşıoglu, D., Demiriz, S. and Türkoglu, D. (2016) Fixed points of non-Newtonian contraction mappings on non-Newtonian metric spaces, *J. Fixed Point Theory Appl.*, V. 18, pp. 213–224.

Binney, J. J., Dowrick, N. J., Fisher, A. J. and Newman, M. E. J. (1993) *The Theory of Critical Phenomena*, Oxford University Press, New York.

Birkhoff, G. (1937) An extended arithmetic, *Duke Math. J.*, V. 3, No. 2, pp. 311–316.

Birkhoff, G. (1938) On product integration, *J. Math. Phys.*, V. 16, pp. 104–132.

Birkhoff, G. (1942) Generalized arithmetic, *Duke Math. J.* V. 9, No. 2, pp. 283–302.

Birkhoff, G. and Bartee, T. C. (1967) *Modern Applied Algebra*, McGraw Hill, New York.

Biro D. and Matsuzawa, T. (2001) Use of numerical symbols by the chimpanzee (Pan troglodytes): Cardinals, ordinals, and the introduction of zero, *Anim. Cogn.*, V. 4, pp. 193–199.

Al-Biruni, R. (2004) *The Book of Instruction in the Elements of the Art of Astrology*, Kessinger Publishing.

Bishop, E. (1970) Mathematics as a numerical language, in *Intuitionism and Proof Theory*, North-Holland, Amsterdam, New York, pp. 53–71.

Bjarnadóttir, K. (2014) History of teaching arithmetic, in *Handbook on the History of Mathematics Education*. Springer, New York, pp. 431–457.

Black, M. (1934) Vagueness: An exercise in logical analysis, *Philos. Sci.*, V. 4, pp. 427–455.

Black, M. (1951) Review: Gottlob Frege, J. L. Austin, The foundations of arithmetic. A logico-mathematical enquiry Into the concept of number, *J. Symbolic Logic*, V. 16, No. 1, p. 67.

Black, M. (1963) Reasoning with loose concepts, *Dialogue*, V. 2, pp. 1–12.

Blake, D. (2006) *Pension Finance*, Willey, New York.

Blass, A., Di Nasso, M. and Forti, M. (2011) *Quasi-selective Ultrafilters and Asymptotic Numerosities*, preprint, arXiv:1011.2089v2.

Błaszczyk P. (2005) On the mode of existence of the real numbers, in *Logos of Phenomenology and Phenomenology of the Logos*, Book One, Analecta Husserliana (The Yearbook of Phenomenological Research), V. 88, Springer, Dordrecht, pp. 137–155.

Blatner, D. (1999) *The Joy of Pi*, Walker & Company, PaloAlto, CA.

Blehman, I. I., Myshkis, A. D. and Panovko, Ya.G. (1983) *Mech. Appl. Logic*, Nauka, Moscow, 1983 (in Russian).

Blizard, W. (1989) Multiset theory, *Notre Dame J. Formal Logic,* V. 30, pp. 36–66.

Blizard, W. D. (1989a) Real-valued multisets and fuzzy sets, *Fuzzy Sets Sys.*, V. 33, pp. 77–97.

Blizard, W. (1990) Negative membership, *Notre Dame J. Formal Logic*, V. 31, No. 1, pp. 346–368.

Blizard, W. D. (1991) The development of multiset theory, *Modern Logic*, V. 1, No. 4, pp. 319–352.

Block, N. (1986) Advertisement for a semantics for psychology, in *Midwest Studies in Philosophy*, University of Minnesota, Minneapolis, pp. 615–678.

Bloom, P. and Wynn, K. (1997) Linguistic cues in the acquistion of number words, *J. Child Lang.*, V. 24, No. 3, pp. 511–533.

Blum, L., Cucker, F., Shub, M. and Smale, S. (1998) *Complexity of Real Computation*, Springer, New York.

Blum, L., Shub, M. and Smale, S. (1989) On a theory of computation and complexity over the real numbers: NP-completeness, recursive functions and universal machines, *Bull. Amer. Math. Soc.*, V. 21, No. 1, pp. 1–46.

Blyth, D. (2000) Platonic number in the parmenides and metaphysics XIII, *Int. J. Philos. Studies*, V. 8, No. 1, pp. 23–45.

Bobis, J. (1991) The effect of instruction on the development of computation estimation strategies, *Math. Education Res. J.*, V. 3, pp. 7–29.

Bobis, J. (1996) Visualisation and the development of number sense with kindergarten children, in *Children's Number Learning*, A Research Monograph of the Mathematics Education Group of Australasia and the Australian Association of Mathematics Teachers, AAMT, Adelaide.

Boccaletti, D., Catoni, F., Cannata, R., Catoni, V., Nichelatti, E. and Zampetti, P. (2008) *The Mathematics of Minkowski Space-Time and An Introduction to Commutative Hypercomplex Numbers*, BirkhäuserVerlag, Basel.

Boccuto, A. and Candeloro, D. (2011) Differential calculus in Riesz spaces and applications to *g*-calculus, *Mediterranean J. Math.*, V. 8, No. 3, pp. 315–329.

Bockstaele, P. (1960). Notes on the first arithmetics printed in Dutch and English, *Isis*, V. 51, No. 3, pp. 315–321.

Boethius (1983) *Boethian Number Theory: A Translation of the "De Institutione Arithmetica,"* Trans. Michael Masi. (Studies in Classical Antiquity, 6.), Rodopi, Amsterdam.

Boissoneault, L. (2017) *How Humans Invented Numbers — And How Numbers Reshaped Our World*: Anthropologist Caleb Everett explores the subject in his new book, *Numbers and the Making of Us*, smithsonian.com, (https://www.smithsonianmag.com/innovation/how-humans-invented-numbersand-how-numbers-reshaped-our-world-180962485/).

Boksic, B. (2017) Getting to 1+1=3: The 3 Types of Relationships in Your Life, (https://www.goalcast.com/2017/06/26/getting-113-3-types-relation ships-life/).

Bolzano, B. (1817) *Rein analytischer Beweis des Lehrsatzes, das zwischen zwey Werthen, die ein entgegengesetztes Resultat gewähren, wenigstens eine reelle Wurzel der Gleichung liege*, Gottlieb Haase, Prag.

Bolzano, B. (1837/1973) *Theory of Science*, Reidel, Dordrecht.

Bolzano, B. (1851) *Paradoxien des Unendlichen*, C. H. Reclam Sen., Leipzig.

Bolzano, B. (1975) Einleitung zur Grössenlehre. Erste Begriffe der allgemeinen Grössenlehre, in *Gesamtausgabe*, II A 7, Friedrich Fromann Verlag, Stuttgart-Bad Cannstatt, Germany.

Bondecka-Krzykowska, I. (2004) Strukturalizm jako alternatywa dla platonizmu w filozofii matematyki, *Filozofia Nauki*, No. 1.

Boncompagni Ludovisi, B. (1862) Opuscoli di Leonardo Pisano secondo un codice della Biblioteca Ambrosiana di Milano contrassegnato E.75. Parte Superiore, in Id., *Scritti di Leonardo Pisano matematico del secolo decimoterzo*, v. II, Roma, Tipografia delle scienze matematiche e fisiche, pp. 253–283.

Boniface, J. (1999) Kronecker. Sur le concept de nombre, *La Gazette des mathématiciens*, V. 81, pp. 49–70.

Boniface, J. (2002) *Les constructions des nombres reels dans le movement d'arithmétisation de l'analyse*, Ellipses, Paris.

Boniface J. (2007) *The Concept of Number from Gauss to Kronecker,* in *The Shaping of Arithmetic after C. F. Gauss's Disquisitiones Arithmeticae,* Springer, Berlin, Heidelberg, pp. 314–342.

Boniface, J. and Schappacher, N. (2002) Sur le concept de nombre dans la mathématique. Cours inédit de Leopold Kronecker à Berlin (1891), *Revue d'Histoire des mathématiques,* V. 7, No. 2, pp. 207–277.

Boole, G. (1860) *Calculus of Finite Differences,* Chelsea Publishing Company.

Boole, G. (1880) *A Treatise on the Calculus of Finite Differences,* Macmillan, London.

Boolos, G. (1987) The consistency of Frege's Foundations of Arithmetic, in *On Being and Saying,* Essays in Honor of Richard Cartwright, The MIT Press, Cambridge, pp. 3–20.

Boolos, G. (1990) The standard of equality of numbers, in *Meaning and Method,* Essays in Honor of Hilary Putnam, Cambridge University Press, Cambridge, pp. 261–277.

Borel, E. (1903) Contribution a l'analyse arithmetique du continu, *J. Math. Appl.,* 5e série, tome IX, pp. 329–375.

Borel, E. (1950) *Probabilité et Certitude,* Presses Universitaires de France, Paris.

Borges, E. (2004) A possible deformed algebra and calculus inspired in nonextensive thermo-statistics, *Physica A,* V. 340, pp. 95–101.

Borghi, P. (1484) Qui comenza la nobel opera de arithmetica (Aritmetica mercantile), Erhard Ratdolt, Vénétie.

Borisov, A. N., Alekseev, A. V., Merkurjeva, G. V., Slyadz, N. N. and Glushkov, V. I. (1989) *Processing Fuzzy Information in Decision-making Systems,* Radio and Svyaz, Moscow (in Russian).

Born, C. (2020) *The Secret of Ishango: On the helix structure of prime numbers,* Books on Demand.

Borodin, A. I. (1986) *From the History of Arithmetic,* Vyscha Shkola, Kiev (in Russian).

Borovik, A. V. (2010) *Mathematics under the Microscope: Notes on Cognitive Aspects of Mathematical Practice,* American Mathematical Society, Providence, RI.

Bortolan, G. and Degani, R. (1985) A review of some methods for ranking fuzzy numbers, *Fuzzy Sets Sys.,* V. 15, pp. 1–19.

Boruah, K. and Hazarika, B. (2016) *Some Basic Properties of G-Calculus and Its Applications in Numerical Analysis,* Preprint in General Mathematics [math.GM] (arXiv:1607.07749).

Borwein, P. B. (2002) *Computational Excursions in Analysis and Number Theory,* Springer-Verlag, New York.

Borwein, J. and Bailey, D. (2008) *Mathematics by Experiment: Plausible Reasoning in the 21st Century,* A K Peters/CRC Press, Natick, MA.

Borwein, J., Bailey, D. and Girgensohn, R. (2004) *Experimentation in Mathematics: Computational Paths to Discovery.* A K Peters/CRC Press, Natick, MA.

Borwein, J. M., Borwein, P. B., Girgensohn, R. and Parnes, S. (1996) Making sense of experimental mathematics, *Math. Intell.,* V. 18, No. 4, pp. 12–18.

Boscarino, G. (2018) An interpretation of Plato's idea and Plato's criticism of Parmenides according to Peano's ideography, *Athens J. Humanities and Arts*, V. 5, No. 1, pp. 7–28.

Bosman, L. (2005) *Meaning and Philosophy of Numbers* (Ibis Western Mystery Tradition), Hays (Nicolas) Ltd.

Bostock, D. (1974) *Logic and Arithmetic*, V. 1, Clarendon Press, Oxford.

Bostock, D. (1979) *Logic and Arithmetic*, V. 2, Clarendon Press, Oxford.

Boucon, J.-L. (2019) *The Ontology of Knowledge, Logic, Arithmetic, Sets Theory and Geometry*, PhilArchive (https://philpapers.org/rec/BOUTOO-2).

Bouhennache, R. (2014) *Truly Hypercomplex Numbers: Unification of Numbers and Vectors*, preprint in General Mathematics (arXiv:1409.2757).

Bourbaki, N. (1939) *Éléments de Mathématiques, Volume 1: Théorie des Ensembles*, Fascicule de Résultats Hermann, Paris (English translation: Bourbaki, N. *Theory of Sets*, Hermann, Paris, 1968).

Bourbaki, N. (1948) L'architecture des mathématiques, Legrands courants de la pensée mathématiques, *Cahiers Sud*, pp. 35–47.

Bourbaki, N. (1950) The architecture of mathematics, *Amer. Math. Monthly*, V. 57, pp. 221–232

Bourbaki, N. (1957) *Structures*, Hermann, Paris.

Bourbaki, N. (1960) *Theorie des Ensembles*, Hermann, Paris.

Bourbaki, N. (1972) *Univers*, in Séminaire de Géométrie Algébrique du Bois Marie — 1963–64 — Théorie des topos et cohomologie étale des schémas — (SGA 4) — vol. 1 (Lecture Notes in Mathematics, V. 269) Springer-Verlag, Berlin/New York, pp. 185–217 (in French).

Boumans, M. (2005) *How Economists Model the World into Numbers*, Routledge, New York.

Boyer, C. B. (1943) An early reference to division by zero, *Ameri. Math. Monthly*, V. 50, No. 8, pp. 487–491.

Boyer, C. B. (1985) *A History of Mathematics*, Princeton University Press, Princeton.

Boyer, C. B. and Merzbach, U. C. (2010) *A History of Mathematics*, Willey, New York.

Boysen, S. T. and Berntson, G. G. (1989) Numerical competence in a chimpanzee (Pan troglodytes), *J. Comp. Psychol.*, V. 103, No. 1, pp. 23–31.

Boysen, S. T., Berntson G. G., Shreyer T. A. and Hannan M. B. (1995) Indicating acts during counting by a chimpanzee (Pan troglodytes), *J. Comp. Psychol.*, V. 109, No. 1, pp. 47–51.

Brahmagupta and Bhaskara, (1817) *Algebra with Arithmetics and Mesuration from the Saskrit of Brahmagupta and Bhaskara*, (Colebrooke, H. T., Transl.), London.

Brannon, E. M. (2002) The development of ordinal numerical knowledge in infancy, *Cognition*, V. 83, pp. 223–240.

Brannon, E. M. and Terrace, H. S. (1998) Ordering of the numerosities 1 to 9 by monkeys, *Science*, V. 282, pp. 746–749.

Brannon, E. M., Abbott, S. and Lutz, D. J. (2004) Number bias for the discrimination of large visual sets in infancy, *Cognition*, V. 93, pp. B59–B68.

Braßel, B., Fischer, S. and Huch, F. (2007) Declaring numbers, in *Workshop on Functional and (Constraint) Logic Programming*, pp. 23–36.

Breger, H. (2008) Natural numbers and infinite cardinal number, in *Kosmos und Zahl*, Steiner, Stuttgart, Germany, pp. 309–318.

Brekke, L. and Freund, P. (1993) p-adic numbers in physics, *Phys. Rep*, V. 231, pp. 1–66.

Breukelaar J. W. C. and Dalrymple-Alford J. C. (1998) Timing ability and numerical competence in rats, *J. Exp. Psychol. Anim. Behav. Process.*, V. 24, pp. 84–97.

Brindle, R. S. (1987) *The New Music*, Oxford University Press.

Broadbent, T. A. A. (1971) The Higher Arithmetic, *Nature*, V. 229 No. 6, pp. 187–188.

Brockman, J. (1997) *What Kind of Thing is a Number? A Talk with Reuben Hersh*, *Edge* (https://www.edge.org/conversation/reuben_hersh-what-kind-of-thing-is-a-number).

Brodsky, A. E., Rogers Senuta, K., Weiss, C. L. A. Marx, C. M., Loomis, C., Arteaga, S. S., Moore, H., Benhorin, R. and Castagnera-Fletcher, A. (2004) When one plus one equals three: the role of relationships and context in community, *Amer. J. Commun. Psychol.*, V. 33, No. 3–4, pp. 229–241.

Brouwer, L. E. J. (1975) *Collected Works*, V. 1, North-Holland, Amsterdam.

Brown, G. (2000) Leibniz on wholes, unities, and infinite number, *Leibniz Rev.*, V. 10, pp. 21–51.

Brown, J. R. (2017) Proofs and guarantees, *Math. Intelli.*, V. 39, No. 4, pp. 47–50.

Brown, N. (2015) *One Plus One Equals Three*: *Combining Email + Facebook Ads To Reach More Customers*, (https://socialmediaweek.org/blog/2015/02/one-plus-one-equals-three-combining-email-facebook-ads-reach-customers/).

Brueckler, F. M. and Stilinović, V. (2013) Teaching arithmetic in the Habsburg Empire at the end of the 18th century — A textbook example, *Historia Mathematica*, V. 40, No. 3, pp. 309–323.

Bruno, A. and Yasaki, D. (2008) *The Arithmetic of Trees*, Preprint in Combinatorics (math.CO), (arXiv:0809.4448).

Bucur, I. and Deleanu, A. (1968) *Introduction to the Theory of Categories and Functors*, Wiley, London.

Bueti, D. and Walsh, V. (2009) The parietal cortex and the representation of time, space, number and other magnitudes, *Philos. Trans. R. Soc.*, B V. 364, pp. 1831–1840.

Buium, A. (2005) *Arithmetic Differential Equations*, Mathematical Surveys and Monographs, V. 118, American Mathematical Society, Providence, RI.

Buium, A. (2017) *Foundations of Arithmetic Differential Geometry*, Mathematical Surveys and Monographs, V. 222, American Mathematical Society, Providence, RI.

Buium, A. and Simanca, S. R. (2010) Arithmetic partial differential equations, I, *Adv. Math.*, V. 225, No. 2, pp. 689–793.

Bulmer-Thomas, I. (1983) Plato's theory of numbers, *The Classical Quarterly*, V. 33, No. 2, pp. 375–384.

Bunn, R. (1977) Quantitative relations between infinite sets, *Ann. Sci.*, V. 34, pp. 177–191.

Burdik, C., Frougny, C., Gazeau, J.-P. and Krejcar, R. (1998) Beta-integers as natural counting systems for quasicrystals, *J. Phys., A: Math. Gen.*, V. 31, No. 30, pp. 6449–6472.

Burdik, C., Frougny, C., Gazeau, J.-P. and Krejcar, R. (2000) Beta-integers as group, in *Dynamical Systems: From Crystals to Chaos*, World Scientific, New York/London/Singapore, pp. 125–136.

Burgess, J. P. and Rosen, G. (1997) *A Subject With No Object: Strategies for Nominalistic Interpretation of Mathematics*, Clarendon Press, Oxford, UK.

Burgin, M. (1977) Non-classical models of natural numbers, *Russian Math. Surv.*, V. 32, No. 6, pp. 209–210 (in Russian).

Burgin, M. (1980) Dual arithmetics, *Abstr. Amer. Math. Soc.*, V. 1, No. 6.

Burgin, M. (1982) Products of operators in a multidimensional structured model of systems, *Math. Social Sci.*, No. 2, pp. 335–343.

Burgin, M. (1985) Abstract theory of properties, in *Non-classical Logics*, Institute of Philosophy, Moscow, pp. 109–118 (in Russian).

Burgin, M. (1986) Quantifiers in the theory of properties, in *Nonstandard Semantics of Non-classical Logics*, Institute of Philosophy, Moscow, pp. 99–107 (in Russian).

Burgin, M. (1989) Numbers as properties, *Abstr. Amer. Mathe. Soci.*, V. 10, No. 1.

Burgin, M. (1989a) *Named Sets, General Theory of Properties, and Logic*, Institute of Philosophy, Kiev, (in Russian).

Burgin, M. (1990) Theory of named sets as a foundational basis for mathematics, in *Structures in Mathematical Theories*, San Sebastian, Spain, pp. 417–420.

Burgin, M. (1990a) Abstract theory of properties and sociological scaling, in *Expert Evaluation in Sociological Studies*, Kiev, pp. 243–264 (in Russian).

Burgin, M. (1990b) Hypermeasures and hyperintegration, *Notices Natl. Acade. Scie. Ukraine*, No. 6, pp. 10–13 (in Russian and Ukrainian).

Burgin, M. (1991) Logical methods in artificial intelligent systems, *Vestnik Computer Soc.*, No. 2, pp. 66–78 (in Russian).

Burgin, M. (1992) Infinite in finite or metaphysics and dialectics of scientific abstractions, *Philoso. Soc. Thought*, No. 8, pp. 21–32 (in Russian and Ukrainian).

Burgin, M. (1992a) Algebraic structures of multicardinal numbers, in *Problems of group theory and homological algebra*, Yaroslavl, pp. 3–20 (in Russian).

Burgin, M. (1994) Is it possible that mathematics gives new knowledge about reality, *Philos. Soc. Thought*, No. 1, pp. 240–249 (in Russian and Ukrainian).

Burgin, M. (1995) Named sets as a basic tool in epistemology, *Epistemologia*, v. XVIII, pp. 87–110.

Burgin, M. (1997) *Non-Diophantine Arithmetics or What Number is 2+2*, Ukrainian Academy of Information Sciences, Kiev (in Russian, English summary).

Burgin, M. (1997a) Mathematical theory of technology, in *Methodological and Theoretical Problems of Mathematics and Informatics*, Ukrainian Academy of Information Sciences, Kiev, pp. 91–100 (in Russian).

Burgin, M. (1997b) *Fundamental Structures of Knowledge and Information*, Ukrainian Academy of Information Sciences, Kiev (in Russian).

Burgin, M. (1997c) Scientific foundations of structuralism, in *Language and Culture*, Ukrainian Academy of Information Sciences, Kiev, V. 1, pp. 24–25 (in Russian).

Burgin, M. (1997d) Logical varieties and covarieties, in *Methodological and Theoretical Problems of Mathematics and Information and Computer Sciences*, Ukrainian Academy of Information Sciences, Kiev, pp. 18–34 (in Russian).

Burgin, M. (1998) *On the Nature and Essence of Mathematics*, Ukrainian Academy of Information Sciences, Kiev (in Russian).

Burgin, M. (1998a) Finite and Infinite, in *On the Nature and Essence of Mathematics*, Ukrainian Academy of Information Sciences, Kiev, Appendix, pp. 97–108 (in Russian).

Burgin, M. (1970/1998) Mathematics as an experimental science, in *On the Nature and Essence of Mathematics*, Ukrainian Academy of Information Sciences, Kiev, Appendix, pp. 97–108 (in Russian).

Burgin, M. (2001) *Diophantine and Non-Diophantine Aritmetics: Operations with Numbers in Science and Everyday Life*, LANL, Preprint Mathematics GM/0108149, 27 p. (electronic edition: http://arXiv.org).

Burgin, M. (2001)*How we Count or is it Possible that Two Times Two is not Equal to Four*, Science Direct Working Paper No S1574-0358(04)70635-8, 12 pp. (electronic edition: http://www.sciencedirect.com/preprintarchive).

Burgin, M. (2002) Theory of hypernumbers and extrafunctions: functional spaces and differentiation, *Discrete Dynamics Nature Soc.*, V. 7, No. 3, pp. 201–212.

Burgin, M. (2003) Levels of System Functioning Description: From Algorithm to Program to Technology, in *Proceedings of the Business and Industry Simulation Symposium*, Society for Modeling and Simulation International, Orlando, FL, pp. 3–7.

Burgin, M. (2003a) Nonlinear phenomena in spaces of algorithms, *International J. Compu. Math.*, V. 80, No. 12, pp. 1449–1476.

Burgin, M. (2004) Hyperfunctionals and Generalized Distributions, in *Stochastic Processes and Functional Analysis*, A Dekker Series of Lecture Notes in Pure and Applied Mathematics, V. 238, pp. 81–119.

Burgin, M. (2004a) Logical Tools for Program Integration and Interoperability, in *Proceedings of the IASTED International Conference on Software Engineering and Applications*, MIT, Cambridge, pp. 743–748.

Burgin, M. (2004b) *Unified Foundations of Mathematics*, Preprint in Mathematics, LO/0403186, 39 p. (electronic edition: http://arXiv.org).

Burgin, M. (2004c) *Named Set Theory Axiomatization: TNZ Theory*, (February 2004) Mathematics Preprint Archive pp. 333–344 (Available at SSRN: https://ssrn.com/abstract=3177551).

Burgin, M. (2005) *Super-recursive Algorithms*, Springer, New York.

Burgin, M. (2005a) Hypermeasures in general spaces, *Int. J. Pure Appl. Math.*, V. 24, pp. 299–323.

Burgin, M. (2006) Nonuniform operations on named sets, in *5th Annual International Conference on Statistics, Mathematics and Related Fields*, 2006 Conference Proceedings, Honolulu, Hawaii, pp. 245–271.

Burgin, M. (2006a) Operational and Program Schemas, in *Proceedings of the 15th International Conference on Software Engineering and Data Engineering* (SEDE-2006), ISCA, Los Angeles, California, pp. 74–78.

Burgin, M. (2007) Elements of non-Diophantine arithmetics, in *6th Annual International Conference on Statistics, Mathematics and Related Fields*, 2007 Conference Proceedings, Honolulu, Hawaii, January, pp. 190–203.

Burgin, M. (2007a) Universality, reducibility, and completeness, *Lec. Notes Comput. Sci.*, V. 4664, pp. 24–38.

Burgin, M. (2008) *Neoclassical Analysis: Calculus closer to the Real World*, Nova Science Publishers, New York.

Burgin, M. (2008a) Hyperintegration approach to the Feynman integral, *Integration: Math. Theory Appl.*, V. 1, No. 1, pp. 59–104.

Burgin, M. (2008b) Inequalities in series and summation in hypernumbers, in *Advances in Inequalities for Series*, Nova Science Publishers, New York, pp. 89–120.

Burgin, M. (2009) Mathematical Theory of Information Technology, in *Proceedings of the 8th WSEAS International Conference on Data Networks, Communications, Computers* (DNCOCO'09), Baltimore, Maryland, USA, November, pp. 42–47.

Burgin, M. (2009a) Structures in mathematics and beyond, in Proceedings of the 8th *Annual International Conference on Statistics, Mathematics and Related Fields*, Honolulu, Hawaii, pp. 449–469.

Burgin, M. (2010) *Introduction to Projective Arithmetics*, Preprint in Mathematics, math.GM/1010.3287, 21 pp. (electronic edition: http://arXiv.org).

Burgin, M. (2010a) Integration in bundles with a hyperspace base: indefinite integration, *Integration: Math. Theory Appl.*, V. 2, pp. 395–435.

Burgin, M. (2010b) *Theory of Information: Fundamentality, Diversity and Unification*, World Scientific, New York/London/Singapore.

Burgin, M. (2010c) *Measuring Power of Algorithms, Computer Programs, and Information Automata*, Nova Science Publishers, New York.

Burgin, M. (2010d) Algorithmic complexity of computational problems, *International Journal of Computing & Information Technology*, V. 2, No. 1, pp. 149–187.

Burgin, M. (2011) *Theory of Named Sets*, Mathematics Research Developments, Nova Science Publishers, New York.

Burgin, M. (2011a) *Differentiation in Bundles with a Hyperspace Base*, Preprint in Mathematics, math.CA/1112.3421, (electronic edition: http://arXiv.org).

Burgin, M. (2012) *Hypernumbers and Extrafunctions: Extending the Classical Calculus*, Springer, New York.

Burgin, M. (2012a) *Structural Reality*, Nova Science Publishers, New York.

Burgin, M. (2012b) Integration in bundles with a hyperspace base: Definite integration, *Integration: Math. Theory Appl.*, V. 3, No. 1, pp. 1–54.

Burgin, M. (2012c) Fuzzy continuous functions in discrete spaces, *Annals of Fuzzy Sets, Fuzzy Logic Fuzzy Systems*, V. 1, No. 4, pp. 231–252.

Burgin, M. (2013) Named sets and integration of structures, in *Topics in Integration Research*, Nova Science Publishers, New York, pp. 55–98.

Burgin, M. (2015) Operations with Extrafunctions and integration in bundles with a hyperspace base, in *Functional Analysis and Probability*, Chapter 1, Nova Science Publishers, New York, pp. 3–76.

Burgin, M. (2015a) Picturesque diversity of probability, in *Functional Analysis and Probability*, Chapter 14, Nova Science Publishers, New York, pp. 301–354.

Burgin, M. (2016) Probability theory in relational structures, *Journal of Advanced Research in Applied Mathematics and Statistics*, V. 1, No. 3&4, pp. 19–29.

Burgin, M. (2016a) *Theory of Knowledge*: *Structures and Processes*, World Scientific, New York/London/Singapore.

Burgin, M. (2016b) Inductive Complexity and Shannon Entropy, in *Information and Complexity*, World Scientific, New York/London/Singapore, pp. 16–32.

Burgin, M. (2017) *Functional Algebra and Hypercalculus in Infinite Dimensions: Hyperintegrals, Hyperfunctionals* and Hyperderivatives, Nova Science Publishers, New York.

Burgin, M. (2017a) Bidirectional named sets as structural models of interpersonal communication, *Proceedings*, V. 1, No. 3, 58.

Burgin, M. (2017b) Ideas of Plato in the context of contemporary science and mathematics, *Athens J. Humanities Arts*, V. 4, No. 3, pp. 161–182.

Burgin, M. (2017c) *Mathematical Knowledge and the Role of an Observer*: *Ontological and epistemological aspects*, Preprint in Mathematics History and Overview (math.HO), 1709.06884, 15 pp. (electronic edition: http://arXiv.org).

Burgin, M. (2017d) *Semitopological Vector Spaces: Hypernorms, Hyperseminorms and Operators*, Apple Academic Press, Toronto, Canada.

Burgin, M. (2018) *Mathematics as an Interconnected Whole*, preprint in General Mathematics 1801.0135, 15 pp. (electronic edition: http://viXra.org/postprints/).

Burgin, M. (2018a) Mathematical Analysis of the Concept Equality, *Research and Reports on Mathematics*, V. 2, No. 2, 10.4172/RRM.1000e102.

Burgin, M. (2018b) Triadic Structures in Interpersonal Communication, *Information*, V. 9, No. 11, 283.

Burgin, M. (2018c) Introduction to non-Diophantine number theory, *Theory Appl. Math. Comput. Sci.*, V. 8, No. 2, pp. 91–134.

Burgin, M. (2018d) Inductive turing machines, in *Unconventional Computing* (Adamatzky, A. (Ed), A volume in the Encyclopedia of Complexity and Systems Science, Springer, Berlin/Heidelberg, pp. 675–688.

Burgin, M. (2019) On weak projectivity in arithmetic, *European J. Pure Appl. Math.*, V. 12, No. 4, pp. 1787–1810.

Burgin, M. (2019a) Information-oriented analysis of discovery and invention in mathematics, in *Philosophy and Methodology of Information*: *The Study of Information in the Transdisciplinary Perspective,* World Scientific, New York/London/Singapore, pp. 171–199.

Burgin, M. S. and Baranovich, T. M. (1975) Linear Ω-algebras, *Russian Mathematical Surveys*, V. 30, No. 4, pp. 61–106 (translated from Russian).

Burgin, M. and Buzaglo, M. (1994) Cardinal typology of sets, *Abstr. Amer. Math. Soc.*, V. 15, No. 4.

Burgin, M. and Czachor, M. (2020) *Non-Diophantine Arithmetics in Mathematics, Physics and Psychology*, World Scientific, New York/London/Singapore.

Burgin, M. and Dantsker, A. M. (1995) A method of solving operator equations of mechanics with the theory of Hypernumbers, *Notices Natl. Acade. Sciences of Ukraine*, No. 8, pp. 27–30 (in Russian).

Burgin, M. and Dantsker, A. M. (2015) Real-time inverse modeling of control systems using hypernumbers, in *Functional Analysis and Probability*, Chapter 17, Nova Science Publishers, New York, pp. 439–456.

Burgin, M., Dantsker, A. M. and Esterhuysen, K. (2012) Lithium battery temperature prediction, *Integration: Math. Theory Appl.*, V. 3, No. 4, pp. 319–331.

Burgin, M. and Karasik, A. (1976) Operators of multidimensional structured model of parallel computations, *Automation Remote Control*, V. 37, No. 8, pp. 1295–1300.

Burgin, M. and Krinik, A. C. (2009) Probabilities and hyperprobabilities, 8th *Annual International Conference on Statistics, Mathematics and Related Fields*, Conference Proceedings, Honolulu, Hawaii, pp. 351–367.

Burgin, M. and Krinik, A. C. (2010) Introduction to conditional hyperprobabilities, *Integration: Math. Theory Appl.*, V. 2, pp. 285–304.

Burgin, M. and Krinik, A. C. (2015) Hyperexpectation in axiomatic and constructive settings, in *Functional Analysis and Probability*, Chapter 12, Nova Science Publishers, New York, pp. 259–288.

Burgin, M. and Meissner, G. (2017) $1 + 1 = 3$: Synergy arithmetic in economics, *Appl. Math.*, V. 8, No. 2, pp. 133–144.

Burgin, M. and Milov, Yu. (1998) Grammatical aspects of language in the context of the existential triad concept, in *On the Nature and Essence of Mathematics*, Ukrainian Academy of Information Sciences, Kiev, pp. 136–142.

Burgin, M. and de Vey Mestdagh, C. N. J. (2011) The representation of inconsistent Knowledge in Advanced knowledge based systems, *Lecture Notes Comput. Sci., Knowledge-Based Intelligent Inform. Eng. Syst.*, V. 6882, pp. 524–537.

Burgin, M. and de Vey Mestdagh, C. N. J. (2015) Consistent structuring of inconsistent knowledge, *J. Intelligent Inform. Syst.*, V. 45, No. 1, pp. 5–28.

Burgin, M. and Schumann, J. (2006) Three levels of the symbolosphere, *Semiotica*, V. 160, No. 1/4, pp. 185–202.

Burke, M. (2015) *Frege, Hilbert, and Structuralism*, Dissertation, Department of Philosophy, Faculty of Arts, University of Ottawa.

Burkert, W. (1972) *Lore and Science in Ancient Pythagoreanism*, Trans. E. L. Minar, Jr., Cambridge, MA.

Burr, S. A. (Ed.) (1992) The unreasonable effectiveness of number theory, in *Proc. Symposia in Applied Mathematics*, Orono, Maine, V. 46.

Burr, D. and Ross, J. (2008) A visual sense of number, *Curr. Biol.*, V. 18, pp. 425–428.

Burnett, C. (2010) *Numerals and arithmetic in the Middle Ages*, Farnham: Ashgate Variorum.

Burton, D. M. (1997) *The History of Mathematics*, McGraw-Hill Co., New York.

Busch, P., Grabowski, M. and Lahti, P. J. (1995) *Operational Quantum Physics*, Springer, Berlin.

Buss, S. R. (1986) *Bounded Arithmetic*, Bibliopolis, Napoli.

Buss, S. R. (1986a) The polynomial hierarchy and intuitionistic bounded arithmetic, in *Structure in Complexity Theory*, Lecture Notes in Computer Science, V. 223 (Springer, Berlin), pp. 125–143.

Buss, S. R. (1994), On Gödel's theorems on lengths of proofs — I: Number of lines and speedup for arithmetics, *J. Symbolic Logic*, V. 59, pp. 737–775.

Buss, S. R. (1995) Relating the bounded arithmetic polynomial time hierarchies, *Ann. Pure Appl. Logic*, V. 75, pp. 67–77.

Buss, S. R. (2013) The computational power of bounded arithmetic from the predicative viewpoint, in *New Computational Paradigms*: *Changing Conceptions of What is Computable*, Elsevier, Amsterdam, pp. 213–222.

Bussmann, J. B. (2013) One plus one equals three (or more . . .): combining the assessment of movement behavior and subjective states in everyday life, *Front. Psychol.*; https://doi.org/10.3389/fpsyg.2013.00216.

Butkovič. P. (2010) *Max-linear Systems: Theory and Algorithms*, Springer Monographs in Mathematics, Springer-Verlag London Ltd., London.

Butterworth, B. (2005) The development of arithmetical abilities, *J. Child. Psychol. Psychiatry*, V. 46, pp. 3–18.

Butterworth, B. Cappelletti, M. and Kopelman, M. (2001) Category specificity in reading and writing: The case of number words, *Nature Neurosci.*, V. 4, pp. 784–786.

Buzaglo, M. (1992) Cardinal numbers: A Wittgensteinian approach, in *Proc. XV International Wittgenstein Symposium*, Kirchberg am Wechesel, Austria.

Byers, W. (2007) *How Mathematicians Think: Using Ambiguity, Contradiction, and Paradox to Create Mathematics*, Princeton University Press, Princeton, NJ.

Byers, W. (2011) *The Blind Spot: Science and the Crisis of Uncertainty*, Princeton University Press, Princeton, NJ.

Byron, G. G. (1974) [*Written 1813–1814*] *Alas! the Love of Women: 1813–1814*, Belknap Press.

Caesar, L. K. and Cech, N. B. (2019) Synergy and antagonism in natural product extracts: when $1 + 1$ does not equal 2, *Natural Products Rep.*, V. 36, pp. 869–888.

Cain, F. (2008/2015) Venus, the morning star and evening star, *Universe Today*; https://www.universetoday.com/22570/venus-the-morning-star/.

Cajori, F. (1919) The controversy on the origin of our numerals, *Scientific Monthly*, V. 9, No. 5, pp. 458–464.

Cajori, F. (2011) *A History of Mathematical Notations*, Dover Books on Mathematics, Dover, New York.

Çakır, Z. (2013) Spaces of continuous and bounded functions over the field of geometric complex numbers, *J. Inequalities Appl.*, V. 2013, Article 363.

Çakmak, A. F. and Başar, F. (2014) Certain spaces of functions over the field of non-Newtonian complex numbers, *Abstract Appl. Anal.*, V. 2014, Article ID 236124, 12 p.

Calandri, F. (1491) *De Arithmetica*, Lorenzo da Morgiani and Giovanni Thedesco da Maganza, Florence.

Call, J. and Tomasello, M. (2008). Does the chimpanzee have a theory of mind? 30 years later, *Trends Cognitive Sci.*, V. 12, pp. 187–192.

Calude, A. S. (2021) The history of number words in the world's languages — what have we learnt so far? *Philos. Transact. of Royal Soc., B*, V. 376 //doi.org/10.1098/rstb.2020.0206.

Cambridge Business English Dictionary, Cambridge University Press, London, 2011.

Camos, V. and Tillmann, B. (2008). Discontinuity in the enumeration of sequentially presented auditory and visual stimuli, *Cognition*, V. 107, No. 3, pp. 1135–1143.

Campbell, J. I. and Alberts, N. M. (2009) Operation-specific effects of numerical surface form on arithmetic strategy, *J. Exp. Psychol. Learn. Mem. Cogn.*, V. 35, No. 4, pp. 999–1011.

Canfora, G. and Troiano, L. (2004) Fuzzy ordering of fuzzy numbers, in *Proc. IEEE Conf. Fuzzy Systems*, pp. 669–674.

Cantlon, J. F. (2012) Math, monkeys, and the developing brain, *Proc. Natl. Acad. Sci.*, V. 109 (Supplement 1), pp. 10725–10732.

Cantlon, J. F. and Brannon, E. M. (2006) Shared system for ordering small and large numbers in monkeys and humans, *Psychol. Sci.*, V. 17, pp. 401–406.

Cantlon, J. F. and Brannon, E. M. (2007) Basic math in monkeys and college students, *PLoS Biol.*, V. 5:e328.

Cantlon, J. F. and Brannon, E. M. (2007a) How much does number matter to a monkey (Macaca mulatta)? *J. Exp. Psychol. Anim. Behav. Process*, V. 33, pp. 32–41.

Cantlon, J. F., Merritt, D. J. and Brannon, E. M. (2016) Monkeys display classic signatures of human symbolic arithmetic, *Anim. Cogn.*, V. 19, pp. 405–415.

Cantlon, J. F., Libertus, M. E., Pinel, P., Dehaene, S., Brannon, E. M. and Pelphrey, K. A. (2009) The neural development of an abstract concept of number, *J. Cognitive Neurosci.*, V. 21, No. 11, pp. 2217–2229.

Cantor, G. (1869) Über die einfachen Zahlensysterme, *Zeits. Math. Phys.* V. 14, pp. 121–128.

Cantor, G. (1872) Über die Ausdehnung eines Satzes aus der Theorie der trigonometrischen Reihen, *Math. Anna.*, V. 5, No. 1, 123–132.

Cantor, G. (1874) Über eine Eigenschaft des Inbegriffes aller reelen algebraischen Zahlen, *J. Reine Angew. Math.*, b. 77, s. 258–262.

Cantor, G. (1878) Ein Beitrag zur Mannigfaltigkeitslehre, *J. Reine Angew. Math.*, b. 84, s. 242–258.

Cantor, G. (1883) *Grundlagen einer allgemeinen Mannigfaltigkeitslehre: Ein mathematisch-philosophischer Versuch in der Lehre des Unendlichen*, Teubner, Leipzig.

Cantor, G. (1895/1897) Beiträge zur Begründung der transfiniten Mengenlehre, *Math. Ann.*, b. 46, s. 481–512; b. 49, s. 207–246.

Cantor, G. (1932) *Gesammelte Abhandlungen Mathematischen und Philosophischen Inhalts*, Springer, Berlin.

Cantù, P. (2018) The epistemological question of the applicability of mathematics, *J. History Analy. Philos.*, V. 6, No. 3, pp. 96–114.

Carbone, A. (1999) Cycling in proofs, feasibility and no speed-up for nonstandard arithmetic, *Trans. Amer. Math. Soc.*, V. 352, No. 5, pp. 2049–2075.

Cardano, G. (1539) *Practica arithmetice et mensurandi singularis*, Antonius Castellioneus, Mediolani.

Cardano, G. (1545) *Artis magnae, sive de regulis algebraicis* (known as *Ars magna*), Johannes Petreius, Nuremberg.

Cardano, G. (2006) *Liber de ludoaleae*, FrancoAngeli, Milan.

Carey S. (1998) Knowledge of number: its evolution and ontogeny, *Science*, V. 282, pp. 641–642.

Carey, S. (2001) Cognitive foundations of arithmetic: evolution and ontogenesis, *Mind Lang.*, V. 16, pp. 37–55.

Carlitz, L. (1931) The arithmetic of polynomials in a Galois field, *Proc. Natl Acad. Sci. USA*, V. 17, No. 2, pp. 120–122.

Carlitz, L. (1968) Bernoulli numbers, *Fibonacci Quart.*, V. 6, pp. 71–85.

Carnap, R. (1937) *The Logical Syntax of Language*, Routledge & Kegan, London.

Carroll, M. L. (2001) *The natural chain of binary arithmetic operations and generalized derivatives*, preprint in Mathematics, [math-HO] (arXiv: 0112050).

Carroll, L. (2009) *Euclid and His Modern Rivals*, Barnes and Noble, New York.

Carroll, P. and Mui, C. (2009) Billion dollar lessons: what you can learn from the most inexcusable business failures, *Portfolio*.

Carruth, P. W. (1942) Arithmetic of ordinals with applications to the theory of ordered abelian groups, *Bull. Amer. Math. Soc*, V. 48, pp. 262–271.

Carter, J. (2008) Structuralism as a philosophy of mathematical practice, *Synthese*, V. 163, No. 2, pp. 119–131.

Cartwright, S. and Schoenberg, R. (2006) 30 years of mergers and acquisitions research, *British J. Manag.*, V. 15, No. 51, pp. 51–55.

Cartwrite, J. H. E. and Piro, O. (1992) The dynamics of Runge–Kutt methods, *Int. J. Bifurcation Chaos*, V. 2, No. 3, pp. 427–450.

Case, J. (1971) Enumeration reducibility and partial degrees, *Ann. Math. Logic*, V. 2, No. 4, pp. 419–439.

Cassani, O. and Conway, J. H. (2018) Neumbering, *Math. Intelligencer*, V. 40, No.1, pp. 91–92.

Cassels, J. W. S. (1957) *An Introduction to Diophantine Approximation*, Cambridge Tracts in Mathematics and Mathematical Physics, 45, Cambridge University Press, Cambridge.

Castaneda, H. (1959) Arithmetic and reality, *Australasian J. Philos.*, V. 37, No. 2, pp. 91–107.

Castaneda, H. (1974) Thinking and the structure of the world: Discourse d'Ontologie, *Critica*, V. 6, pp. 3–40.

Castano-Bernard, R., Catanese, F., Kontsevich, M., Pantev, T., Soibelman, Y., and Zharkov, I. (Eds.) (2014) *Homological Mirror Symmetry and Tropical Geometry*, Lecture Notes of the Unione Matematica Italiana, Springer.

Castonguay, C. (1972) *Meaning and Existence in Mathematics*, Springer Vienna and New York.

Catoni, F., Cannata, R. and Nichelatti, E. (2004) The parabolic analytic functions and the derivative of real functions, *Adv. Appl. Clifford Algebras*, V. 14, No. 2, pp. 185–190.

Cauchy A.-L. (1821) *Course d'Analyse de l'Ecole Royale Politechnique; Analyse Algébrique*, L'imprimerie Royale, Paris.

Caveing, M. (1997) *La figure et le nombre. Recherches sur les premières mathématiques des Grecs. La constitution du type mathématique de l'idéalié dans la pensée grecque*, Septentrion, Paris.

Cayley, A. (1843) Chapters in the analytical geometry of (n) dimensions, *Cambridge Math. J.*, V. 4, pp. 119–127.

Cayley, A. (1845) On Jacobi's elliptic functions and on quaternions, *Philos. Magazine*, V. 26, pp. 208–211.

Cerf, V., Gostelow, K., Estrin, G. and Volanski, S. (1971) *Proper Termination of Flow of Control in Programs Involving Concurrent Processes*, Report ENG-7178, Computer Science Department, UCLA.

Cerveny, K. P. (2004) *Somewhere Between the One and The Zero: The Philosophy of Number 1=0*, Trafford Publishing, Bloomington, IN.

Chahal J. S. (1988) *Algebraic number fields.* in *Topics in Number Theory*, The University Series in Mathematics. Springer, Boston, MA, pp. 63–91.

Chakraverty S., Sahoo D. M., and Mahato N. R. (2019) Fuzzy numbers, in *Concepts of Soft Computing*, Springer, Singapore, pp. 53–69.

Chambers, E. (1728) *Cyclopaedia; Or, An Universal Dictionary of Arts and Sciences*, Volume the First, James and John Knapton *et al.*, London.

Chapin, E. W. (1974) Set-valued set theory, I, *Notre Dame J. Formal Logic*, V. 15, pp. 619–634.

Chapin, E. W. (1975) Set-valued set theory, II, *Notre Dame J. Formal Logic*, V. 16, pp. 255–267.

Chapman, C. H. (1892) Weierstrass and Dedekind on general complex numbers, *Bull. New York Math. Soc.*, V. 1, No. 7, pp. 150–156.

Chauvenet, R. (2012) *Chemical Arithmetic and Calculation of Furnace Charges*, Forgotten Books.

Chemla, K. C. (2015) East Asian mathematics, *Encyclopedia Britannica*, https://www.britannica.com/science/East-Asian-mathematics.

Chen, J. R. (1973) On the representation of a larger even integer as the sum of a prime and the product of at most two primes, *Sci. Sinica*, V. 16, pp. 157–176.

Chen, S.-M. (1994) Fuzzy system reliability analysis using fuzzy number arithmetic operations, *Fuzzy Sets Syst.*, V. 64, No. 1, pp. 31–38.

Chen, Y. and Skien, S. (2016) The books of numbers: Quantifying historical trends in numeracy, *Math. Intelligencer*, V. 38, No. 1, pp. 67–73.

Cheng, Y. (2019) *Incompleteness for Higher-Order Arithmetic: An Example Based on Harrington's Principle*, Springer, New York.

Cheng, Y. and Schindler, F. (2019) Harrington's principle in higher order arithmetic, *Bulletin of Symbolic Logic*, v. 80, No. 2, pp. 477–489.

Chester, R. (1915) *Algebra of al-Khowarizmi*, Macmillan, London.

Cheyne, C. (1997) Getting in touch with numbers: intuition and mathematical platonism, *Philo. and Phenom. Res.*, V. 57, No. 1, pp. 111–125.

Chihara, C. (1982) A Gödelian thesis regarding mathematical objects: Do they exist? And can we perceive them? *Philos. Rev.*, V. 91, pp. 211–227.

Chin, G. and Culotta, E. (2014) What the numbers tell us, *Science*, V. 344, No. 6186, pp. 818–821.

Chindea, G. (2007) Le nombre est-il une réalité parfaitement intelligible? Une analyse de l'intelligibilité du nombre chez Plotin, *Chôra*, V. 5, pp. 97–109.

Chrisman, N. R. (1998) Rethinking levels of measurement for cartography, *Cartography Geographic Inform. Sci.*, V. 25, No. 4, pp. 231–242.

Chrisomalis, S. (2010) *Numerical Notation: A Comparative History*, Cambridge University Press. Cambridge.

Christie, J. A. (1865) *The Constructive Arithmetic*, Virtue Brothers and Co., London.

Christov, C. I. and Todorov, T. D. (1974) Asymptotic numbers: algebraic operations with them, *Serdica, Bulgaricae Math. Publ.*, V. 2, pp. 87–102.

Chudakov, N. G. (1937) On the Goldbach problem, *Dok. Akad. Nauk USSR*, V. 17, pp. 335–338.

Chung, W. and Hassanabadi, H. (2019) Deformed classical mechanics with $\alpha-$deformed translation symmetry and anomalous diffusion, *Mod. Phys. Lett. B*, V. 33, p. 950368.

Chung, W. S. and Hassanabadi, H. (2020) Possible non-additive entropy based on the α-deformed addition, *European Phys. J. Plus*, V. 135, No. 1, DOI:10.1140/epjp/s13360-019-00047-6.

Chung, W. S. and Hounkonnou, M. N. (2020) *Deformed Special Relativity Based on α-Deformed Binary Operations*, preprint in Physics [physics.gen-ph], arXiv:2005.11155.

Chung, W. S., Hassanabadi, H. and Lütfüoğlu, B. C. (2021) α-Boson gas model based on α-additive entropy, *J. Statis. Mech. Theory Exp.*, No. 5, 053101 DOI: 10.1088/1742-5468/abf5d6.

Chunikhin, A. (1997) Fuzzy sets and multisets: Unity in diversity, in *Methodological and Theoretical Problems of Mathematics and Information and Computer Sciences*, Kiev, pp. 9–17.

Chunikhin, A. (2012) *Polymultisets, Multisuccessors, and Multidimensional Peano Arithmetics*, preprint in Mathematics (electronic edition: arXiv:1201.1820).

Chunikhin, A. *Introduction to Multidimensional Number Systems. Theoretcal Foundations and Applications*, LAP LAMBERT Academic Publishing, 2012a, 140 p. (in Russian).

Chunikhin A. (2018) Multidimensional numbers and semantic numeration systems: Theoretical foundation and application, *JITA – J. Inform. Technol. Appl., PanEuropien University APEIRON*, V. 8, No. 2, p. 49–53.

Church, A. (1932) A set of postulates for the foundation of logic, *Ann. of Math.*, (2), V. 33, No. 2, pp. 346–366.

Church, A. (1936) An unsolvable problem of elementary number theory, *Amer. J. Math.*, V. 58, pp. 345–363.

Church, A. (1956) *Introduction to Mathematical Logic*, Princeton University Press, Princeton.

Church, R. M. and Meck, W. H. (1984) The numerical attribute of stimuli, in *Animal Cognition*, Erlbaum, Hillsdale, pp. 445–464.

Chwistek, L. (1935) *The Limits of Science: Outline of Logic and Methodology of Science*, Ksiaznica-Atlas, Lwów-Warszawa.

Clark, P. (1995) *Hellman Geoffrey. Mathematics Without Numbers. Towards a Modal-structural Interpretation.* Clarendon Press, Oxford University Press, Oxford and New York 1989, xi + 154 pp [Book Review], *J. Symbolic Logic*, V. 60, No. 4, pp. 1310–1312.

Clarke-Doane, J. (2008) Multiple reductions revisited, *Philos. Math.*, V. 16, No. 2, pp. 244–255.

Clawson, C. C. (1994) *The Mathematical Traveler: Exploring the Grand History of Numbers* (Language of Science), Plenum Press, New York.

Cleary, J. J. (1998) Mathematics as Paideia in Proclus, *The Paideia Archive: Twentieth World Congress of Philosophy*, V. 3, pp. 79–84.

Clement, L. (1908) *The Ancient Science of Numbers: The Practical Application of Its Principles in the Attainment of Health, Success, and Happiness*, Roger Brothers, New York.

Cleveland, A. (2008) Circadian Math: One plus one doesn't always equal two, *RPI News*, June 6, 2008, https://news.rpi.edu/luwakkey/2456.

Clifford, W. K. (1873) Preliminary sketch of bi-quaternions, *Proc. London Math. Soc.*, V. 4, pp. 381–395.

Clifford, A. H. (1954) Naturally totally ordered commutative semigroups, *Amer. J. Math.*, V. 76, pp. 631–646.

Close, F. (2011) *The Infinity Puzzle: Quantum Field Theory and the Hunt for an Orderly Universe*, Basic Books, New York.

Cloud, M. J., Moore, R. E. and Kearfott, R. B. (2009) *Introduction to Interval Analysis*, Society for Industrial and Applied Mathematics (SIAM), Philadelphia.

Cocker, E. (1678) *Cocker's Arithmetick*, John Hawkins, London.

Cockle, J. (1848) On certain functions resembling quaternions and on a new imaginary in algebra, *London–Dublin–Edinburgh Philos. Magazine*, V. 33, pp. 435–439.

Cockle, J. (1849) On a new imaginary in algebra, *London–Dublin–Edinburgh Philos. Magazine* (3), V. 34, pp. 37–47.

Cockle, J. (1849a) On the symbols of algebra and on the theory of tessarines, *London–Dublin–Edinburgh Philos. Magazine* (3), V. 34, pp. 406–410.

Cohen, P. J. (1963) The independence of the continuum hypothesis I, *Proc. Natl. Acad. Sci. USA*, V. 50, pp. 1143–1148.

Cohen, L. B. (2011) Making $1 + 1 = 3$: improving sedation through drug synergy, *Gastrointest. Endosc.*, V. 73, No. 2, pp. 215–217.

Cohen, L. W. and Ehrlich, G. (1977) *The Structure of the Real Number System*, Robert E. Krieger P. C., Huntington, New York.

Cohen L. W. and Goffman C. (1949) Theory of transfinite Convergence, *Tran. Amer. Math. Soc.* V. 66, pp. 65–74.

Cohen, G., Gaubert, S. and Quadrat, J. P. (1999) Max-plus algebra and system theory: Where we are and where to go now, *Ann. Rev. Control*, V. 23, pp. 207–219.

Cohen Kadosh, R. and Dowker, A. (Eds.). (2015) *Oxford Library of Psychology, The Oxford Handbook of Numerical Cognition*, Oxford University Press, New York.

Cohen Kadosh, R. and Walsh, V. (2009) Numerical representation in the parietal lobes: abstract or not abstract? *Behav. Brain Sci.*, V. 32 (discussion 328–373), pp. 313–328.

Cohen Kadosh, R., Cohen Kadosh, K., Kaas, A., Henik, A. and Goebel, R. (2007) Notation-dependent and -independent representations of numbers in the parietal lobes, *Neuron*, V. 53, pp. 307–314.

Cohn, P. M. (1965) *Universal Algebra*, Harper & Row, New York/Evanston/London.

Coleman, P. and Pietronero, L. (1992) The fractal structure of the universe, *Phys. Rep.*, V. 213, 311.

Collins, J. C. (1984) *Renormalization*, Cambridge University Press, Cambridge.

Collins, A. W. (1998) On the question 'do numbers exist?' *Philos. Quart.*, V. 48, No. 190, pp. 23–36.

Colombeau, J.-F. (1984) *New Generalized Functions and Multiplication of Distributions*, North-Holland, Amsterdam.

Colombeau, J.-F. (1985) *Elementary Introduction to New Generalized Functions*, North-Holland, Amsterdam.

Colvin, F. H. and Cheney, W. L. (1904) *Engineers Arithmetic: A Pocket Book Containing the Foundation Principles Involved in Making Such Calculations As Comes [!] Into the Practical Work of the Stationary Engineer*, Derry-Collard Company, New York.

Comba, J. L. D. and Stolfi, J. (1993) Affine arithmetic and its applications to computer graphics", in *Proc. SIBGRAPI'93 — VI Simpósio Brasileiro de Computação Gráfica e Processamento de Imagens (Recife, BR)*, pp. 9–18.

Conant, L. L. (1896) *The Number Concept, Its Origin and Development*, Macmillan, New York.

Condry, K. F. and Spelke, E. S. (2008) The development of language and abstract concepts: The case of natural number, *J. Exp. Psychol.*, V. 137, pp. 22–28.

Constantinides, G. A., Cheung, P. Y. K. and Luk, W. (2003) Synthesis of saturation arithmetic architectures, *ACM Tran. Design Automation Electronic Syst.*, V. 8, No. 3, pp. 334–354.

Conway, J. H. (1976) *On Numbers and Games*, Academic Press, London.

Conway, J. H. (1994) The surreals and the reals, in *Real Numbers, Generalizations of the Reals, and Theories of continua*, Springer, The Netherlands, pp. 93–103.

Conway, J. H. and Guy, R. K. (1996) *The Book of Numbers*, Springer-Verlag, New York.

Cook, S. A. and Urquhart, A. (1993) Functional interpretations of feasibly constructive arithmetic, *Ann. Pure Appl. Logic*, V. 63, pp. 103–200.

Cordes, S., Gelman, R. and Gallistel, C. R. (2002) Variability signatures distinguish verbal from nonverbal counting for both large and small numbers, *Psychol. Bull. Rev.*, V. 8, pp. 698–707.

Corfield, D. (2003) *Towards a Philosophy of Real Mathematics*, Cambridge University Press, Cambridge.

Cornell, G. and Silverman, J. H. (Eds.). (1986) *Arithmetic Geometry*, Springer-Verlag, New York.

Cornelis, C., Deschrijver, C. and Kerre, E. E. (2004) Implication in intuitionistic and interval-valued fuzzy set theory: Construction, classification, application, *Int. J. Approxi. Reasoning*, V. 35, pp. 55–95.

Corry, L. (1992) Nicolas Bourbaki and the concept of mathematical structure, *Synthese*, V. 92, pp. 315–348.

Corry, L. (1996) *Modern Algebra and the Rise Mathematical Structures*, Birkhäuser, Basel.

Corry, L. (2015) *A Brief History of Numbers*, Oxford University Press.

Courant, R. and Robbins, H. (1960) *What is Mathematics*, Oxford University Press, London.

Covey, S. R. (2004) *The 7 Habits of Highly Effective People Personal Workbook*, Touchstone.

Crandall, R. and Pomerance, C. (2001) *Prime Numbers: A Computational Perspective*, Springer, New York.

Crippa, D. (2017) Descartes on the Unification of Arithmetic, Algebra and Geometry Via the Theory of Proportions, *Ciências Formais e Filosofia: Lógica Mat.*, V. 73, No. 3–4, pp. 1239–1258.

Crocker, R. L. (1964) Pythagorean mathematics and music, *J. Aesthetics Art Criticism*, V. 22, No. 3, pp. 325–335.

Crollen, V., Grade, S., Pesenti, M. and Dormal, V. (2013) A common metric magnitude system for the perception and production of numerosity, length, and duration, *Front. Psychol.*, V. 4, Article 449.

Crosby, A. W. (1997) *The Measure of Reality: Quantification and Western Society, 1250–1600*, Cambridge University Press, Cambridge.

Crossley, J. N. (1987) *The Emergence of Number*, World Scientific, Singapore.

Crossley J. N. (2007) *Growing Ideas of Number*, Australian Council for Educational Research, Camberwell.

Crump, T. (1990) *The Anthropology of Numbers*, Cambridge University Press, Cambridge.

Cummins, D. D. (1991) Childrens's interpretations of arithmetic word problems, *Cogn. Instr.*, V. 8, No. 3, pp. 261–289.

Cuninghame-Green, R. A. (1979) *Minimax Algebra*, Springer Lecture Notes in Economics and Mathematical Systems, V. 166, Springer, Berlin.

Cuninghame-Green, R. A. (1995) *Minimax Algebra and Applications*, Advances in Imaging and Electron Physics, V. 90, Academic Press, New York, pp. 1–121.

Cunnington, S. (1904) *The Story of Arithmetic, A Short History of Its Origin and Development*, Swan Sonnenschein, London.

Curry, H. B. (1941) A formalization of recursive arithmetic, *Amer. J. Math*, V. 63, pp. 263–283.

Cutland, N. (1988) *Nonstandard Analysis and its Applications*, London Mathematical Society, London.

Czachor, M. (2016) Relativity of arithmetic as a fundamental symmetry of physics, *Quantum Stud.: Math. Found.*, V. 3, pp. 123–133.

Czachor, M. (2016a) *Dark Energy as a Manifestation of Nontrivial Arithmetic*, preprint in Physics, [math-ph gr-qc quant-ph], arXiv: 1604.05738.

Czachor, M. (2017) If gravity is geometry, is dark energy just arithmetic? *Int. J. Theoret. Phy.*, V. 56, pp. 1364–1381.

Czachor, M. (2017a) Information Processing and Fechner's Problem as a Choice of Arithmetic, in *Information Studies and the Quest for Transdisciplinarity: Unity through Diversity*, World Scientific, New York, pp. 363–372.

Czachor, M. (2019) Waves along fractal coastlines: From fractal arithmetic to wave equations, *Acta Phys. Polonica* B, V. 50, No. 4, pp. 813–831 (also: Preprint in Mathematics, [math.DS], arXiv:1707.06225).

Czachor, M. (2020) Non-Newtonian mathematics instead of non-Newtonian physics: Dark matter and dark energy from a mismatch of arithmetics, *Found. Sci.*, V. 26, pp. 75–95 (also: Preprint in Physics, [physics.gen-ph], arXiv:1911.10903).

Czachor, M. (2020a) A loophole of all 'loophole-free' Bell-type theorems, *Found. Sci.*, V. 25, pp. 971–985. DOI: 10.1007/s10699-020-09666-0; (also: Preprint in Quantum Physics, [quant-ph], arXiv:1710.06126v2).

Czachor, M. (2020c) Unifying aspects of generalized calculus, *Entropy*, V. 22, No. 10, 1180.

Czachor, M. (2021) Arithmetic loophole in Bell's theorem: An overlooked threat for entangled-state quantum cryptography, *Acta Phys. Polon. A*, V. 139, pp. 70–83 (also Preprint in Physics, [physics.gen-ph], arXiv:2004.04097).

Czachor, M. and Nalikowski, K. (2021) *Faking Quantum Probabilites: Beyond Bell's Theorem and Tsirelson Bounds*, Preprint in Quantum Physics, arXiv:2105.12728 [quant-ph].

Czachor, M. and Naudts, J. (2002) Thermostatistics based on Kolmogorov–Nagumo averages: Unifying framework for extensive and nonextensive generalizations. *Phys. Lett. A*, V. 298, pp. 369–374.

Czachor, M. and Posiewnik, A. (2016) Wavepacket of the Universe and its spreading, *Int. J. Theor. Phys.*, V. 55, pp. 2001–2011.

Dacke, M. and Srinivasan, M. V. (2008) Evidence for counting in insects, *Anim. Cogn.*, V. 11, pp. 683–689.

Dahan-Dalmedico, A. and Peiffer, J. (2010) *History of Mathematics: Highways and byways*, Spectrum Series, V. 66, Mathematical Association of America.

Dales, H. G. and Woodin, W. H. (1996) *Super-real Fields*, London Mathematical Society Monographs, New Series, V. 14, Oxford University Press, Oxford.

Danielou, A. (1943) *Introduction to the Study of Musical Scales*, MuThe India Society, London.

Dantzig, T. (1930/2007) *Number: The Language of Science*, Macmillan, New York.

Darrigol, O. (2003) Number and measure: Hermann von Helmholtz at the crossroads of mathematics, physics, and psychology, *Studies History Philos. Sci. Part A*, V. 34, No. 3, pp. 515–573.

Datta, B. and Singh, A. N. (1935) *History of Hindu Mathematics: A Source Book, Pt I, Numeral Notation and Arithmetic*, Mortilal Banarsi Das, Lahore.

Dauben, J. W. (1979) *Georg Cantor: His Mathematics and Philosophy of the Infinite*, Harvard University Press, Cambridge, MA.

Dauben, J. (1980) The development of Cantorian set theory, in *From the calculus to set theory, 1630–1910*, Princeton University Press, Princeton, pp. 181–219.

Davenport, C. M. (1978) *An Extension of the Complex Calculus to Four Real Dimensions, with an Application to Special Relativity*, University of Tennessee, Knoxville, TN.

Davenport, H. (1992) *The Higher Arithmetic: An Introduction to the Theory of Numbers*, Cambridge University Press, Cambridge.

Davies, E. B. (2003) *Science in the Looking Glass: What Do Scientists Really Know?* Oxford University Press, New York.

Davis, P. J. (1964) Number, *Sci. American*, V. 323, No. 5, pp. 51–59.

Davis, P. J. (1972) Fidelity in mathematical discourse: Is one and one really two? *Amer. Math. Monthly*, V. 79, No. 3, pp. 252–263.

Davis, P. J. (2000) Four Thousand — or possibly thirty-seven thousand — years of mathematics, *SIAM News*, V. 33, No. 9, p. 6.

Davis, P. J. (2003) Is Mathematics a unified whole, *SIAM News*, V. 36, No. 3, p. 6.

Davis, P. J. and Hersh, R. (1986) *The Mathematical Experience*, Houghton Mifflin Co., Boston, MA.

Davis, P. J. and Hersh, R. (1987) *Descartes' Dream: The World According to Mathematics*, Houghton Mifflin Co., Boston, MA.

Davis, H. and Perusse, R. (1988). Numerical competence in animals: Definitional issues, current evidence, and a new research agenda. *Behav. Brain Sci.*, V. 11, No. 4, pp. 561–579.

Dawood, H. (2011). *Theories of Interval Arithmetic: Mathematical Foundations and Applications*, LAP LAMBERT Academic Publishing, Saarbrücken.

d'Errico, F., Backwell, L., Villa, P., Degano, I., Lucejko, J. J., Bamford, M. K., Higham, T. F. G., Colombini, M. P. and Beaumont, P. B. (2012) Early evidence of San material culture represented by organic artifacts from Border Cave, South Africa, *Proc. Natl. Acad. Sci. USA*, V. 109, No. 33, pp. 13214–13219.

De Cruz, H. (2008) An extended mind perspective on natural number representation, *Philosop. Psychol.*, V. 21, pp. 475–490.

de Finetti, B. (1931) Sulconcetto di media. *Giornale dell' Instituto, Italiano degli Attuarii*, V. 2, pp. 369–396.

De Morgan, A. (1930) *The Elements of Arithmetic*, Taylor, London.

De Morgan, A. (1837) *Elements of Trigonometry*, Taylor & Walton, London.

De Morgan, A. (1837a) *Elements of Algebra*, Taylor & Walton, London.

De Morgan, A. (1847) *Arithmetical Books from the Invention of Printing to the Present Time: Being Brief Notices of a Large Number of Works Drawn Up from Actual Inspection*, Taylor Walton, London.

De Morgan, A. (1872) *A Budget of Paradoxes*, Longmans, Green, London.

De Villiers, M. (1923) *The Numeral-Words, Their Origin, Meaning, History and Lesson*, Witherby, London.

Deco, G. and Rolls, E. T. (2006) Decision-making and Weber's law: A neurophysiological model, *Eur. J. Neurosci.*, V. 24, pp. 901–916.

Decock, L. (2008) The conceptual basis of numerical abilities: One-to-one correspondence versus the successor relation, *Philos. Psychol.*, V. 21, pp. 459–473.

Dedekind, R. (1872) *Stetigkeit und Irrational Zahlen*, Friedr. Viemeg & Sohn, Braunschweig.

Dedekind, R. (1885) Zur Theorie der aus n Haupteinheitengebildeten komplexen Grössen, *Nachrichten Königl. Ges. der Wiss. zu Göttingen*, V. 10, pp. 141–159.

Dedekind, R. (1887) Zur Theorie der aus n Haupteinheitengebildeten komplexen Grössen, *Nachrichten Königl. Ges. der Wiss. zu Göttingen*, V. 12, pp. 1–7.

Dedekind, R. (1888) *Was Sind und was Sollen die Zahlen?* Vieweg, Braunschweig.

Dedekind, R. (1890) Letter to Keferstein, in *From Frege to Gödel: A Source Book in Mathematical Logic*, 1879–1931, pp. 98–103.

Dedekind R. (1963) *Essays on the Theory of Numbers. I: Continuity and Irrational Numbers. II: The Nature and Meaning of Numbers*, Dover Publications, New York.

Dehaene, S. (1992) Varieties of numerical abilities, *Cognition*, V. 44, No. 1–2, pp. 1–42.

Dehaene, S. (1997) *The Number Sense: How the Mind Creates Mathematics*, Oxford University Press, New York.

Dehaene, S. (1997a) *What Are Numbers, Really? A Cerebral Basis for Number Sense*, Edge, http://www.edge.org/3rd_culture/dehaene/index.html.

Dehaene, S. (2001) Precis of the number sense, *Mind Lang.*, V. 16, pp. 16–36.

Dehaene, S. (2002) Single-neuron arithmetic, *Science*, V. 297, pp. 1652–1653.

Dehaene, S. (2003) The neural basis of the Weber–Fechner law: A logarithmic mental number line, *Trends Cognitive Sciences*, V. 7, pp. 145–147.

Dehaene, S. (2009) Origins of mathematical intuitions: The case of arithmetic, *Ann. N. Y. Acad. Sci.*, V. 1156, pp. 232–259.

Dehaene, S. (2020) *How We Learn: Why Brains Learn Better Than Any Machine ...for Now*, Viking, New York.

Dehaene, S. and Brannon, E. (Eds.), (2011) *Space, Time and Number in the Brain: Searching for the Foundations of Mathematical Thought*, Attention and Performance Series, Academic Press, Amsterdam.

Dehaene, S. and Changeux, J. P. (1993) Development of elementary numerical abilities: A neuronal model, *J. Cogn. Neurosci.*, V. 5, pp. 390–407.

Dehaene, S. and Cohen, L. (1994) Dissociable mechanisms of subitizing and counting: Neuropsychological evidence from simultanagnosic patients, *J. Exp. Psychol. Human Perception Perform.*, V. 20, pp. 958–975.

Dehaene, S. and Cohen, L. (1995) Towards an anatomical and functional model of number processing, *Math. Cognition*, V. 1, pp. 83–120.

Dehaene, S., Dehaene-Lambertz, G. and Cohen, L. (1998) Abstract representations of numbers in the animal and human brain, *Trends. Neurosci.*, V. 21, pp. 355–361.

Dehaene, S., Izard, V., Spelke, E. and Pica, P. (2008) Log or linear? Distinct intuitions of the number scale in Western and Amazonian indigene cultures, *Science*, V. 320, pp. 1217–1220.

Dehaene, S., Molko, N., Cohen, L. and Wilson, A. J. (2004) Arithmetic and the brain, *Current Opinion Neurobio.*, V. 14, pp. 218–224.

Dehaene, S., Piazza, M., Pinel, P. and Cohen, L. (2003) Three parietal circuits for number processing, *Cognitive Neuropsycho.*, V. 20, pp. 487–506.

Dejnožka, J. (2007) Are the natural numbers just any progression? Peano, Russell, and Quine, *Rev. Mod. Logic*, V. 10, No. 3 & 4, pp. 91–111.

Delgado, M., Vila, M. A. and Voxman, W. (1998) On a canonical representation of fuzzy numbers, *Fuzzy Sets Syst.*, V. 93, 125–135.

Dembart, L. (1982) Book Review: Should We Count on Mathematics? *The Los Angeles Times*, Tuesday, November 11, p. 92.

Demopoulos, W. (1998) The philosophical basis of our knowledge of number, *Noûs*, V. 32, pp. 481–503.

Demopoulos, W. (2000) On the origin and status of our conception of number, *Notre Dame J. Formal Logic*, V. 41, No. 3, pp. 210–226.

Deninger, C. (2002) A *Note on Arithmetic Topology and Dynamical Systems*, Preprint in Number Theory (arXiv:math/0204274[math.NT]).

Depman, I.Ya. (1959) *History of Arithmetic*, Uchpedgis, Moscow, (in Russian).

Derboven, J. (2011) One plus one equals three: Eye-tracking and semiotics as complementary methods in HCI, in *CCID2: The Second International Symposium on Culture, Creativity, and Interaction Design*, Newcastle, UK. https://ccid2.files.wordpress.com/2011/02/one-plus-one-equals-three_derboven.pdf.

d'Errico, F., Doyon, L., Colagé, I., Queffelec, A., Le Vraux, E., Giacobini, G., Vandermeersch, B. and Maureille, B. (2018) From number sense to number symbols. An archaeological perspective, *Philos. Trans. R. Soc. B*, doi:3732016051820160518.

Deringer, W. (2018) *Calculated Values: Finance, Politics, and the Quantitative Age*, Harvard University Press, London/Cambridge, MA.

Descartes, R. (1637/2006) *A Discourse on the Method of Correctly Conducting One's Reason and Seeking Truth in the Sciences (Transl. I. Maclean)* Oxford University Press, New York.

Descartes, R. (1641) Meditationes de prima philosophia, in qua Dei existential et animae immortalitas demonstrantur, Michel Soly, Paris.

Detlefsen, M. (1980) The arithmetization of metamathematics in a philosophical setting, *Rev. Int. Philos.*, V. 34, No. 131/132, pp. 268–292.

Devito, J. A. (1994) *Human Communication: The Basic Course*, Harper Collins, New York.

Devlin, H. (2017-09-13) Much ado about nothing: ancient Indian text contains earliest zero symbol, *The Guardian*.

Di Giorgio, N. (2010) *Non-standard Models of Arithmetic: A Philosophical and Historical perspective*, MSc Thesis, Universiteit van Amsterdam.

Di Nasso, M. (2010) Fine asymptotic densities for sets of natural numbers, *Proc. Amer. Math. Soc.*, V. 138, pp. 2657–2665.

Di Nasso, M. (1996) Hyperordinals and nonstandard alpha-models, in *Logic and Algebra*, A. Ursini and P. Agliano, (Eds.), Lecture Notes in Pure and Applied Mathematics, V. 180, Marcel Dekker, New York, pp. 457–475.

Di Nasso, M. and Forti, M. (2005) Ultrafilter semirings and nonstandard submodels of the Stone-Cech compactification of the natural numbers, in *Logic and its Applications*, A. Blass and Y. Zhang (Eds.), AMS Contemporary Mathematics, V. 380, pp. 45–51.

Di Nasso, M. and Forti, M. (2010) Numerosities of point sets over the real line, *Trans. Amer. Math. Soc.*, V. 362, pp. 5355–5371.

Dickson, L. E. (1924) Algebras and their arithmetics, *Bull. Amer. Math. Soc.*, V. 30, No. 5-6, pp. 247–257.

Dickson, L. E. (1919/2005) *History of the Theory of Numbers, v. I: Divisibility and primality*, Dover Publications, New York.

Dickson, L. E. (1932) *History of the Theory of Numbers*, Carnegie Institute of Washington, Washington.

Diderot, D. and d'Alembert, J. R. (Eds.). (1993) Encyclopédie ou dictionnaire raisonné des sciences, des arts et des métiers, Editions Flammarion.

Dieudonné, J. (1979) The Difficult Birth of Mathematical Structures (1840–1940), in V. Mathieu and P. Rossi (Eds.), *Scientific Culture in the Contemporary World*, Scientia, Milano, pp. 7–23.

Diester, I. and Nieder, A. (2007) Semantic associations between signs and numerical categories in the prefrontal cortex, *PLoS Bio.*, V. 5, No. 11, e294.

Diewert, W. E. (1987) Index numbers, in *The New Palgrave: A Dictionary of Economics*, V. 2, Palgrave, pp. 767–780.

Diez, J. A. (1997) A hundred years of numbers. An historical introduction to measurement theory 1887–1990, Pt. 1, *Studies History Philos. Sci.*, V. 28, No. 1, pp. 167–185.

Diez, J. A. (1997) A hundred years of numbers. An historical introduction to measurement theory 1887–1990, Pt. 2, *Studies History Philos. Sci.*, V. 28, No. 2, pp. 237–265.

Dijkman, J. G., van Haeringen, H. and De Lange, S. J. (1983) Fuzzy numbers, *J. Math. Anal. Appl.*, V. 92, pp. 301–341.

Dijksterhuis, E. J. (1987) *Archimedes, With a New Bibliographic Essay by Wilbur R. Knorr*, Princeton University Press, New Jersey.

Dijkstra, E. W. (1982) *Why Numbering Should Start at Zero*, https://www.cs.utexas.edu/users/EWD/transcriptions/EWD08xx/EWD831.html.

Dillon, J. (1987) Iamblichus of Chalcis (c. 240–325 A.D.), *Aufstief und Niedergang der Römischen Welt*. Part 2, *Principat*, V. 36, No. 2, pp. 862–909.

Dilworth, T. (1743) *The Schoolmaster's Assistant, Being a Compendium of Arithmetic both Practical and Theoretical*, London.

Diophantus, (1974) *Arithmetic*, (translated by Veselovskii, I. N.), Nauka, Moscow (in Russian).

Dirac, P. A. M. (1978) *Directions in Physics*, John Wiley & Sons, New York.

Dixon, G. M. (1994) *Division Algebras: Octonions, Quaternions, Complex Numbers and the Algebraic Design of Physics*, Kluwer Academic Publishers, Dordrecht.

Docampo Rey, J. (2006) Reading Luca Pacioli's Summa in Catalonia: An early 16th-century Catalan manuscript on algebra and arithmetic, *Historia Mathematica*, V. 33, No. 1, pp. 43–62.

Dold-Samplonius, Y. (1992) The 15th century Timurid mathematician Ghiyath al-Din Jamshid al-Kashi and his computation of the Qubba, in *Amphora: Festschrift for Hans Wussing on the Occasion of His 65th Birthday*, pp. 171–181.

Dollard, J. D. and Friedman, C. N. (1979) *Product Integration, with Applications to Differential Equations*, Addison-Wesley.

Domański, Z. and Błaszak, M. (2017) *Deformation Quantization with Minimal Length*, preprint in Mathematical Physics [math-ph] (arXiv:1706.00980).

Donaldson, T. M. E. (2014) If there were no numbers, what would you think? *Philos.*, V. 3, No. 4, pp. 283–287.

Drake, F. R. (1974) *Set Theory: An Introduction to Large Cardinals*, Studies in Logic and the Foundations of Mathematics, V. 76, Elsevier Science Ltd., Amsterdam.

Drozdyuk, A. and Drozdyuk, D. (2010) *Fibonacci, His Numbers and His Rabbits*, Choven Pub., Toronto.

Drummond, J. (1985) Frege and Husserl: Another look at the issue of influence. *Husserl Studies*, V. 2, pp. 245–265.

Du Bois-Reymond, P. (1870/1871) Sur la grandeur relative des infinis des fonctions, *Ann. Mat. Pura Appli.*, V. 4, pp. 338–353.

Du Bois-Reymond, P. (1875) Über asymptotische Werthe, infinitäre Approximationen und infinitäre Auflösung von Gleichungen, *Math. Ann.*, V. 8, pp. 363–414.

Du Bois-Reymond, P. (1877) Über die Paradoxen des Infinitärcalcüls, *Math. Ann.*, V. 11, pp. 149–167.

Du Bois-Reymond, P. (1882) *Die allgemeine Functionentheorie I: Metaphysik und Theorie der* mathematischen Grundbegriffe: Grösse, Grenze, Argument und Function, Verlag der H. Laupp'schen Buchhandlung, Tübingen.

Duff, M. J. (2004) *Comment on Time-Variation of Fundamental Constants*, preprint in high energy physics, (arXiv: hep-th/0208093v3).

Dukkipati, A. (2010) On Kolmogorov–Nagumo averages and nonextensive entropy, in *Int. Sympo. Information Theory & Its Applications*, Taichung, pp. 446–451.

Dukkipati, A., Murty, M. N. and Bhatnagar, S. (2005) *Nongeneralizability of Tsallis Entropy by Means of Kolmogorov–Nagumo Averages Under Pseudo-additivity*, preprint in Mathematical Physics (math-ph) arXiv: math-ph/0505078.

Dummett, M. (1975) Wang's paradox, *Synthese*, V. 30, No. 3/4, pp. 301–324.

Dummett, M. (1991) *Frege: Philosophy of Mathematics*, Harvard University, Cambridge, MA.

Dunlop, K. (2016) Poincaré on the foundations of arithmetic and geometry. Part 1: Against "dependence-hierarchy" interpretations, *HOPOS: History Philos. Sci.*, V. 6, No. 2, pp. 274–308.

Dunlop, K. (2017) Poincaré on the foundations of arithmetic and geometry. Part 2: intuition and unity in mathematics, *HOPOS: The J. Int. Soc. History Philoso. Sci.*, V. 7, No. 1, pp. 88–107.

Dunnington, G. W. (2004) *Carl Friedrich Gauss: Titan of Science*, Mathematical Association of America, Washington, DC.

Durkheim, E. (1984) *The Division of Labor in Society*, The Free Press, New York.

Dutta, P., Boruah, H. and Ali, T. (2011) Fuzzy arithmetic with and without using α-cut method: A comparative study, *Int. J. Latest Trends Comput.*, V. 2, No. 1, pp. 99–107.

Duyar, C., Sagır, B. and Ogur, O. (2015) Some basic topological properties on non-Newtonian real line, *British J. Math. Comput. Sci.*, V. 9, No. 4, pp. 300–307.

Dwyer, P. S. (1951) *Linear Computations*, Wiley, Oxford.

Dyson, F. J. (2004) *Infinite in All Directions*, Harper Perennial, New York.

Dyson, G. (2012) *Turing's Cathedral: The Origins of the Digital Universe*, Pantheon Books, New York.

Dzhafarov, E. N. and Colonius, H. (2011) The Fechnerian idea, *Amer. J. Psychol.*, V. 124, pp. 127–140.

Edwards, H. M. (1992) Kronecker's arithmetic theory of algebraic quantities, *Jahresbericht der Deutschen Mathematiker Vereinigung*, V. 94, No. 3, pp. 130–139.

Edwards, H. M. (1995) Kronecker on the Foundations of Mathematics, in *From Dedekind to Gödel, Essays on the Development of the Foundations of Mathematics*, Boston, pp. 45–52.

Edwards, H. M. (1996) *Fermat's Last Theorem*, Springer-Verlag, New York.

Edwards, H. M. (2008) *Higher Arithmetic: An Algorithmic Introduction to Number Theory*, American Mathematical Society, Providence, RI.

Eger, E., Michel, V., Thirion, B., Amadon, A., Dehaene, S. and Kleinschmidt, A. (2009) Deciphering cortical number coding from human brain activity patterns, *Curr. Biol.*, V. 19, pp. 1608–1615.

Ehrlich, P. (1992) Universally extending arithmetic continua, in *Le Labyrinthe du Continu*, Colloque de Cerisy, Springer-Verlag France, Paris, pp. 168–178.

Ehrlich, P. (1994) All numbers great and small, in *Real Numbers, Generalizations of the Reals, and Theories of Continua*, Kluwer Academic Publishers, Dordrecht, pp. 239–258.

Ehrlich, P. (Ed) (1994a) *Real Numbers, Generalizations of the Reals, and Theories of Continua*, Kluwer Academic Publishers, Dordrecht, Holland.

Ehrlich, P. (1995) Hahn's "Über die Nichtarchimedischen Grössensysteme" and the origins of the modern theory of magnitudes and numbers to measure them, in *From Dedekind to Gödel: Essays on the Development of the Foundations of Mathematics* (J. Hintikka, Ed.), Kluwer Academic Publishers, pp. 165–213.

Ehrlich, P. (2001) Number systems with simplicity hierarchies: A generalization of Conway's theory of surreal numbers, *J. Symbol. Logic*, V. 66, pp. 1231–1258.

Ehrlich, P. (2006) The rise of non-Archimedean mathematics and the roots of a misconception. I. The emergence of non-Archimedean systems of magnitudes, *Arch. Hist. Exact Sci.*, V. 60, No. 1, pp. 1–121.

Ehrlich, P. (2011) Conway names, the simplicity hierarchy and the surreal number tree, *J. Logic Anal.*, V. 3, No. 1, pp. 1–26.

Ehrlich, P. (2012) The absolute arithmetic, continuum and the unification of all numbers great and small, *Bull. Symbol. Logic*, V. 18, pp. 1–45.

Eilenberg, S. and Mac Lane, S. (1942) Natural isomorphisms in group theory, *Proc. Natl. Acad. Sci. USA*, V. 28, pp. 537–543.

Eilenberg, S. and Mac Lane, S. (1945) General theory of natural equivalence, *Trans. Amer. Math. Soc.*, V. 58, pp. 231–294.

Einstein, A. (1905) Zur elektrodynamik bewegter körper, *Ann. der Phy.*, V. 17, pp. 891–921.

Eisenstein, G. (1847) *Mathematische Abhandlungen. Besonders aus dem Gebiete der höheren Arithmetik und der elliptischen Funktionen*, Reimer, Berlin.

Eisler, H. (1963) Magnitude scales, category scales, and Fechnerian integration, *Psychol. Rev.*, V. 70, pp. 243–253.

Ekman, G. (1964) Is the power law a special case of Fechner's law? *Perceptual Motor Skills*, V. 19, pp. 730.

Ekman, G. and Hosman, B. (1965) Note on subjective scales of number, *Perceptual Motor Skills*, V. 21, pp. 101–102.

El Naschie, M. S. (2008) Quantum gravity unification via transfinite arithmetic and geometrical averaging, *Chaos, Solitons Fractals*, V. 35, No. 2, pp. 252–256.

Elden, S. (2005) *Speaking Against Number: Heidegger, Language and the Politics of Calculation* (Taking on the Political EUP), Edinburgh University Press, Edinburgh.

Elekes, G. (1997) On the number of sums and products, *Acta Arith.*, V. 81, pp. 365–367.

Elliott, P. D. T. A. (1985) Information and Arithmetic, in *Arithmetic Functions and Integer Products*, Grundlehren der mathematischen Wissenschaften (A Series of Comprehensive Studies in Mathematics), V. 272, Springer, New York, NY, pp. 343–355.

Elsas, A. (1894). Husserls philosophie der arithmetik, *Philos. Monatshefte*, V. 30, pp. 437–440.

Endler, O. (1972) *Valuation Theory*, Springer, New York.

Enge, E. (2017) *SEO and social*: $1 + 1 = 3$, SearchEngineLand, https://searchengineland.com/seo-social-1-1-3-271978.

Engelhard, K. and Mittelstaedt, P. (2008) Kant's theory of arithmetic: A constructive approach? *Zeit. Allgemeine Wissenschaftstheorie*, V. 39, No. 2, pp. 245–271.

Enriques, F. (1911) Sui numeri non archimedei e su alcune loro interpretazioni, *Boll. Math., Soc. Italiana Mat.* IIIa, pp. 87–105.

Epstein, D. and Levy, S. (1995) Experimentation and proof in mathematics. *Notices Amer. Math. Soc.* V. 42, pp. 670–674.

Erdös, P. and Szemerédi, E. (1983) On Sums and Products of Integers, *Studies in Pure Mathematics*, Birkhäuser, Basel, pp. 213–218.

Ershov, Y. L. (1972) The theory of enumerations, in *Actes du Congrès International des Mathématiciens* 1970, V. 1, Gauthier-Villars, Paris, pp. 223–227.

Ershov, Y. L. (1977) *theory of enumerations*, Monographs in Mathematical Logic and Foundations of Mathematics, Nauka, Moscow (in Russian).

Ershov, Yu. A. and Palyutin, E. A. (1986) Mathematical Logic, Mir, Moscow.

Ershov, Y. L. (1999) Theory of numberings, in *Handbook of Computability Theory*, Stud. Logic Found. Math., V. 140, North-Holland, Amsterdam, pp. 473–503.

Escobedo, J. (2010) Mercantile arithmetic and the incunable Catalan printing. Suma de la art de arismètica, by Francesc Santcliment (1482), *Contributions Sci.*, V. 6, No. 1, pp. 59–73.

Ésénine-Volpin (1961) Le programme ultra-intuitionniste des fondements des mathématiques, in *Intuionitistic Methods, Proc. Symp. Foundations of Mathematics*, PWN Warsaw, pp. 201–223.

Esenin-Volpin, A. S. (1970) The ultra-intuitionistic criticism and the antitraditional program for foundations of mathematics, in *Intuitionism and Proof Theory*, North-Holland, pp. 3–45.

Estermann, T. (1938) On Goldbach's problem: Proof that almost all even positive integers are sums of two primes, *Proc. London Math. Soc.*, V. 44, pp. 307–314, doi:10.1112/plms/s2-44.4.307.

Estrada, R. and Kanwal, R. P. (1994) *Asymptotic Analysis, A Distributional Approach*, Birkhauser, Boston.

Euclid, (1956) *The Thirteen Books of Euclid's Elements, with Introduction and Commentary by T. L. Heath*, Dover, New York.

Euler, L. (1780) De infinities infinitis gradibus tam infinite magnorum quam infinite parvorum, *Acta Acade. Sci. Imperialis Petropolitanae*, V. 2, No. 1, pp. 102–118.

Euler, L. (1748) *Introductio in Analys Ininfinitorum*, Springer, Heidelberg.

Euler, L. (1988) *Introduction to Analysis of the Infinite*, Book 1, Springer, New York.

Euler, L. (1990) *Introduction to Analysis of the infinite*, Book 2, Springer, New York.

Evans, M. G. (1955) Aristotle, Newton, and the Theory of Continuous Magnitude, *J. History of Ideas*, V. 16, No. 4, pp. 548–557.

Everett, C. (2013) Linguistic relativity and numeric cognition: New light on a prominent test case, in *Proc. 37th Annual Meeting of the Berkeley Linguistics Society*, pp. 91–103.

Everett, C. (2017) *Numbers and the Making of Us: Counting and the Course of Human Cultures*, Harvard University Press, Cambridge, MA.

Eves, H. (1988) *Return to Mathematical Circles*, Prindle, Weber and Schmidt, Boston.

Eves, H. (1990) *An Introduction to the History of Mathematics*, Saunders College Publishing, Philadelphia.

Eves, H. (1990a) *Foundations and Fundamental Concepts of Mathematics*, Dover, New York.

Faith, C. (2004) *Rings and Things and a Fine Array of Twentieth Century Associative Algebra*, Mathematical Surveys & Monographs, American Mathematical Society, Providence, RI.

Falmagne, J. C. (1971) The generalized Fechner problem and discrimination, *J. Math. Psych.*, V. 8, pp. 22–43.

Falmagne, J. C. (1985) *Elements of Psychophysical Theory*, Oxford University Press, Oxford.

Falmagne J.-C. and Doble, C. (2015) *On Meaningful Scientific Laws*, Springer, New York.

Faltin F., Metropolis N., Ross B., Rota G.-C. (1975) The real numbers as a wreath product, *Advances in Mathematics*, V. 16, pp. 278–304.

Farmelo, G. (2019) *The Universe Speaks in Numbers: How Modern Math Reveals Nature's Deepest Secrets*, Basic Books, New York.

Fayek, A. R. (Ed.) (2018) *Fuzzy Hybrid Computing in Construction Engineering and Management*, Emerald Publishing Limited.

Fechner, G. T. (1860) *Elemente der Psychophysik*, Breitkopf und Hartel, Leipzig.

Feferman, S. (1960) Arithmetization of metamathematics in a general setting, *Fund. Math.*, V. 49, pp. 35–92.

Feferman, S. (1974) *The Number Systems: Foundations of Algebra and Analysis*, Addison-Wesley, Reading, MA.

Feferman, S. and Hellman, G. (1995) Foundations of predicative arithmetic, *Philos. Logic*, V. 24, pp. 1–17.

Feferman, S. and Strahm, T. (2010) Unfolding finitist arithmetic, *Rev. of Symbolic Logic*, V. 3, No. 4, pp. 665–689.

Feigenson, L., Dehaene, S. and Spelke, E. (2004) Core systems of number, *Trends. Cogn. Sci.*, V. 8, pp. 307–314.

Feigenson, L. and Carey, S. (2005) On the limits of infants' quantification of small object arrays. *Cognition.* V. 97, No. 3, pp. 295–313.

Feigenson, L., Carey, S. and Spelke, E. (2002) Infants' discrimination of number vs. continuous extent. *Cognitive Psychol.*, V. 44, pp. 33–66.

Felka, K. (2014) Number words and reference to numbers, *Philos. Studies*, V. 168, No. 1, pp. 261–282.

Fellows, M. R., Gaspers, S. and Rosamond, F. A. *Parameterizing by the Number of Numbers*, preprint in Data Structures and Algorithms (cs.DS), 2010 (arXiv:1007.2021).

Felscher, W. (1974) *Logic and Arithmetic*, Oxford University at the Clarendon Press, London.

Felscher, W. (1978) *Naive Mengen und abstrakte Zahlen, I*, Bibliographisches Institut, Mannheim.

Felscher, W. (1978) *Naive Mengen und abstrakte Zahlen, II, Algebraische und reelle Zahlen*, Bibliographisches Institut, Mannheim.

Felscher, W. (1979) *Naive Mengen und abstrakte Zahlen, III, Transfinite Methoden*, Bibliographisches Institut, Mannheim.

Felscher, W. (2000) *Lectures on Mathematical Logic, V. 3: Logic of Arithmetic*, Gordon and Breach Science Amsterdam.

Fenstad, J. E. (2015) On what there is — infinitesimals and the nature of numbers, *Inquiry: Interdisciplinary J. Philos.*, V. 58, No. 1, pp. 57–79.

Fenster, D. D. (1998) Leonard Eugene Dickson and his work in the Arithmetics of Algebras, *Arch. History Exact Sci.*, V. 52, No. 2, pp. 119–159.

Fenster, D. D. (2007) Beyond class field theory: Helmut Hasse's arithmetic in the theory of algebras in Early 1931, *Arch. History Exact Sci.*, V. 61, No. 5, pp. 425–456.

Fernandes, D. P. (2017) Do Conceito de número e magnitude na matemática Grega Antiga, *Rev. Humanidades de Valparaíso*, V. 5, No. 9, pp. 7–23.

Ferreirós, J. (2005) Richard Dedekind (1888) and Giuseppe Peano (1889), Booklets on the foundations of arithmetic, in *Landmark Writings in Western Mathematics, 1640–1940*, (I. Grattan-Guinness, Ed.), Elsevier, Amsterdam, Chapter 47, pp. 613–626.

Ferreirós, J. (2007) The rise of pure mathematics, as arithmetic in Gauss, in *The Shaping of Arithmetic after C.F. Gauss's Disquisitiones Arithmeticae*, C. Goldstein, N. Schappacher & J. Schwermer (Eds.), Springer, Berlin, pp. 235–268.

Fesenko, I. (2003) Analysis on arithmetic schemes, I, *Docum. Math.*, Extra volume, pp. 261–284.

Fetterman, J. G. (1993) Numerosity discrimination: both time and number matter, *J. Exp. Psychol. Anim. Behav. Process.*, V. 19, pp. 149–164.

Fetters, M. D. and Freshwater, D. (2015) The $1 + 1 = 3$ integration challenge, *J. Mixed Methods Res. (JMMR)*, V. 9, No. 2, pp. 115–117.

Feynman, R. P. (1948) Space–time approach to non–relativistic quantum mechanics, *Rev. Mod. Phys.*, V. 20, pp. 367–387.

Feynman, R. P. and Hibbs, A. R. (1965) *Quantum Mechanics and Path Integrals*, McGraw-Hill Companies, New York.

Fias, W. (2001) Two routes for the processing of verbal numbers: evidence from the SNARC effect, *Psychol. Res.*, V. 65, pp. 250–259.

Fias, W., Brysbaert, M., Geypens, F. and d'Ydewalle, G. (1996) The importance of magnitude information in numerical processing: evidence from the SNARC effect, *Math. Cogn.*, V. 2, No. 1, pp. 95–110.

Fibonacci, (1202/2002) *Liber Abaci*, A book on calculations (Translation by Laurence Sigler).

Field, H. (1980) *Science without Numbers: The Defense of Nominalism*, Princeton Legacy Library, Princeton University Press, Princeton.

Field, J. V. (1997) *The Invention of Infinity: Mathematics and Art in the Renaissance*, Oxford University Press, Oxford/New York.

de Figueiredo, L. H. and Stolfi, J. (1996) Adaptive enumeration of implicit surfaces with affine arithmetic, *Comput. Graphics Forum*, V. 15, No. 5, pp. 287–296.

de Figueiredo, L. H. and Stolfi, J. (2004) Affine arithmetic: concepts and applications, *Numer. Algorithms*, V. 37, No. 1–4, pp. 147–158.

Filip, D. A. and Piatecki, C. (2014) A non-Newtonian examination of the theory of exogenous economic growth, *Math. Aetherna*, V. 4, No. 2, pp. 101–117.

Fine, H. B. (1891) *The Number System of Algebra Treated Theoretically and Historically,* Leach, Shewell & Sanborn, Boston.

Fischbein, E., Tirosh, D. and Hess, P. (1979) The intuition of infinity, *Educational Studies Math.,* V. 10, pp. 3–40.

Fisher, G. (1994) Veronese's non-Archimedean linear continuum, in *Real Numbers, Generalizations of the Reals, and Theories of continua,* Springer, The Netherlands, pp. 107–145.

Fischer, M. H. (2018) Why numbers are embodied concepts? *Front. Psychol.,* V. 8, Article 2347.

Fischer, M. J. and Rabin, M. O. (1974) Super-exponential complexity of Presburger arithmetic, in *Proc. SIAM-AMS Sympos. Applied Mathematics,* V. 7, pp. 27–41.

Fitting, M. (2015) Intensional Logic, in *The Stanford Encyclopedia of Philosophy,* E. N. Zalta, (Ed.); https://plato.stanford.edu/archives/sum2015/entries/logic-intensional/.

Flannery, S. and Flannery, D. (2000) *In Code: A Mathematical Journey,* Profile Books, London.

Flegenheimer, M. (2012) When the Calculator Says $1 + 1 = 4$, *The New York Times,* April 13, 2012.

Flegg, G. (Ed.) (2002) *Numbers: Their History and Meaning,* Dover Publications, Mineola, NY.

Fleming, W. H. (2002) Max-plus stochastic control, in *Stochastic Theory and Control,* Lecture Notes in Control and Information Science, V. 280, pp. 111–119.

Fleming, W. H. (2004) Max-plus stochastic processes, *Applied Math. Optim.,* V. 48, pp. 159–181.

Flombaum, J. I., Junge, J. A., Hauser, M. D. Rhesus monkeys (Macacamulatta) spontaneously compute addition operations over large numbers. *Cognition,* V. 97, No. 3, pp. 315–325.

Florack, L. (2012) Regularization of positive definite matrix fields based on multiplicative calculus, in *Scale Space and Variational Methods in Computer Vision,* Lecture Notes in Computer Science, V. 6667, Springer, pp. 786–796.

Florack, L. and van Assen, H. (2012) Multiplicative calculus in biomedical image analysis, *J. Math. Imaging Vis.,* vol. 42, No. 1, pp. 64–75.

Flynn, M. J. and Oberman, S. S. (2001) *Advanced Computer Arithmetic Design,* Wiley, New York.

von Foerster, H. (2003) *Understanding Understanding: Essays on Cybernetics and Cognition,* Springer, New York.

Fogel, R. W., Fogel, E. M., Guglielmo, M. and Grotte, N. (2013) *Political Arithmetic: Simon Kuznets and the Empirical Tradition in Economics,* National Bureau of Economic Research Series on Long-Term Factors in Economic Development, University of Chicago Press, Chicago.

Fomin, F. V., Gaspers, S., Pyatkin, A. V. and Razgon, I. (2008) On the minimum feedback vertex set problem: Exact and enumeration algorithms, *Algorithmica,* V. 52, No. 2, pp. 293–307.

Forrest, P. and Armstrong, D. M. (1987) The nature of number, *Philos. Studies* V. 16, pp. 165–186.

Fortin, J., Dubois, D. and Fargier, H. (2008) Gradual numbers and their application to fuzzy interval analysis, *IEEE Trans. on Fuzzy Syst.*, V. 16, No. 2, pp. 388–402.

Fraenkel, A. A. (1922) Zur den Grundlagen der Cantor-Zermeloschen Mengenlehre, *Math. Ann.*, V. 86, pp. 230–237.

Fraenkel, A. A., Bar-Hillel, Y. and Levy, A. (1973) *Foundations of Set Theory*, North-Holland Amsterdam.

Frame, A. and Meredith, P. (2008) One plus one equals three: Legal hybridity in Aotearoa/New Zealand, in *Hybrid Identities*, pp. 313–332.

Franci, R. (2002). Il Liber Abaci di Leonardo Fibonacci 1202-2002, *Boll. Unione Mat. Italiana*, V. 5, No. 2, pp. 293–328.

Frank, M., Everett, D. L., Fedorenko, E. and Gibson, E. (2008) Number as a cognitive technology: Evidence from Pirahã language and cognition, *Cognition*, V. 108, pp. 819–824.

Franklin, J. (2014) *An Aristotelian Realist Philosophy of Mathematics: Mathematics as the Science of Quantity and Structure*, Palgrave Macmillan.

Franklin, J. (2014) Quantity and number, in *Neo-Aristotelian Perspectives in Metaphysics*, Routledge, New York, pp. 221–244.

Frappat, L., Sciarrino, A. and Sorba, P. (2000) *Dictionary on Lie Algebras and Superalgebras*, Academic Press, Burlington, MA.

Frege, G. (1879) *Begriffsschrift, eine der arithmetischen nachgebildete Formalsprache des reinen Denkens*, L. Nebert, Halle.

Frege, G. (1884) *Die Grundlagen der Arithmetik: Eine logisch-mathematische* Untersuchung über den Begriff der Zahl, Verlage Wilhelm Koebner, Breslau.

Frege, G. (1892) Über Sinn und Bedutung, *Zeits. Philos. Philosophische Kritik*, V. 100, pp. 25–50.

Frege, G. (1893) *Grundgesetze der Arithmetic*, Begriffsschriftlich abgeleitet, Bd. I, Verlag Hermann Pohle, Jena.

Frege, G. (1903) *Grundgesetze der Arithmetik*, Begriffsschriftlich abgeleitet, Bd. II, Verlag Hermann Pohle, Jena.

Frege, G. (1984) *Collected Papers on Mathematics, Logic, and Philosophy*, Blackwell, Oxford.

Frege, G. (1988) Begriffsschrift, eine der arithmetischen Nachgebildete Formelsprache des reinen Denkens, in *Begriffsschrift und andere Aufsätze*, Georg Olms Verlag, Hildesheim.

Freiberg, U. (2003) A survey on measure geometric Laplacians on Cantor like sets, *Arabian J. Sci. Eng.*, V. 28, pp. 189–198.

Freiberg, U. and Seifert, C. (2015) Dirichlet forms for singular diffusion in higher dimensions, *J. Evolution Equations*, V. 15, pp. 869–878.

Friedberg, R. M. (1958) Three theorems on recursive enumeration, *J. Symbolic Logic*, V. 23, pp. 309–316.

Friedberg, R. (1968) *An Adventurer's Guide to Number Theory*, McGraw-Hill Book Company, New York.

Friedman, H. (1976) Systems of second order arithmetic with restricted induction, I, II, *J. Symbolic Logic*, V. 41, pp. 557–559.

Friedrich, B. (2019) *Multiplikative Euklidische Vektorräume als Grundlage für das Rechnen mit positiv-reellen Größen*, Logos Verlag, Berlin.

Fuchs, L. (1963) *Partially Ordered Algebraic System*, Pergamon Press, Oxford.

Fugal, D. L. and Lyons, R. G. (2014) *The Essential Guide to Digital Signal Processing*, Prentice-Hall, Englewood Cliffs, N.J.

Fuson, K. C. (1988) *Children's Counting and Concepts of Number*, Springer-Verlag, New York.

Galaugher, J. (2013) Why there is no Frege-Russell definition of number? in *The Palgrave Centenary Companion to Principia Mathematica. History of Analytic Philosophy*, Palgrave Macmillan, London, pp. 130–160.

Galileo, (1638) *Discorsi e dimostrazioni matematiche, intorno à due nuove scienze*, V. 213, Leida, Appresso gli Elsevirii (Louis Elsevier, Leiden) (Mathematical discourses and demonstrations, relating to Two New Sciences, English translation by Henry Crew and Alfonso de Salvio, 1914).

Gallace, A., Tan, H. Z. and Spence, C. (2008) Can tactile stimuli be subitised? An unresolved controversy within the literature on numerosity judgments, *Perception*, V. 37, No. 5, pp. 782–800.

Gallistel, C. R. (1988) Counting versus subitizing versus the sense of number. (Commentary on Davis and Pérusse: Animal counting), *Behav. Brain Sci.*, V. 11, pp. 585–586.

Gallistel, C. R. (1990) *The Organization of Learning*, MIT Press, Cambridge, MA.

Gallistel, C. R. and Gelman, R. (1990) The what and how of counting, *Cognition*, V. 44, pp. 43–74.

Gallistell, R. C. and Gellman, R. (1992) Preverbal and verbal counting and computation. *Cognition*, V. 44, pp. 43–74.

Gallistell, R. C. and Gellman, R. (2000) Non-verbal numerical cognition: from reals to integers, *Trends Cogn. Sci.*, V. 4, pp. 59–65.

Gallistel, C. R., and King, A. P. (2010) *Memory and the computational* Brain: Why Cognitive Science Will Transform Neuroscience, Wiley/Blackwell, New York.

Galovich, S. (1989) *Introduction to Mathematical Structures*, Harcourt Brace Jovanovich, San Diego.

Gandz, S. (1926) The origin of the Term "Algebra", *Amer. Math. Monthly*, V. 33, No. 9, pp. 437–440.

Gandz, S. (1936) The sources of al-Khowārizmī's algebra, *Osiris*, V. 1, No. 1, pp. 263–277.

Gann, K. (2019) *The Arithmetic of Listening: Tuning Theory and History for the Impractical Musician*, University of Illinois Press, Champaign, IL.

Ganor-Stern, D., Pinhas, M., Kallai, A. and Tzelgov, J. (2010) Holistic representation of negative numbers is formed when needed for the task, *Q. J. Exp. Psychol.*, V. 63, No. 10, pp. 1969–1981.

Gardner, M. (2005) Review of Science in the looking glass: what do scientists really know? By E. Brian Davies (Oxford University Press, 2003), *Notices of the American Mathematical Society*, V. 52, No. 11, http://www.ams.org/notices/200511/rev-gardner.pdf.

Garai, T., Chakraborty, D. and Roy, T. K. (2017) Expected value of exponential fuzzy number and its application to multi-item deterministic inventory model for deteriorating items, *J. Uncertain. Anal. Appl.*, V. 5, p. 8.

Garrido, A. (2012) Axiomatic of fuzzy complex numbers, *Axioms*, V. 1, pp. 21–32.

Gaskin, D. (2007) Denise Gaskins' let's play math, https://denisegaskins.com/2007/11/07/how-to-read-a-fraction/.

Gasking, D. A. T. (1940) Mathematics and the world, *Australasian J. Philoso.*, V. 18, No. 2, pp. 97–116.

Gathmann, A. and Markwig, H. (2005) The Caporaso-Harris Formula and Plane relative Gromov-Witten invariants in Tropical Geometry, Preprint in Mathematics (arXiv: math.AG/0504392).

Gaubert, S. (1997) *Methods and Applications of* (max, +) *Linear Algebra*, Tech. Report 3088, Institut National de Recherche en Informatique et en Automatique.

Gaukroger, S. (1982) The one and the many: Aristotle on the individuation of numbers, *Classical Quart.*, V. 32, No. 2, pp. 312–322.

Gauss, C. F. (1801) *Disquisitiones Arithmeticae*, Fleischer, Leipzig.

Gauss, C. F. (1808) Theorematis arithmetici demonstratio nova, *Comment. Soc. Regiae Sci. Göttingen*, XVI, 69; Werke II, pp. 1–8.

Gauss, C. F. (1831) Theoria residuorum biquadraticorum: Commentatio secunda, *Göttingische gelehrte Anzeigen*, pp. 625–638.

Gauss, C. F. (1900) *Arithmetik und Algebra*: Nachträgezu Band 1–3, v. VIII.

Gauthier, Y. (1989) Finite arithmetic with infinite descent, *Dialectica*, V. 43, No. 4, pp. 329–337.

Gauthier, Y. (2000) The internal consistency of arithmetic with infinite descent, *Modern Logic*, V. 8, No. 1-2, pp. 47–87.

Gauthier, Y. (2013) Kronecker in contemporary mathematics, general arithmetic as a foundational programme, *Rep. Math. Logic*, No. 48, pp. 37–65.

Gauthier, Y. (2015) *Towards an Arithmetical Logic: The Arithmetical Foundations of Logic*, Studies in Universal Logic, Springer.

Gauthier, Y. (2015) Arithmetization of logic, in *Towards an Arithmetical Logic*, Studies in Universal Logic, Birkhäuser, pp. 25–53.

Geary, D. C., Berch, D. B. and Koepke, K. M. (Eds.) (2015) *Evolutionary Origins and Early Development of Number Processing*, Academic Press, London.

Geary, D. C. and Widaman, K. F. (1987) Individual differences in cognitive arithmetic, *J. Exp. Psychol. Gen.*, V. 116, No. 2, pp. 154–171.

de Gelder, J. (1824) *Allereerste Gronden der Cijferkunst [Introduction to Numeracy]*, de Gebroeders van Cleef's, Gravenhage and Amsterdam (in Dutch).

Gelman, R. (1972) The nature and development of early number concepts, *Adv. Child Development Behav.*, V. 7, pp. 115–167.

Gelman, R. (1990) First principles organize attention to and learning about relevant data: Number and animate–inanimate distinction as examples, *Cognitive Sci.*, V. 14, pp. 79–106.

Gelman, R. and Butterworth, B. (2005) Number and language: how are they related? *Trends Cogn. Sci.*, V. 9, pp. 6–10.

Gelman, R. and Gallistel, C. R. (1978) *The Child's Understanding of Number*, Harvard University Press, Cambridge, MA.

Gelman, R. and Gallistel, C. R. (2015) Foreword, in *Evolutionary Origins and Early Development of Number Processing*, Academic Press, London/, pp. XIII–XIX.

Gentzen, G. (1936) Die Widerspruchsfreiheit der reinen Zahlentheorie, *Math. Ann.*, V. 112, pp. 493–565.

Gentzen, G. (1938) Neue Fassung des Widerspruchsfreiheitsbeweises für die reine Zahlentheorie, *Forschungen zur Logik und zur Grundlegung der exakten Wissenschaften*, V. 4, pp. 19–44.

Gentzen, G. (1943) Beweisbarkeit und Unbeweisbarkeit der Anfangsfällen der transfiniten Induktion in der reinen Zahlentheorie, *Math. Ann.*, V. 120, pp. 140–161.

Gershaw, D. A. (2015) *Two Plus Two Equals Four, But Not Always*, electronic edition: http://virgil.azwestern.edu/~dag/lol/TwoPlusTwo.html.

Gersten, R. and Chard, D. (1999) Number sense: rethinking arithmetic instruction for students with mathematical disabilities, *J. Special Education*, V. 33, No. 1, pp. 18–28.

Ghosh, S. (2014) Spontaneous generation of a crystalline ground state in a higher derivative theory, *Physica* A, V. 407, p. 245.

Giachetti, R. E. and Young, R. E. (1997) A parametric representation of fuzzy numbers and their arithmetic operators, *Fuzzy Sets Sys.*, V. 91, No. 2, pp. 185–202.

Giaquinto, M. (1993) Visualizing in Arithmetic, *Philos. Phenomenological Res.*, V. 53, pp. 385–396.

Giaquinto, M. (2001) Knowing numbers, *J. Philos.*, V. 98, No. 1, pp. 5–18.

Giaquinto, M. (2001a) What cognitive systems underlie arithmetical abilities? *Mind Lang.*, V. 16, pp. 56–68.

Giaquinto, M. (2015) Philosophy of number, in *The Oxford Handbook of Numerical Cognition*, Oxford University Press, New York, pp. 17–31.

Giaquinto, M. (2017) Cognitive access to numbers: The philosophical significance of empirical findings about basic number abilities, *Philos. Trans. Royal Soc. B: Biol. Sci.*, V. 373(1740), p. 20160520.

Giaquinto, M. The epistemology of visual thinking in mathematics, *The Stanford Encyclopedia of Philosophy* (Spring 2020 Edition), Edward N. Zalta (ed.), URL = https://plato.stanford.edu/archives/spr2020/entries/epistemology-visual-thinking.

Gibbon, J. (1977) Scalar expectancy theory and Weber's law in animal timing, *Psychol. Rev.*, V. 84, pp. 279-335.

Gibbs, W. W. (2003) A digital slice of Pi. The new way to do pure math: experimentally, *Sci. Amer.*, V. 288, pp. 23–24.

Gibson, J. (1966) *The Senses Considered as Perceptual Systems*, George Allen and Unwin, London

Gill, R. D. and Johansen, S. (1990) A survey of product-integration with a view toward application in survival analysis, *Ann. Statist.*, V. 18, pp. 1501–1555.

Gillies, D. A. (1982) *Frege, Dedekind, and Peano on the Foundations of Arithmetic*, Van Gorcum, Assen.

Gioia, A. (2001) *Number Theory, An Introduction*, Dover Publications, New York.

Giordano, P. (2010) The ring of Fermat reals, *Adv. Math.*, V. 225, No. 4, pp. 2050–2075.

Giordano, P. and Katz, M. (2011) *Two ways of Obtaining Infinitesimals by Refining Cantor's Completion of the Reals*, preprint in mathematical logic, http://arxiv.org/1109.3553[math.LO].

Girard, D. (1629) *Invention Nouvelle en L'Algèbre*, Guillaume Jansson Blaeuw, Amsterdam.

Girard, P. R. (1984) The quaternion group and modern physics, *Eur. J. Phys.*, V. 5, pp. 25–32.

Gisin, N. (2017) *Indeterminism in Physics, Classical Chaos and Bohmian Mechanics. Are Real Numbers Really Real?* Preprint in Quantum Physics, arXiv:1803.06824 [quant-ph].

Glaser, A. (1981) *History of Binary and Other Non-decimal Numeration*, Tomash Publishers.

Gleick, J. (1989) *Chaos: Making a New Science*, Cardinal, London.

Gleyzal, A. (1937) Transfinite real numbers, *Proc. Acad. Sci.*, V. 23, pp. 581–587.

Glyn, A. (2017) *One plus one equals three — the power of Data Combinations*, Luciad, 30 Nov, http://www.luciad.com/blog/one-plus-one-equals-three-the-power-of-data-combinations.

Gödel, K. (1931/1932) Über formal unentscheidbare Sätze der Principia Mathematica und verwandter Systeme I, *Monatsh. Math. Phys.*, V. 38, No. 1, pp. 173–198.

Gödel, K. (1934) Besprechung von Über die Unmöglichkeit einer vollstandigen Charakterisierung der Zahlenreihe mittels eines endlichen Axiomensystems, *Zentralblatt für Math. Grenzgebiete*, V. 2, No. 3.

Gödel, K. (1947) What is Cantor's continuum problem? *Amer. Math. Monthly*, V. 54, pp. 515–525.

Gödel, K. (1964) Russell's mathematical logic, in *Philosophy of Mathematics: Selected Readings*, P. Benacerraf and H. Putnam, (Eds.), Prentice-Hall, Englewood Cliffs, NJ.

Goetzmann, W. N. (2003) *Fibonacci and the Financial Revolution*, Yale School of Management International Center for Finance Working Paper No. 03–28.

Goetzmann, W. N. and Rouwenhorst, K. G. (2005) *The Origins of Value: The Financial Innovations that Created Modern Capital Markets*, Oxford University Press Inc, New York.

Gogen, J. A. (1967) L-fuzzy sets, *J. Math. Anal. Appl.*, V. 18, pp. 145–174.

Goldberg, D. (1991) What every computer scientist should know about floating-point arithmetic, *ACM Comput. Sur.*, V. 2, pp. 1 5–48.

Goldblatt, R. (1984) *Topoi: The Categorical analysis of Logic*, North-Holland, Amsterdam.

Goldman, J. R. (1997) The Queen of Mathematics: A Historically Motivated Guide to Number Theory, A K Peters/CRC Press, New York.

Goldman, J. G. (2010) What are the origins of (large) number representation? *Scientific American*, August https://blogs.scientificamerican.com/thoughtful-animal/what-are-the-origins-of-large-number-representation/.

Goldstein, A. B. and Avenevoli, S. (2018) Strength in numbers, *Prev. Sci.*, V. 19, pp. 109–111.

Goldthwaite, R. A. (1972) Schools and teachers of commercial arithmetic in renaissance florence, *J. European Econ. History*, V. 1, pp. 418–433.

Gollob, H. F., Rossman, B. B. and Abelson, R. P. (1973) Social inference as a function of the number of instances and consistency of information presented, *J. Personality Social Psychol.*, V. 27, pp. 19–33.

Gómez-Torrente, M. (2015) On the essence and Iientity of numbers, *Theoria: Rev. Teoría, Historia Fundamentos de la Ciencia*, V. 30, No. 3, pp. 317–329.

Gondran, M. and Minoux, M. (1979) *Graphes et algorithmes*, Editions Eyrolles, Paris.

Gonshor, H. (1980) Number theory for the ordinals with a new definition for multiplication, *Notre Dame J. Formal Logic*, V. 21, pp. 708–710.

Gonshor, H. (1986) *An Introduction to the Theory of Surreal Numbers*, Cambridge University Press, Cambridge.

Gontar, V. (1993) New theoretical approach for physicochemical reactions dynamics with chaotic behavior, in *Chaos in Chemistry and Biochemistry*, World Scientific, London, pp. 225–247.

Gontar, V. (1997) Theoretical foundation for the discrete dynamics of physicochemical systems: chaos, self-organization, time and space in complex systems, *Discrete Dynamics Nature Soc.*, V. 1, No. 1, pp. 31–43.

Gontar, V. and Ilin, I. (1991) New mathematical model of physicochemical dynamics, *Contrib. Plasma Phys.*, V. 31, No. 6, pp.681–690.

Gonthier, G. (2008) Formal proof — the four-color theorem, *Notices*, V. 55, No. 11, pp. 1382–1393.

González, S. (Ed.) (1994) *Non-Associative Algebra and Its Applications*, Springer, New York.

Goodman, N. (1968) *Languages of Art: An Approach to a Theory of Symbols*, Bobbs-Merrill, IN.

Goodman, N. D. (1979) Mathematics as an objective science, *Amer. Math. Monthly*, V. 88, pp. 540–551.

Goodrich, M. T. and Tamassia, R. (2015) *Algorithm Design and Applications*, Wiley, New York.

Goodstein, R. L. (1954) Logic-free formalizations of recursive arithmetic, *Mat. Scand.*, V. 2, pp. 247–261.

Goodstein, R. L. (1965) Multiple successor arithmetics, in *Formal Systems and Recursive Functions*, North-Holland, pp. 265–271.

Gordon, P. A. (2004) Numerical cognition without words: Evidence from Amazonia, *Science*, V. 306, pp. 496–499.

Gottlieb, A. (2013) '1 + 1 = 3': The synergy between the NEW key technologies, Next-generation Enterprise WANs, *Network World*, July 15, 2013, http://www.networkworld.com/article/2224950/cisco-subnet/-1---1---3---the-synergy-between-the-new-key-technologies.html.

Gottwald, S. (2010) An early approach toward graded identity and graded membership in set theory, *Fuzzy Sets Syst.*, V. 161, No. 18, pp. 2369–2379.

Gowers, T. (1998) A new proof of Szemerédi's theorem for arithmetic progressions of length four, *Geom. Funct. Anal.*, V. 8, No. 3, pp. 529–551.

Grabiner, J. V. (1974) Is mathematical truth time-dependent? *Amer. Math. Monthly*, V. 81, pp. 354–365.

Grabisch, M., Marichal, J.-L., Mesiar, R. and Pap, E. (2009) *Aggregation Functions*, Cambridge University Press, New York.

Grant, M. and Johnston, C. (2013) $1 + 1 = 3$: *CMO & CIO Collaboration Best Practices That Drive Growth*, Canadian Marketing Association, Don Mills, Canada.

Grant, H. and Kleiner, I. (2015) Hypercomplex numbers: from algebra to algebras, in *Turning Points in the History of Mathematics*, Birkhäuser, New York, NY, pp. 67–73.

Grassmann, H. (1844) *Die lineale Ausdehnungslehre, ein neuer Zweig der Mathematik*, dargestellt und durch Anwendungen auf die übrigen Zweige der Mathematik, wie auch die Statistik, Mechanik, die Lehre vom Magnetismus und die Krystallonomie erläutert, Wiegand, Leipzig.

Grassmann, H. (1861) *Lehrbuch der Arithmetik für höhere Lehranstalten, bd 1*, Enslin, Berlin.

Grassmann, H. (1862) *Die Ausdehnungslehre*. Vollständig und in strenger Form *bearbietet*, Enslin, Berlin.

Grassmann, H. (1894–1911) *Gesammelte mathematische und physikalische Werke*, 3 vols., (Engel, F. Ed.) B. G. Teubner, Leipzig.

Grassmann, R. (1872) *Die Formenlehre oder Mathematik*, Georg Olms Verlagsbuchhandlung, Hildesheim.

Grassmann, R. (1872a) *Die Begriffslehre oder Logik*. Zweites Buch der Formenlehre oder Mathematik, Verlag von R. Graßmann, Stettin.

Grassmann, R. (1891) *Die Zahlenlehre oder Arithmetik — streng wissenschaftlich in strenger Formelentwicklung*, Verlag von R. Graßmann, Stettin.

Grattan-Guinness, I. (1975) Fuzzy membership mapped onto interval and many-valued quantities, *Z. Math. Logik. Grundladen Math.*, V. 22, pp. 149–160.

Grattan-Guinness, I. (1996) Numbers, magnitudes, ratios, and proportions in Euclid's Elements: How did he handle them? *Historia Math.*, V. 23, pp. 355–375.

Grattan-Guinness, I. (1998) *The Norton History of the Mathematical Sciences*: *The Rainbow of Mathematics*, W. W. Norton, New York.

Grattan-Guinness, I. (1998a) Some neglected niches in the understanding and teaching of numbers and number systems, *ZentralblattfürDidaktik Math.*, V. 30, pp. 12–19.

Grattan-Guinness, I. (2000) *The Search for Mathematical Roots, 1870–1940*: *Logic, Set Theories and the Foundations of Mathematics from Cantor through Russel to Gödel*, Princeton University Press, Princeton.

Grattan-Guinness, I. (2008) Solving Wigner's mystery: The reasonable (though perhaps limited) effectiveness of mathematics in the natural sciences, *Math. Intelligencer*, V. 30, No. 3, pp. 7–17.

Grattan-Guinness, I. (2011) Numbers as moments of multisets: A new-old formulation of arithmetic. *Math Intelligencer*, V. 33, pp. 19–29.

Grätzer, G. (1996) *General Lattice Theory*, Birkhäuser, Boston.

Graves, C. (1845) On algebraic triplets, *Proc. Royal Irish Acad.*, V. 3, pp. 51–54; 57–64; 80–84; 105–108.

Graves-Gregory, N. (2014) Historical changes in the concepts of number, mathematics and number theory, *Proce. Alternative Natural Philosophy Associ.*, No. 34, pp. 25–52.

Gray, J. (1979) Non-Euclidean geometry: a re-interpretation, *Historia Math.*, V. 6, pp. 236–258.

Gray, J. (1998) Mathematicians as Philosophers of Mathematics: Part 1, Learning Math., V. 18, No. 3, pp. 20–24.

Gray, J. (1999) Mathematicians as Philosophers of Mathematics: Part 2, Learning Math., V. 19, No. 1, pp. 28–31.

Gray, J. (2008) *Plato's Ghost: The Modernist Transformation of Mathematics*, Princeton University Press, Princeton.

Green, B. (2009) Book reviews: "Terence C. Tao and Van H. Vu, Additive combinatorics," *Bull. Amer. Math. Soc.*, V. 46, No. 3, pp. 489–497.

Greene, D. H. and Knuth, D. E. (1990) *Mathematics for the Analysis of Algorithms*, Birkhäuser, Boston, MA.

Gregg, D. G. (2010) Designing for collective intelligence, *Commun. ACM*, V. 53, No. 4, pp. 134–138.

Gregory, F. H. (1996) *Arithmetic and Reality: A Development of Popper's Ideas*, Working Paper Series No. WP96/01, Department of Information Systems, City University of Hong Kong, Hong Kong.

Griffin, M. (2008) Looking behind the symbol: Mythic algebra, numbers, and the illusion of linear sequence, *Semiotica*, V. 171, pp. 1–13.

Griffiths, P. (1990) *Equality in language: Aspects of the theory of linguistic equality*, Durham theses, Durham University, Available at Durham E-Theses.

Grigoriev, D. and Shpilrain, V. (2014) Tropical cryptography, *Commun. Algebra*, V. 42, No. 6, pp. 2624–2632.

Grimsley, S. (2018) *Synergy in Business: Definition & Examples*, https://study.com/academy/lesson/synergy-in-business-definition-examples-quiz.html.

Gromov, N. A. and Kuratov, V. V. (2005) Noncommutative space-time models. *Czechoslovak J. Phys.*, V. 55, No. 11, pp. 1421–1426.

Gross, M. (2011) *Tropical Geometry and Mirror Symmetry, Regional Conference Series in Mathematics*, American Mathematical Society, Providence, RI.

Gross, H. J., Pahl, M., Si, A., Zhu, H., Tautz, J. and Zhang, S. (2009) Number-based visual generalisation in the honeybee, *PLoS One*, V. 4, e4263.

Grossman, J. (1981) *Meta-Calculus: Differential and Integral*, Archimedes Foundation, Rockport, MA.

Grossman, M. (1979) *The First Non-linear System of Differential and Integral Calculus*, Mathco, Rockport, MA.

Grossman, M. (1979a) An introduction to non-Newtonian calculus, *Int. J. Math. Education Science Technol.*, V. 10, No. 4, pp. 525–528.

Grossman, M. (1983) *Bigeometric Calculus: A System with Scale-Free Derivative*, Archimedes Foundation, Rockport.

Grossman, M. (1988) Calculus and discontinuous phenomena, *Int. J. Math. Education Sci. Technol.*, V. 19, No. 5, pp. 777–779.

Grossman, M. and Katz, R. (1972) *Non-Newtonian Calculus*, Lee Press, Pigeon Cove, MA.

Grossman, M. and Katz, R. (1984) Isomorphic calculi, *Int. J. Math. Education Sci. Technol.*, V. 15, No. 2, pp. 253–263.

Grossman, M. and Katz, R. (1986) A new approach to means of two positive numbers, *Int. J. Math. Education Science Technol.*, V. 17, No. 2, pp. 205–208.

Grossman, J., Grossman, M. and Katz, R. (1980) *The First Systems of Weighted Differential and Integral Calculus*, Archimedes Foundation, Rockport, MA.

Grossman, J., Grossman, M. and Katz, R. (1983) *Averages: A New Approach*, Archimedes Foundation, Rockport, MA.

Grossman, J., Grossman, M. and Katz, R. (1987) Which growth rate? *Int. J. Math. Education Science Technology*, V. 18, No. 1, pp. 151–154.

Grotheer, M., Herrmann, K.-H. and Kovaćs, G. (2016) Neuroimaging evidence of a bilateral representation for visually presented numbers, *J. Neurosci.*, V. 36, No. 1, pp. 88–97.

Gouvêa, F. Q. (1997) *p-Adic Numbers: An Introduction*, Springer-Verlag, New York.

Grünwald, V. (1885) Intorno al l'aritmetica dei sistemi numerici a base negativa con particolare riguardo al sistema numerico a base negativo-decimale per lo studio delle sue analogie col l'aritmetica ordinaria (decimale), *Giornale di Matematiche di Battaglini*, pp. 203–221, 367.

Grzegorczyk, A. (1957) On the definition of computable real continuous functions, *Fundam. Mat.*, V. 44, pp. 61–71.

Grzegorzewski, P. (2003) Distance and orderings in a family of intuitionistic fuzzy numbers, in *Proc. Third Conf. Fuzzy Logic and Technology (EUSFLAT 03)*, pp. 223–227.

Grzymala-Busse, J. (1987) Learning from examples based on rough multisets, in *Proc. 2nd Int. Symp. Methodologies for Intelligent Systems*, Charlotte, NC, USA, pp. 325–332.

Guaspari, D. (1979) Partially conservative extensions of arithmetic, *Trans. Amer. Math. Soc.*, V. 254, pp. 47–68.

Guedj, D. (1996) *L'empire of Nombres*, Editions Gallimard, Paris.

Guénard, F. and Lemberg, H. (2001) *La méthode expérimentale en mathématiques*, Springer-Verlag, Heidelberg, Germany.

Guenther, R. A. (1983) Product integrals and sum integrals, *Int. J. Math. Education Sci. Technol.*, V. 14, pp. 243–249.

Guerra, M. L. and Stefanini, L. (2005) Approximate fuzzy arithmetic operations using monotonic interpolations, *Fuzzy Sets Syst.*, V. 150 (1), pp. 5–33.

Gugoiu, L. and Gugoiu, T. (2006) *The Book of Fractions*, La Citadelle, Mississauga, Ontario, Canada.

Gullberg, J. (1997) *Mathematics: From the Birth of Numbers*, W. W. Norton, New York.

Gvozdanovic, J. (Ed.) (1992) *Indo-European Numerals.* deGruyter, Berlin-York.

Gwiazda, J. (2010) *Infinite Numbers Are Large Finite Numbers*, https://philarchive.org/rec/GWIINA?all_versions=1.

Gwiazda, J. (2012) On infinite number and distance, *Constructivist Foundations*, V. 7, No. 2, pp. 126–130.

Haaparanta, L. (Ed.) (1994) *Mind, Meaning and Mathematics. Essays on the Philosophical Views of Husserl and Frege*, Number 237 in Synthese Library, Kluwer Academic Publishers, Dordrecht/.

Hachtman, S. (2017) Determinacy in third order arithmetic, *Annals of Pure and Applied Logic*, v. 168, No. 11.

Hahn, H. (1907) Über die nichtarchimedischen Grössensysteme, *Sitzungsberichte der Kaiserlichen Akademie der Wissenschaften*, Wien, Mathematisch-Naturwissenschaftliche Klasse 116 (Abteilung IIa), pp. 601–655.

Hajek, P. and Pudlak, P. (1993) *Metamathematics of First-Order Arithmetic*, Springer, New York.

Hájek, P. (2005) On arithmetic in the Cantor–Lukasiewicz fuzzy set theory, *Archive Math. Logic*, V. 44, pp. 763–782.

Hájek, P. (2013) On equality and natural numbers in Cantor–Lukasiewicz set theory. *Logic J. IGPL*, V. 21, pp. 91–100.

Hájek, P. and Hájková, M. (1972) On interpretability in theories containing arithmetic, *Fundam. Math.*, V. 76, pp. 131–137.

Hájek, P. and Pudlák, P. (1993) *Metamathematics of First-order Arithmetic*, Perspectives of Mathematical Logic, Springer-Verlag, Berlin.

Hájková, M. (1971) The lattice of bi-numerations of arithmetic, *Comment. Math. Univ. Carolinae*, V. 12, Part I: pp. 81–104, Part II: pp. 281–306.

Halberda, J. and Feigenson, L. (2008) Set representations required for the acquisition of the "natural number" concept, *Behavioral Brain Sci.*, V. 31, pp. 655–656.

Halberda, J., Mazzocco, M. M. M. and Feigenson, L. (2008) Individual differences in non-verbal number acuity correlate with maths achievement, *Nature*, V. 455, pp. 665–668.

Hale, B. (2000) Reals by abstractio, *Philosophia Mathematica*, V. 8, pp. 100–123.

Hale, B. (2002) Real numbers, quantities and measurement, *Philos. Math.*, V. 10, pp. 304–323.

Hale, B. (2004) Real numbers and set theory — extending the Neo-Fregean programme Beyond Arithmetic, *Synthese*, V. 147, pp. 21–41.

Halmos, P. R. (1974) *Naive Set Theory*, Springer, New York.

Hamilton, W. R. (1833) *Introductory Lecture on Astronomy*, Dublin University Review and Quarterly Magazine, v. I, Trinity College., January 1833.

Hamilton, W. R. (1843) On Quaternions; or on a new System of Imaginaries in Algebra, in a *letter to John T. Graves*, dated October 17, 1843.

Hamilton, W. R. (1844) On quaternions, or on a new system of imaginaries in algebra, *Philos. Magazine*, V. 25, No. 3, pp. 489–495.

Hamilton (1848) Note, by Sir W. R. Hamilton, respecting the researches of John T. Graves, Esq., *Trans. Royal Irish Acad.*, V. 21, pp. 338–341.

Hamilton, W. R. (1866) *Elements of Quaternions*, Longmans, Green, London.

Hamming, R. W. (1980) The unreasonable effectiveness of mathematics, *Amer. Math. Monthly*, V. 87, No. 2, pp. 81–90.

Hankel, H. (1867) *Vorlesungenüber die complexen Zahlen und ihre Functionen*, L. Voss, Leipzig.

Hanna, R. (2002) Mathematics for humans: Kant's philosophy of arithmetic revisited, *European J. Philos.*, V. 10, No. 3, pp. 328–352.

Hansen, E. R. (1975) A generalized interval arithmetic, in *Interval Mathematics*, Lecture Notes in Computer Science, V. 29, Springer, Berlin, Heidelberg, pp. 7–18.

Hansen, H. K. and Porter, T. (2012) What do numbers do in transnational governance? *Int. Political Soc.*, V. 6, No. 4, pp. 409–426.

Hanss, M. (2005) *Applied fuzzy arithmetic: An Introduction with Engineering Applications*, Springer, New York.

Hardy, G. H. (1910) *Orders of Infinity, The "Infinitärcalcül" of Paul Du Bois–Reymond*, Cambridge University Press, Cambridge.

Hardy, G. H. (1929) Mathematical proof, *Mind*, V. 38, 149.

Hardy, G. H., Littlewood, J. E. and Pólya, G. (1934) *Inequalities*, Cambridge University Press, Cambridge.

Harmer, A. (2014) Leibniz on infinite numbers, infinite wholes, and composite substances, *British J. History Philos.*, V. 22, No. 2.

Harrow, K. (1978) The bounded arithmetic hierarchy, *Inform. Control*, V. 36, pp. 102–117.

Hartle, J. B. (2003), *Gravity: An Introduction to Einstein's General Relativity*, Pearson, Boston/.

Hartnett, P. M. (1992) The development of mathematical insight: From one, two, three to infinity, *Dissertation Abstr. Int.*, V. 52, p. 3921.

Hartnett, P. and Gelman, R. (1998) Early understandings of numbers: paths or barriers to the construction of new understandings? *Learning and Instruction*, V. 8, No. 4, pp. 341–374.

Harvey, B. M., Fracasso, A., Petridou, N. and Dumoulin, S. O. (2015) Topographic representations of object size and relationships with numerosity reveal generalized quantity processing in human parietal cortex, *Proc. Natl Acad. Sci. USA*, V. 112, No. 44, pp. 13525–13530.

Hatami, R. R. and Pejlare, J. (2019) Rizanesander's Recknekonsten or "The art of arithmetic" – the oldest known textbook of mathematics in Swedish, in *Proc. Eleventh Congr. European Society for Research in Mathematics Education*.

Hatcher, W. S. (2000) *Foundations of Mathematics: An Overview at the Close of the Second Millenium*, Landegg Academy, Switzerland.

Hatcher, A. (2014) *Topology of numbers*, Cornell University.

Hattangadi, V. (2017) *What is the meaning of synergy in business?* http:// drvidyahattangadi.com/what-is-the-meaning-of-synergy-in-business/.

Haven, T., Anderson, D. and Keller, J. (2010) A fuzzy choquet integral with an interval type 2 fuzzy number-valued integrand, in *Proc. 2010 IEEE Int. Conf. Fuzzy Systems*, Barcelona, Spain, 18–23 July 2010, pp. 1–8.

Hauser, M., Tsao, F., Garcia, P. and Spelke, E. (2003) Evolutionary foundations of number: spontaneous representation of numerical magnitudes by cotton-top tamarins. *Proc. Roy. Soc. B: Biol. Sci.*, V. 270, No. 1523, pp. 1441–1446.

Hawking, S. (Ed.) (2005) *God Created the Integers: The Mathematical Breakthroughs that Changed History*, Running Press, London.

Hayashi, T. (2008) Bakhshālī manuscript, in *Encyclopaedia of the History of Science, Technology, and Medicine in Non-Western Cultures*, V. 1, Springer, pp. B1–B3.

Hayes, B. (2009) The higher arithmetic, *Amer. Sci.*, V. 97, No. 5, pp. 364–367.

Heath, T. L. (1921) *A History of Greek Mathematics*, Clarendon Press, Oxford.

Heaton, H. (1898) Infinity, the Infinitesimal, and Zero, *Amer. Math. Monthly*, V. 5, No. 10, pp. 224–226.

Heaviside, O. (1893) On operators in physical mathematics, I, *Proc. Roy. Soc., London*, V. 52, pp. 504–529.

Heaviside, O. (1894) On operators in physical mathematics, II, *Proc. Roy. Soc., London*, V. 54, pp. 105–143.

Heck, R. (1993) The development of arithmetic in *Frege's Grundgesetze der Arithmetik*, *J. Symbolic Logic*, V. 58, pp. 579–601.

Heck, R. G. (1995) Definition by induction in Frege's *Grundgesetze der Arithmetik*, in *Frege's Philosophy of Mathematics*, Harvard University Press, Cambridge, MA, pp. 295–333.

Heck, R. G. (1996) The consistency of predicative fragments of Frege's *Grundgesetze der Arithmetik*, *History Philos. Logic*, V. 17, pp. 209–220.

Heck, R. G., Jr. (1999) *Grundgesetze der Arithmetik* I §10, *Philoso. Math.*, vol. 7, pp. 258–292.

Heck, R. (2000) Cardinality, counting, and equinumerosity, *Notre Dame J. Formal Logic*, The George Boolos Memorial Symposium, Notre Dame, IN, 1998, V. 41, No. 3, pp. 187–209.

Heeffer, A. (2017) Arithmetic in the renaissance, in *Encyclopedia of Renaissance Philosophy*, Springer, Cham.

Heidelberger, M. (1993) *Nature from Within: Gustav Theodore Fechner and His Psychophysical Worldview*, C. Klohr (trans.), University of Pittsburgh Press, Pittsburgh, 2004.

Heidelberger, M. (1993a) Fechner's impact for measurement theory, commentary on D. J. Murray, "A perspective for viewing the history of psychophysics", *Behav. Brain Sci.*, V. 16, No. 1, pp. 146–148.

Heilpern, S. (1992) The expected value of a fuzzy number, *Fuzzy Sets Syst.*, V. 47, pp. 81–86.

Heine, E. (1872) Die elemente der functionenlehre,*J. Reine Angew. Math.*, V. 74, pp. 172–188.

Heine, E. (1880) Essay of introduction to the theory of analytic functions according to the principles of Prof. Weierstrass, *Giornale Mat.*, V. 18.

Heis, J. (2015) Arithmetic and number in the philosophy of symbolic forms, in *The Philosophy of Ernst Cassirer: A Novel Assessment*, pp. 123–140.

Heisenberg, W. (1998) *Philosophie — Le manuscrit de 1942*, Seuil, Paris.

Helfgott, H. A. (2013) *The Ternary Goldbach Conjecture is True*, preprint, math.NT, arXiv: 1312.7748.

Hellendoorn, H. (1991) Fuzzy numbers and approximate reasoning. *Ann. Univ. Sci. Budapest, Sect. Comp.* V. 12, pp. 113–119.

Hellman, G. (1989) *Mathematics Without Numbers*, Oxford University Press, New York.

Hellman, G. (2001) Three varieties of mathematical structuralism, *Philos. Math.*, V. 9, pp. 184–211.

Helmut, W. C. (2010) Frege, the complex numbers, and the identity of indiscernibles, *Logique Anal.*, V. 53, No. 209, pp. 51–60.

Hempel, C. G. (1939) Vagueness and logic. *Philos. Sci.*, V. 6, pp. 163–180.

Henle, J. M. (1999) Non-nonstandard analysis: Real infinitesimals, *Math. Intelligencer*, V. 21, No. 1, pp. 67–73.

Henle, J. M. (2003) Second-order non-nonstandard analysis, *Studia Logica*, V. 74, No. 3, pp. 399–426.

Hensel, K. (1897) Übereineneue Begründung der Theorie der algebraischen Zahlen, *Jahresbericht der Deutschen Mathematiker-Vereinigung* V. 6, No. 3, pp. 83–88.

Hensel, K. (1908) *Theorie der algebraischen Zahlen*, Teubner, Leipzig.

Hensel, K. (1913) *Zahlentheorie*, Göschen, Berlin.

Henson, C. W., Kaufmann, M. and Keisler, H. J. (1984) The strength of nonstandard methods in arithmetic, *J. Symbolic Logic*, V. 49, No. 4, pp. 1039–1058.

Herbert, N. (1987) *Quantum Reality: Beyond the New Physics*, Anchor Books, New York.

Herbrand, J. (1931) Sur la non-contradiction de l'arithmétiqué, *J. Reine Angew. Math.*, V. 166, pp. 1–8.

Hermite, C. (1873) Sur le fonctionexponentielle, *Comptes Rendus Acad. Sci.*, V. 77, pp. 18–24.

Herrmann, E., Hernández-Lloreda, M. V., Call, J., Hare, B. and Tomasello, M. (2010) The structure of individual differences in the cognitive abilities of children and chimpanzees, *Psychol Sci.*, V. 21, No. 1, pp. 102–110.

Hernández, P., Cubillo, S., Torres-Blanc, C. and Guerrero, J. A. (2017) New Order on Type 2 Fuzzy Numbers, *Axioms*, 6, 22.

Herranz, F. J. and Santander, M. (2002) Conformal compactification of spacetimes. *J. Phys. A*, V. 35, No. 31, pp. 6619–6629.

Herrlich, H. and Strecker, G. E. (1973) *Category Theory*, Allyn and Bacon Inc., Boston.

Hersh, R. (1999) *What is Mathematics, Really?* Oxford University Press, New York.

Herscovics, N. and Linchevski, L. (1994) A cognitive gap between arithmetic and algebra, *Educational Studies Math.*, V. 27, No. 1, pp. 59–78.

Hessenberg, G. (1906) Grundbegriffe der Mengenlehre, *Abhandlungen der Friesschen Schule. Neue Folge*, V. 1, pp. 478–706.

Hewitt, E. and Hewitt, R. E. (1979) The Gibbs–Wilbraham phenomenon: An episode in Fourier analysis, *Arch. History Exact Sci.*, V. 21, No. 2, pp. 129–160.

Heyting, A. (1930) Die formalin Regeln der intuitionistischen Mathematik II, *Sitz. Preuss. Akad. Wiss., Phys.-Math. Kl.*, pp. 57–71.

Hickey, T., Ju, Q. and van Emden, M. H. (2001) Interval arithmetic: From principles to implementation, *J. ACM (JACM)*, V. 48, No. 5, pp. 1038–1068.

Hickman, J. L. (1980) A note on the concept of multiset, *Bull. Australian Math. Soc.*, V. 22, pp. 211–217.

Hickman, J. (1983) A note on Conway multiplication of ordinals, *Notre Dame J. Formal Logic*, V. 24, pp. 143–145.

Higgins, P. M. (2008) *Number Story: From Counting to Cryptography*, Copernicus, London.

Hilbert, D. (1897) Die Theorie der algebraischen Zahlkörper. *Jahresbericht der Deutschen Mathematiker-Vereinigung*, V. 4 ("1894–1895"), pp. 177–546.

Hilbert, D. (1899) *Die Grundlagen der Geometrie*, Festschrift zur Feier der Enthüllung des Gauss-Weber Denkmals in Göttingen, Teubner, Leipzig.

Hilbert, D. (1900) Über der Zahlbegriff, *Jaresbericht der Deutschen Mathematiker-Vereinigung*, bd. 8, pp. 180–184.

Hilbert, D. (1902) Mathematical problems, *Bull. Amer. Math. Soc.*, V. 8, No. 10, pp. 437–479.

Hilbert, D. (1926) Uber das Unendliche, *Math. Ann.*, V. 95, pp. 161–190.

Hilbert, D. (1931) Die Grundlegung der elementaren Zahlentheorie, *Math. Ann.*, V. 104, pp. 485–94.

Hilbert, D. (2013) *David Hilbert's Lectures on the Foundations of Arithmetics and Logic 1917-1933*, Springer Heidelberg.

Hilbert, D. and Bernays, P. (1968) *Grundlagen der Mathematik, I*, Springer-Verlag, Berlin.

Hill, G. F. (1915) *The Development of Arabic Numerals in Europe*. Clarendon Press, Oxford.

Hill, C. O. (2000) Frege's attack on Husserl and Cantor, in *Husserl or Frege?*, pages 95–107. Open Court, Chicago and La Salle, pp. 95–107.

Hill, C. O. (2010) Husserl on Axiomatization and Arithmetic, in *Phenomenology and Mathematic s*, Springer, New York.

Hinrichs, J., Yurko, D. and Hu, J. (1981) Two-digit number comparison: use of place information. *J. Exper. Psych.: Human Perception Perf.*, V. 7, pp. 890–901.

Hintikka, J. (1970) The semantics of modal notions and the indeterminacy of ontology. *Synthese*, V. 21, pp. 408–424.

Hintikka, J. (1970a) Propositional attitudes *De Dicto* and *De Re*. *J. Philos.*, V. 67, pp. 869–883.

Hirsch, M. W. (1994) *Differential Topology*, Springer, New York.

Hitch, G. J. (1978) The role of short-term working memory in mental arithmetic, *Cognitive Psychol.*, V. 10, No. 3, pp. 302–323.

Hjelmslev, J. (1950) Eudoxus' axiom and Archimedes' Lemma, *Centaurus*, V. 1, pp. 2–11.

Hlodovskii, I. (1959) A new proof of the consistency of arithmetic, *Uspehi Mat. Nauk*, V. 14, No. 6, pp. 105–140 (in Russian).

Hoborski, A. (1923/1938) Aus der theoretischen arithmetik, *Opusc. math. Kraków*, V. 2, No. 11–12.

Hodes, H. (1984) Logicism and the ontological commitments of arithmetic, *Journal of Philosophy*, V. 81, pp. 123–149.

Hölder, O. (1901) Die Axiome der Quantität und die Lehre vom Mass, *Berichteüber die Verhandlungen der Königlich Sachsischen Gesellschaft der Wissenschaften zu Leipzig, Mathematische-Physicke Klasse*, V. 53, pp. 1–64.

Hölder, O. (1996) The axioms of quantity and the theory of measurement, *J. Math. Psychol.* V. 40, pp. 235–252.

Aczel, P., *et al.* (2013) *Homotopy Type Theory: Univalent Foundations of Mathematics*, The Univalent Foundations Program, Institute for Advanced Study.

Hopcroft, J. E., Motwani, R. and Ullman, J. D. (2007) *Introduction to Automata Theory, Languages, and Computation*, Addison-Wesley, Boston.

Hopkins, B. (2002) Authentic and symbolic numbers in Husserl's philosophy of arithmetic, in *The New Yearbook for Phenomenology and Phenomenological Philosophy II*, pp. 39–71.

Horne, J. (2020) *The Ontology of Number*, preprint, ResrarchGate, https://www.researchgate.net/publication/339416068.

Horsten, L. (2019) Philosophy of Mathematics, in *The Stanford Encyclopedia of Philosophy*, Spring 2019 Edition), URL= https://plato.stanford.edu/archives/spr2019/entries/philosophy-mathematics/.

Hossack, K. (2020) *Knowledge and the Philosophy of Number: What Numbers are and How they Are Known* (Mind, Meaning and Metaphysics), Bloomsbury Academic.

Houston, E. (1990) The history of $2 + 2 = 5$, *Math. Magazine*, V. 63, No. 5, pp. 338–339.

Howard, S. R., Avarguès-Weber, A., Garcia, J. E., Greentree, A. D. and Dyer, A. G. (2018) Numerical ordering of zero in honey bees, *Science*, V. 360, pp. 1124–1126.

Hsu, Y. F. and Szücs, D. (2011). Arithmetic mismatch negativity and numerical magnitude processing in number matching. *BMC Neuroscience*, V. 12, 83, https://doi.org/10.1186/1471-2202-12-83.

Hudson, P. and Ishizu, M. (2016) *History by Numbers: An Introduction to Quantitative Approaches*, Bloomsbury Academic.

Huizing, C., Kuiper, R. and Verhoeff, T. (2012) Generalizations of Rice's theorem, applicable to executable and non-executable formalisms, in *Turing-100, The Alan Turing Centenary*, EPiC Series, V. 10, pp. 168–180.

Hurford, J. R. (1975) *The linguistic theory of numerals*, Cambridge University Press, Cambridge.

Hurford, J. (1987) *Language and Number: The Emergence of a Cognitive System*, Blackwell, Oxford.

Hurwitz, A. (1923) Über die Komposition der quadratischenFormen, *Math. Ann.*, V. 88, No. 1–2, pp. 1–25.

Husemöller, D. (1994) *Fibre Bundles*, Springer Verlag, Berlin.

Husserl, E. (1887) *Über den Begriff der Zahl (Psychologische Analysen)*. Heynemann'sche Buchdruckerei (F. Beyer), Halle a.d. Saale.

Husserl, E. (1891) *Philosophie der Arithmetik (Psychologische und Logische Untersuchungen)*, C. E. M. Pfeffer, Halle.

Husserl, E. (1970) *Philosophie der Arithmetik. Mit Ergänzenden Texten (1890–1901)*, Number XII in Husserliana, Nijhoff, Den Haag.

Husserl, E. (1983) *Contributions à la Theorie du Calcul des Variations*. Queen's Papers in Pure and Applied Mathematics. Queen's University, V. 65, Kingston, Ontario.

Husserl, E. (1983a) *Studien zur Arithmetik und Geometrie*, Husserliana, v. XXI, M. Nijhoff, The Hague.

Husserl, E. (1981) On the concept of number, in *Husserl: Shorter Works*, P. Mc Cormick and F. Elliston (Eds), University of Notre Dame Press, Notre Dame, pp. 92–120.

Husserl, E. (2005) Lecture on the concept of number (WS 1889/90), in *The New Yearbook for Phenomenology and Phenomenological Philosophy*, V. 5, pp. 279–309.

Ierna, C. (2005) The beginnings of Husserl's philosophy. Part 1: From Über den Begriff der Zahl to Philosophie der Arithmetik, in *The New Yearbook for Phenomenology and Phenomenological Philosophy*, V. 5, pp. 1–56.

Ierna, C. (2006) The beginnings of Husserl's philosophy. Part 2: Mathematical and Philosophical Background, in *The New Yearbook for Phenomenology and Phenomenological Philosophy*, V. 6, pp. 33–81.

Ierna, C. (2008) Review of Edmund Husserl *Philosophy of Arithmetic*, trans. Dallas Willard (Husserliana Collected Works X). *Husserl Studies*, V. 24, No. 1, pp. 53–58.

Ierna, C. (2009) Husserl et Stumpf sur la Gestalt et la fusion. *Philosophiques*, V. 36, No. 2, pp. 489–510.

Ierna, C. (2011a) Brentano and Mathematics. *Revue Roumaine de Philosophie*, V. 55, No. 1, pp. 149–167.

Ierna, C. (2011b) Edmund Husserl's philosophie der arithmetik in reviews, in *The New Yearbook for Phenomenology and Phenomenological Philosophy*, XI.

Ifould, R. (2013) Do the maths: the science behind the numbers that govern our lives, *The Guardian*, Sat 16 November 2013, https://www.theguardian.com/lifeandstyle/2013/nov/16/science-behind-numbers-govern-lives-maths.

Ifrah, G. (1985) *From One to Zero: A Universal History of Numbers*, Viking, New York.

Ifrah, G. (2000) *The Universal History of Numbers: From Prehistory to the Invention of the Computer*, Willey, New York.

Imaeda, K. and Imaeda, M. (2000) Sedenions: algebra and analysis, *Appl. Math. Comput.*, V. 115, No. 2, pp. 77–88.

Inönü, E. and Wigner, E. P. (1953) On the contraction of groups and their representations, *Proc. Natl. Acad. Sci.*, V. 39, No. 6, pp. 510–524.

Ishango bone, *Wikipedia*, https://en.wikipedia.org/wiki/Ishango_bone.

Itenberg, I. and Mikhalkin, G. (2012) Geometry in the tropical limit. *Math. Semesterberichte*, V. 59, No. 1, pp. 57–73.

Itenberg, I., Mikhalkin, G. and Shustin, E. I. (2009) *Tropical Algebraic Geometry*, V. 35, Springer Science & Business Media, New York.

Iverson, K. E. (1962) *A Programming Language*, Wiley, New York.

Izard, V., Streri, A. and Spelke, E. S. (2014) Toward exact number: Young children use one-to-one correspondence to measure set identity but not numerical equality, *Cognitive Psychology*, V. 72, pp. 27–53.

Izard, V., Pica, P., Spelke, E. S. and Dehaene, S. (2008) Understanding exact numbers, *Philos. Psychol.*, V. 21, No. 4, pp. 491–505.

Jackson, B. B. (2013) Defusing easy arguments for numbers, *Linguistics Philos.*, V. 36, No. 6, pp. 447–461.

Jackson, J. D. (1975) *Classical Electrodynamics*, Wiley, New York.

Jackson, T. (2018) *Mathematics: An Illustrated History of Numbers* (*Ponderables: 100 Breakthroughs that Changed History*), Shelter Harbor Press.

Jacobsthal, E. (1909) Zur Arithmetik der transfiniten Zahlen, *Math. Ann.*, V. 67, pp. 130–143.

Jaffe, K. (2020) *The Thermodynamic Roots of Synergy and Its Impact on Society*, preprint, https://arxiv.org/ftp/arxiv/papers/1707/1707.06662.pdf.

Jahn, K. U. (1975) Intervall-wertige Mengen, *Math.Nach.*, V. 68, pp. 115–132.

Jánossy, L. (1955) Remarks on the foundation of probability calculus, *Acta Physica*, V. 4, 333.

Jay, C. B. (1989) A note on natural numbers Objects in monoidal categories, *Studia Logica*, V. 48, No. 3, pp. 389–393.

Jeans, J. (1968) *Science and Music*, Dover Publications, New York.

Jech, T. (2002) *Set Theory*, Springer, New York.

Jeřábek, E. (2020) Induction rules in bounded arithmetic, *Archive Math. Logic*, V. 59 No. 3, pp. 461–501.

Jizba, P. and Korbel, J. (2020) When Shannon and Khinchin meet Shore and Johnson: equivalence of information theory and statistical inference axiomatics, *Phys. Rev. E*, V. 101, p. 042126.

Jørgensen, K. F. (2005) *Kant's Schematism and the Foundations of Mathematics*, PhD thesis, Roskilde University, Roskilde.

Jørgensen, K. F. (2005) *Kant and the Natural Numbers*, https://www.scribd.com/document/116697225/Kant-and-the-Natural-Numbers.

Jude, B. (2014) *Synergy* - 1 + 1 = 3, Aug 20, 2014, https://www.linkedin.com/pulse/20140820054514-115081853-synergy-1-1-3.

Julia, B. (1990) Statistical theory of numbers, in *Number Theory and Physics*, Proceedings in Physics, V. 47, pp. 276–293.

Jung, C. G. (1969) *The Structure and Dynamics of the Psyche*, Princeton University Press, Princeton.

Kaczmarz, S. (1932) Axioms for arithmetic, *J. London Math. Soc.*, V. 7, No. 3, pp. 179–182.

Kadak, U. (2015) Non-Newtonian fuzzy numbers and related applications, *Iranian J. Fuzzy Syst.*, V. 12, No. 5, pp. 117–137.

Kahn, C. (2001) *Pythagoras and the Pythagoreans*, Hackett, Indianapolis.

Kalderon, M. E. (1996) What numbers could be (and, hence, necessarily are), *Philos. Math.*, V. 4, No. 3, pp. 238–255.

Kalinowski, J. (2002) On rank equivalence and rank preserving operators, *Novi Sad J. Math.*, V. 32, No. 1, pp. 133–139.

Kalvesmaki, J. (2013) *The Theology of Arithmetic: Number Symbolism in Platonism and Early Christianity*, Hellenic Studies Series 59, Center for Hellenic Studies, Washington, DC.

Kanamori, A. (2003) *The Higher Infinite: Large Cardinals in Set Theory from Their Beginnings*, Springer, New York.

Kanamori, A. and Magidor, M. (1978) *The Evolution of Large Cardinal Axioms in Set Theory. Higher Set Theory*. Lecture Notes in Mathematics, V. 669, Springer, Berlin.

Kang, J., Wu, J., Smerieri, A. and Feng, J. (2010) Weber's law implies neural discharge more regular thana Poisson process, *European J. Neurosci.*, V. 31, pp. 1006–1018.

Kaniadakis, G. (2001) Nonlinear kinetics underlying generalized statistics, *Physica A*, V. 296, p. 405.

Kaniadikis, G. (2002) Statistical mechanics in the context of special relativity, *Phys. Rev. E*, V. 66, p. 056125.

Kaniadikis, G. (2005) Statistical mechanics in the context of special relativity (II), *Phys. Rev. E*, V. 72, p. 036108.

Kaniadakis, G. (2006) Towards a relativistic statistical theory, *Physica A*, V. 365, pp. 17–23.

Kaniadikis, G. (2011) Power-law tailed statistical distributions and Lorentz transformations, *Phys. Lett. A*, V. 375, pp. 356–359.

Kaniadikis, G. (2012) Physical origin of the power-law tailed statistical distributions, *Mod. Phys. Lett. B*, V. 26, p. 1250061.

Kant, I. (1763) *Versuch den Begriff der negative Größen in die Weltweisheitein zu führen*, Kanter, Königsberg.

Kant, I. (1770/1929) *Inaugural Dissertation*, The Open Court, Chicago.

Kant, I. (1786/2004) *Metaphisical Foundations of Natural Science*, Cambridge University Press, Cambridge.

Kant, I. (1787/1958) *The Critique of Pure Reason*, Macmillan, & Co, London.

Kantor, I. L. and Solodovnikov, A. S. (1989) *Hypercomplex Numbers: An Elementary Introduction to Algebras*, Springer-Verlag, New York.

Kaplan R. (1999) *The Nothing That Is: A Natural History of Zero*, Oxford University Press, New York.

Kapranov, M. (1995) Analogies between the Laglands correspondence and topological quantum field theory, *Progress in Math.*, V. 131, pp. 119–151.

Karp, A. and Schubring, G. (Eds.) (2014) *Handbook on the History of Mathematics Education*, Springer Science & Business Media.

Karpinski, L. C. (1925) *The History of Arithmetic*, Rand McNally & company, Chicago.

Kasatkin, V. N. (1982) *New About Numerical Systems*, Vyshcha Shkola, Kiev (in Russian).

Kästner, A. (1786) *Mathematische Anfangsgründe*, Vandenhoeck, Gottingen.

Katasonov, V. N. (1993) *Metaphysical Mathematics of the XVII Century*, Nauka, Moscow.

Kato, K. and Sh. Saito, S. (1986) Global class field theory of arithmetic schemes, *Contemp. Math.*, V. 55, part I, pp. 255–331.

Katz, E. (1957) The two-step flow of communication, *Public Opinion Quart.*, V. 21, pp. 61–78.

Katz, F. M. (1981) *Sets and Their Sizes*. Ph.D. Dissertation, MIT.

Katz, E. (2017) What is Tropical Geometry? *Notices. Amer. Math. Soc.*, V. 64, No. 4, pp. 380–382.

Katz, K. and Katz, M. (2010) Zooming in on infinitesimal $1 - .9\ldots$ in a post-triumvirate era, *Educational Studies Math.*, V. 74, No. 3, pp. 259–273.

Katz, K. U. and Katz, M. G. (2012) Stevin numbers and reality, *Found. Sci.*, V. 17, No. 2, pp. 109–123.

Katz, V. J. and Barton, B. (2007) Stages in the history algebra with implications for teaching, *Educational Studies Math.*, V. 66, No. 2, pp. 185–201.

Kaufmann, A. (1977) *Introduction à la théorie des sous-ensembles flous: A l'usage des ingénieurs*, Masson, Paris.

Kaufmann, A. (1985) *Introduction to Fuzzy Arithmetic: Theory and Applications*, Van Nostrand Reinhold Electrical/Computer science and Engineering Series, Van Nostrand Reinhold Co.

Kauffman, L. H. (1995) Arithmetic in the form, *Cybernetics Syst.*, V. 26, pp. 1–57.

Kauffman, L. H. (1999) What is a number? *Cybernetics Sys.*, V. 30, No. 2, pp. 113–130.

Kaufmann, A. and Gupta, M. M. (1985) *Introduction to Fuzzy Arithmetic: Theory and Applications*, Van Nostrand, New York.

Kaufman, E. L., Lord, M. W., Reese, T. W. and Volkmann, J. (1949) The discrimination of visual number, *Amer. J. Psychol.*, V. 62, No. 4, pp. 498–525.

Kaufmann, R. M. and Yeomans, C. (2017) Math by pure thinking: R first and the divergence of measures in Hegel's philosophy of mathematics, *European J. Philos.*, V. 25, No. 4, pp. 985–1020.

Kaye, G. R. (1919) Indian Mathematics, *Isis*, V. 2, No. 2, pp. 326–356.

Kaye, G. R. (1917/2004) *The Bakhshālī Manuscripts: A Study in Medieval Mathematics*, Aditya Prakashan, New Delhi.

Kaye, R. (1991) *Models of Peano Arithmetic*, Oxford Logic Guides, Oxford University Press, New York.

Kaye, R. (1993) Using Herbrand-type theorems to separate strong fragments of arithmetic, in *Arithmetic, Proof Theory, and Computational Complexity*, Oxford Logic Guides, V. 23, Oxford University Press, New York, pp. 238–246.

Keenan, A., Schweller, R., Michael Sherman, M. and Zhong, X. (2013) *Fast Arithmetic in Algorithmic Self-Assembly*, preprint on Data Structures and Algorithms (arXiv:1303.2416 [cs.DS])

Keisler, H. J. (1994) The hyperreal line. Real numbers, generalizations of the reals, and theories of continua, *Synthese Library*, V. 242, pp. 207–237.

Kelly, J. L. (1955) *General Topology*, Van Nostrand Co., Princeton/New York.

Kennedy, H. C. (1974) Peano's concept of number, *Historia Math.*, V. 1, pp. 387–408.

Kern, I. (1964). *Husserl und Kant.* Phaenomenologica 16. Nijhoff, Den Haag.

Kernighan, B. W. (2018) *Millions, Billions, Zillions: Defending Yourself in a World of too Many Numbers*, Princeton University Press, Princeton.

Kesseböhmer, M., Samuel, T. and Weyer, H. (2014) *A Note on Measure-Geometric Laplacians*, preprint in Mathematics, math.FA, (arXiv:1411.2491).

Khrennikov, A. (1995) p-adic probability interpretation of Bell's inequality, *Phys. Lett. A* v.200.

Khrennikov, A. (1997) *Non-Archimedean Analysis: Quantum Paradoxes, Dynamical Systems and Biological Models*, Kluwer, Boston.

Khrennikov, A. and Segre, G. (2007) Hyperbolic quantization, in L. Accardi, W. Freudenberg, and M. Schurman (Eds.) Quantum probability and infinite dimensional analysis, World Scientific Publishing, Hackensack, NJ, pp. 282–287.

Kim, J. (2014) Euclid strikes back at Frege, *Philos. Quart.*, V. 64, No. 254, pp. 20–38.

Kim, M. (2017) *Arithmetic Gauge Theory: A Brief Introduction*, preprint in Mathematical Physics (math-ph) (arXiv:1712.07602).

Kirby, A. (2000) *Water Arithmetic doesn't Adds*, BBC News, http://news.bbc.co.uk/hi/english/sci/tech/newsid_671000/671800.stm.

Kircher, A. (1650) *Musurgia Universalis*, Corbelletti, Rome.

Kirillov, A. N. (2001) Introduction to tropical combinatorics, in *Physics and Combinatorics* 2000, *Proc. the Nagoya 2000 Intern. Workshop*, World Scientific, pp. 82–150.

Kiselyov, O., Byrd, W. E., Friedman, D. P. and Shan, C.-C. (2008) Pure, declarative, and constructive arithmetic relations (declarative pearl), in *Proc. 9th Int. Conf. Functional and Logic programming* (FLOPS'08), pp. 64–80.

Kisil, V. V. (2007) Two-dimensional conformal models of space-time and their compactification, *J. Math. Phys.*, V. 48, No. 7, 073506.

Kisil, V. V. (2012) Hypercomplex representations of the Heisenberg group and mechanics, *Internat. J. Theoret. Phys.*, V. 51, No. 3, pp. 964–984.

Kisil, V. V. (2012a) Is commutativity of observables the main feature, which separate classical mechanics from quantum? *Izvestiya Komi Nauchnogo Centra UrO RAN*, V. 3, No. 11, pp. 4–9.

Kisil, V. V. (2013) Induced representations and hypercomplex numbers, *Adv. Appl. Clifford Algebras*, V. 23, No. 2, pp. 417–440.

Kitcher, P. (1975) Kant and the Foundations of Mathematics, *Philos. Rev.*, V. 84, pp. 23–50.

Kitcher, P. (1983) *The Nature of Mathematical Knowledge*, Oxford University Press, Oxford.

Klaua, D. (1959/60) Transfinite reelle Zahlenräume, *Wiss. Zeitschr. Humboldt-Universität, Berlin, Math.-Nat. R.*, V. 9, pp. 169–172.

Klaua, D. (1960) Zur Struktur der reellen Ordinalzahlen, *Z. math. Logik Grundlagen Math.*, V. 6, pp. 279–302.

Klaua, D. (1965) Über einen Ansatz zurmehrwertigen Mengenlehre. *Monatsb. Deutsch. Akad. Wiss. Berlin,* V. 7, pp. 859–876.

Klaua, D. (1966) Über einen zweiten Ansatz zur mehrwertigen Mengenlehre, *Monatsber. Deutsch. Akad. Wiss. Berlin,* V. 8, pp. 161–177.

Klaua, D. (1966a) Grundbegriffe einer mehrwertigen Mengenlehre. *Monatsber. Deutsch. Akad. Wiss. Berlin,* V. 8, pp. 782–802.

Klaua, D. (1981) Inhomogene Operationen reeller Ordinalzahlen, *Arch. Math. Logik,* V. 21, No. 1, pp. 149–167.

Klaua, D. (1994) Rational and real ordinal numbers, in *Real Numbers, Generalizations of the Reals, and Theories of Continua,* Kluwer Academic Publishers, pp. 259–276.

Kleene, S. C. (1945) On the interpretation of intuitionistic number theory, *J. Symbolic Logic,* V. 10, pp. 109–124.

Kleene, S. C. (1956) Representation of events in nerve sets and finite automata, in *Automata Studies,* Princeton University Press, Princeton, pp. 3–40.

Kleene, S. C. (2002) *Mathematical Logic,* Courier Dover Publications, New York.

Klees, E. (2006) *One Plus One Equals Three — Pairing Man/Woman Strengths: Role Models of Teamwork* (The Role Models of Human Values Series, V. 1) Cameo Press, New York.

Klein, F. (1872) *Programm zum Eintritt in die philosophische Facultät und den Senat der k. Friedrich-Alexanders-Universität zu Erlangen,* A. Deichert, Erlangen.

Klein, F. (1924) *Elementarmathematik vom Hoheren Standpunkte,* Springer, Berlin.

Kline, M. (1967) *Mathematics for Nonmathematicians,* Dover Publications, New York.

Kline, M. (1990) *Mathematics: The Loss of Certainty,* Oxford University Press, New York.

Kline, M. (1990a) *Mathematical Thought from Ancient to Modern Times,* Oxford University Press, New York.

Kleinberg, J. and Tardos, E. (2006) *Algorithm Design,* Pearson – Addison-Wesley, Boston/.

Kleiner, I. and Movshovitz-Hadar, N. (1994) The role of paradoxes in the evolution of mathematics, *Amer. Math. Monthly,* V. 101, No. 10, pp. 963–974.

Klement, E. P., Mesiar, R. and Pap, E. (2000) *Triangular Norms,* Springer-Science+Business Media, B.V.

Klement, E. P., Mesiar, R. and Pap, E. (2010) A universal integral as common frame for Choquet and Sugeno integral, *IEEE Trans. Fuzzy Sys.,* V. 18, No. 1, pp. 178–187.

Klir, G. J. (1997) Fuzzy arithmetic with requisite constraints, *Fuzzy Sets Syst.,* V. 91, pp. 165–175.

Klir, G. (1997) The role of constrained fuzzy arithmetic in engineering, in *Uncertainty Analysis in Engineering and Sciences: Fuzzy Logic, Statistics, and Neural Network Approach,* Kluwer, Dordrecht, The Netherlands, pp. 1–19.

Klir, G. J. and Pan, Y. (1998) Constrained fuzzy arithmetic: Basic questions and some answers, *Soft Comput.*, V. 2, No. 2, pp. 100–108.

Klir, G. J. and Yuan, B. (1995) *Fuzzy Sets and Fuzzy Logic: Theory and Applications*, Prentice Hall, Upper Saddle River, NJ.

Klotz, I. M. (1995) Number mysticism in scientific thinking, *Math. Intelligencer*, V. 17, pp. 43–51.

Knapp, T. R. (2009) *Percentages: The Most Useful Statistics Ever Invented*, http://www.statlit.org/pdf/2009Knapp-Percentages.pdf.

Kneusel, R. T. (2015) *Numbers and Computers*, Springer International Publishing, Switzerland.

Knight, J., Case, J. and Berman, K. (2006) *Financial Intelligence: A Manager's Guide to Knowing What the Numbers Really Mean*, Harvard Business Review Press.

Knill, O. (2017) *On the Arithmetic of Graphs*, preprint in Combinatorics (math.CO), arXiv:1706.05767.

Knopp, K. (1952) *Elements of the Theory of Functions*, Dover, New York.

Knuth, D. E. (1974) *Surreal Numbers: How Two Ex-students Turned on to Pure Mathematics and Found Total Happiness*, Addison-Wesley, Reading, MA.

Knuth, D. E. (1976) Mathematics and computer science: coping with finiteness, *Science*, V. 194, No. 4271, pp. 1235–1242.

Knuth, D. (1997) *The Art of Computer Programming*, V. 2: *Seminumerical Algorithms*, Addison-Wesley, Reading, MA.

Koblitz, N. (1977) *p-Adic Numbers, p-adic Analysis, and Zeta-Functions*, Springer-Verlag, New York.

Koeplinger, J. and Shuster, J. A. (2012) "1 + 1 = 2" *A Step in the Wrong Direction?* http://fqxi.org/community/forum/topic/1449.

Kolmogorov, A. N. (1930) Sur la notion de la moyenne, *Atti Accad. Naz. Lincei*, V. 12, pp. 388–391.

Kolmogorov, A. N. (1961) Automata and Life, *Tekhnika Molodezhi*, No. 10, pp. 16–19; No. 11, pp. 30–33 (in Russian).

Kolmogorov, A. N. (1979) Automata and life, in *Kibernetika — neogranichennye vozmozhnosti i vozmozhnye ogranichenija. Itogi razvitija.* Moscow, Nauka, pp. 10–29 (in Russian).

Kolokoltsov, V. N. (1998) Nonexpansive maps and option pricing theory, *Kybernetika*, V. 34, No. 6, pp. 713–724.

Kolokoltsov, V. N. and Maslov, V. P. (1997) *Idempotent Analysis and Its Applications*, Kluwer Academic Publishers, Dordrecht.

Korner, S. (1968) *The Philosophy of Mathematics*, Dover Publications, New York.

Kossak, R. and Schmerl, J. H. (2006) *The Structure of Models of Peano Arithmetic*, Clarendon Press, Oxford.

Kozen, D. (1997) *Automata and Computability*, Springer-Verlag, New York.

Kra, B. and Schmeling, J. (2010) *Diophantine Numbers, Dimension and Denjoy Maps*, preprint, https://sites.math.northwestern.edu)~kra⟩papers⟩denjoy.

Krajiček, J. (1995) *Bounded Arithmetic, Propositional Logic and Complexity Theory*, Cambridge University Press, Cambridge.

Krantz, S. G. (2003) *Calculus demystified*, McGraw-Hill, New York.

Krantz, D. H. (1971) Integration of just-noticeable differences, *J. Math. Psych.*, V. 8, pp. 591–599.

Krantz, D. H., Luce, R. D, Suppes, P. and Tversky, A. (1971) *Foundations of Measurement*, V. 1, Academic Press, San Diego, CA.

Krause, E. F. (1987) *Taxicab Geometry*, Dover Publications, New York.

Krause, G. M. (1981) *A Strengthening of Ling's Theorem on Representation of Associative Functions*, Ph.D. Thesis in Mathematics, Illinois Institute of Technology.

Kreisel, G. (1965) *Mathematical Logic*, Lectures on Modern Mathematics, John Wiley & Sons, Inc., New York.

Kress, S. (2015) *Synergy: When One Plus One Equals Three*, http://www.summitteambuilding.com/synergy-when-one-plus-one-equals-three.

Kreuzer, A.P. (2015) Measure Theory and Higher Order Arithmetic, *Proceedings of the American Mathematical Society*, V. 143, No. 12, pp. 5411–5425.

Kripke, S. (1992) *Logicism, Wittgenstein, and De Re Beliefs about Numbers*, Unpublished lectures.

Kroiss, M., Fischer, U. and Schultz, J. (2009) When one plus one equals three: Biochemistry and bioinformatics combine to answer complex questions, *Fly*, V. 3, No. 3, pp. 212–214.

Kronecker, L. (1887) Über den Zahlbegriff, *Crelle J. Reine Angew. Math.*, V. 101, pp. 337–355.

Kronecker, L. (1887/1889) Ein Fundamentalsatz der allgemeinen Arithmetik, in *Werke*, V. II, Teubner, Leipzig, pp. 211–240.

Kronecker, L. (1901) *Vorlesungen Uber Zahlentheorie*, Leipzig, Druck und Verlag von B.G.Teubner.

Kronz, F. and Lupher, T. (2012) Quantum Theory: von Neumann vs. Dirac, *The Stanford Encyclopedia of Philosophy*, E. N. Zalta, (Ed.), http://plato.stanford.edu/archives/sum2012/entries/qt-nvd/.

Krysztofiak, W. (2008) Modalna arytmetyka indeksowanych liczb naturalnych: możliwe światy liczb, *Przegląd Filozoficzny — Nowa Seria*, V. 17, pp. 79–107 (in Polish).

Krysztofiak, W. (2012) Indexed Natural numbers in mind: a formal model of the basic nature number competence, *Axiomathes*, V. 22, No. 4, pp. 433–456.

Krysztofiak, W. (2016) Representational structures of arithmetical thinking: Part I, *Axiomathes*, V. 26, No. 1, pp. 1–40.

Kuczma, M., Choczewski, B. and Ger, R. (1990) *Iterative Functional Equations*, Cambridge University Press.

Kuleshov, V. A. (1969) *Theory of Hyperfields*, Nauka i Tehnika, Minsk, (in Russian).

Kulisch, U. W. (1982) Computer arithmetic and programming languages, in *APL'82 Proc. of the Int. Con. APL*, Heidelberg, Germany, pp. 176–182.

Kulsariyeva, A. T. and Zhumashova, Z. (2015) Numbers as cultural significant, *Procedia — Social Behav. Sci.*, V. 191, pp. 1660–1664.

Kul'vetsas, L. L. (1989) The status of the concept of quantity in physics theory and H. Helmholtz's book "Zählen und Messen", in *Studies in the History of Physics and Mechanics,* Nauka, Moscow, pp. 170–186 (Russian).

Kumar, A. and Kaur, M. (2013) A ranking approach for intuitionistic fuzzy numbers and its application, *J. Appl. Res. Technol.*, V. 11, pp. 381–396.

Kumari, G. (1980) Some significant results of algebra of pre-Aryabhata era, *Math. Ed.*, V. 14, No. 1, pp. B5–B13.

Kuratowski, K. (1966) *Topology*, V. 1, Academic Press, Warszawa.

Kuratowski, K. (1968) *Topology*, V. 2, Academic Press, Warszawa.

Kuratowski, K. and Mostowski, A. (1967) *Set Theory*, North-Holland, Amsterdam.

Kurosh, A. G. (1963) *Lectures on General Algebra*, Chelsea P. C., New York.

Kusraev, A. G. and Kutateladze, S. S. (2005) Boolean valued analysis and positivity, in *Positivity IV — Theory and Applications*, Supplements, Technische Universitat Dresden, pp. 2–5.

Kutter, E. F., Bostroem, J., Elger, C. E., Mormann, F. and Nieder, A. (2018) Single neurons in the human brain encode numbers, *Neuron*, DOI: https://doi.org/10.1016/j.neuron.2018.08.036.

Kvasz, L. (2014) Paradoxes in scientific theories and the boundaries of the language of science, *Organon*, V. 21, pp. 70–87.

Kwaśniewski, A. K. and Czech, R. (1992) On quasi-number algebras, *Rep. Math. Phys.*, V. 31, No. 3, pp. 341–351.

Laba, I. (2008) From harmonic analysis to arithmetic combinatorics, *Bull. Amer. Math. Soc.*, V. 45, No. 1, pp. 77–115.

Lacroix, S. F. (1830) *Traité élémentaire d'arithmétique, à l'usage de l'Ecole Centrale des Quatre-Nations*, Bachelier, Paris.

Ladyman, J. (1998) What is structural realism? *Studies History Philos. Sci.*, V. 29, No. 3, pp. 409–424.

Lagrange, J. L. (1773/1775) *Recherches d'Arithmétique*, Nouveaux mémoires de l'Académieroyale des sciences et belles-lettres de Berlin, pp. 695–795.

Lakoff, G. and Núñez, R. (1997) The metaphorical structure of mathematics: Sketching out cognitive foundations for a mind-based mathematics, in *Mathematical Reasoning: Analogies, Metaphors, and Images*, Erlbaum, Mahwah, NJ, pp. 267–280.

Lakoff, G. and Núñez, R. (2000) *Where Mathematics Comes From: How the Embodied Mind Brings Mathematics Into Being*, Basic Books, New York.

Lakshman, V., Nayagam, G., Jeevaraja, S. and Sivaraman, G. (2016) Complete ranking of intuitionistic fuzzy numbers, *Fuzzy Inform. Eng.*, V. 8, No. 2, pp. 237–254.

Lamandé, P. (2004) 'La conception des nombresen France autour de 1800: L'oeuvre didactique de Sylvestre François Lacroix', *Revue Hist. Math.*, V. 10, pp. 45–106.

Lamb, R. (2010) *How Math Works*, https://science.howstuffworks.com/math-concepts/math1.htm.

Lamb, E. (2016) Modular arithmetic at the music stand, *Scientific Amer.*, V. 314, No. 5, https://blogs.scientificamerican.com/roots-of-unity/modular-arithmetic-at-the-music-stand/.

Lambert, I. H. (1770) Vorläufige Kenntnisse für die, so die Quadratur und Rektifikation des Cirkuls suchen, *Beyträge zum Gebrauche der Mathematik und deren Anwendung II. – Berlin*, pp. 140–169.

Landau, E. (1966) *Foundations of Analysis: The Arithmetic of whole, rational, Irrational and complex Numbers*, A supplement to textbooks on the differential and integral calculus, Chelsea Pub. Co, New York.

Landau, E. (1999) *Elementary Number Theory*, American Mathematical Society, RI.

Lang, S. (1994) *Algebraic Number Theory*, Graduate Texts in Mathematics, V. 110, Springer-Verlag, New York.

Lang, M. (2014) One plus one equals three: multi-line fiber lasers for nonlinear microscopy, *Optik & Photonik*, V. 9, No. 4, pp. 53–56.

de La Roche, E. (1520) *Larismetique nouellement composée*, Constantin Fradin, Lyon.

La Roche, (2019) Estienne De, Encyclopedia.com, https://www.encyclopedia. com/science/dictionaries-thesauruses-pictures-and-press-releases/la-roche-es tienne-de.

Lasswell, H. (1948) The structure and function of communication in society, in *The Communication of Ideas*, Bryson, L. (Ed.) Institute for Religious and Social Studies, New York, pp. 37–51.

Laugwitz, D. (1961) Anwendungen unendlichkleiner Zahlen I. Zur Theorie der Distributionen, *J. Reine Angew. Math.*, V. 207, pp. 53–60.

Laugwitz, D. (1961a) Anwendungen unendlichkleiner Zahlen II. Ein Zugang zur Operatorenrechnung von Mikusinski Distributionen, *J. Reine Angew. Math.*, V. 208, pp. 22–34.

Laugwitz, D. (2001) Curt Schmieden's approach to infinitesimal, in Antipodes — *Constructive and Nonstandard Views of the Continuum. Synthese Library*, Studies in Epistemology, Logic, Methodology, and Philosophy of Science, V. 306, Springer, Dordrecht.

Lautman, A. (1938) *Essai sur les notions de structure et d' existence en mathématique*, Hermann, Paris.

Law, S. (2012) *A Brief History of Numbers and Counting, Pt 1: Mathematics Advanced with Civilization*, Deseret News, https://www.deseretnews.com/ article/865560110/A-brief-history-of-numbers-and-counting-Part-1-Mathem atics-advanced-with-civilization.html.

Law, S. (2012) *A Brief History of Numbers and Counting, Pt 2: Indian Invention of Zero was Huge in Development of Math*, Deseret News, https://www. deseretnews.com/article/865560133/A-brief-history-of-numbers-and-counti ng-Part-2-Indian-invention-of-zero-was-huge-in-development-of.html.

Lawvere, F. W. (1963) *Functorial Semantics of Algebraic Theories*, Ph.D. Thesis, Columbia University, New York.

Lawvere, F. W. (1994) Cohesive toposes and Cantor's "lauter Einsen", *Philos. Math.*, V. 2, No. 1, pp. 5–15.

Lawvere, F. W. and Rosebrugh, R. (2003) *Sets for Mathematics*. Cambridge University Press.

Lawrence, C. (2011) Making $1+1 = 3$: Improving sedation through drug synergy, *Gastrointestinal Endoscopy*, February 2011.

Lawrence, R. (2017) Talking about numbers: easy arguments for mathematical realism, *History Philos. Logic*, V. 38, No. 4, pp. 390–394.

Le Corre, M. and Carey, S. (2007) One, two, three, four, nothing more: an investigation of the conceptual sources of the verbal counting principles, *Cognition*, V. 105, pp. 395–438.

Le Corre, M., Brannon, E. M., Van de Walle, G. A. and Carey, S. (2006) Re-visiting the competence/performance debate in the acquisition of the counting principles, *Cognitive Psychol.*, V. 52, No. 2, pp. 130–169.

Lea, R. (2016) Why one plus one equals three in big analytics, *Forbes*, May 27, 2016, https://www.forbes.com/sites/teradata/2016/05/27/why-one-plus-one-equals-three-in-big-analytics/#1aa2070056d8.

Lebombo bone, *Wikipedia*, https://en.wikipedia.org/wiki/Lebombo_bone.

L'Ecuyer, P. (2012) Random number generation, in *Handbook of Computational Statistics*, Springer-Verlag. pp. 35–71.

Ledford, H. (2013) Plants perform molecular maths, *Nature*, https://www.nature.com/news/plants-perform-molecular-maths-1.13251#:~:text=Computer%2Dgenerated%20models%20published%20in,is%20off%20the%20menu1.

Lee, J. M. (2000) *Introduction to Topological Manifolds*, Graduate Texts in Mathematics, V. 202, Springer, New York.

Lee, E. and Li, R. (1988) Comparison of fuzzy numbers based on the probability measure of fuzzy events, *Comput. Math. Appl.*, V. 15, pp. 887–896.

Lefschetz, S. (1950) The structure of mathematics, *American Sci.*, V. 38, No. 1, pp. 105–111.

Legendre, A. M. (1830) *Théorie des Nombres*, Firmin Didot frères, Paris.

Lejeune Dirichlet, P. G. (1879) *Vorlesungenüber Zahlentheorie*, Druck und Verlag *von* Friedrich Vieweg und Sohn, Braunschweig.

Leibniz, G. (1666) *Disputatio Arithmetica De Complexionibus*, Literis Spörelianis, Universität Leipzig.

Leibniz, G. (1703) Explication de l'arithmétique binaire, Que se sert seul caractéres 0 & 1; avec des Remarques sur son utilité, & sur ce qu'elle donne le sens des anciennes figures Chinoises de Fohy *Memo. Acad. Roy. Sci.*, pp. 85–93.

Leibniz, G. (1860–1875) *Die Mathematischen Schriften von G. W. Leibniz*, C. I. Gerhardt (Ed.) Winter, Berlin.

Leibovich, T. and Ansari, D. (2016) The symbol-grounding problem in numerical cognition: A review of theory, evidence, and outstanding questions, *Can. J. Exp. Psychol.*, V. 70, No. 1, pp. 12–23.

Lemaire, P., Abdi, H. and Fayol, M. (1996) The role of working memory resources in simple cognitive arithmetic, *Euro. J. Cogn. Psychol.*, V. 8, No. 1, pp. 73–103.

Leng, M. (2010) *Mathematics and Reality*, Oxford University Press, Oxford.

Leng, M. (2018) Does $2+3 = 5$? In defence of a near absurdity, *Math Intelligencer*, V. 40, pp. 14–17.

Lesk, A. M. (2000) The unreasonable effectiveness of mathematics in molecular biology, *Math. Intelligencer*, V. 22, No. 2, pp. 28–37.

Leslie, A. M., Gelman, R. and Gallistel, C. R. (2008) The generative basis of natural number concepts. *Trends Cognitive Sci.*, V. 12, No. 6, pp. 213–218.

Leuthesser, L., Kohli, C. and Suri, R. (2003) 2+2=5? A framework for using co-branding to leverage a brand, *J. Brand Manag.*, V. 11, No. 1, pp. 35–47.

Lev, F. M. (1989) Modular representations as a possible basis of finite physics, *J. Math. Phys.*, V. 30, pp. 1985–1998.

Lev, F. M. (1993) Finiteness of physics and its possible consequences, *J. Math. Phys.*, V. 34, pp. 490–527.

Lev, F. M. (2017) Why finite mathematics is the most fundamental and ultimate quantum theory will be based on finite mathematics. *Phys. Part. Nuclei Lett.*, V. 14, pp. 77–82.

Lev, F. M. (2020) *Finite Mathematics as the Foundation of Classical Mathematics and Quantum Theory with Application to Gravity and Particle Theory*, Springer, New York.

Lev, F. M. (2021) *Popular Discussion of Classical and Finite Mathematics*, hal-03006612v4.

Lewis, T. D. (2006) *The Arithmetization of Analysis: From Eudoxus to Dedekind*, Southern University.

Li, P., Le Corre, M., Shui, R., Jia, G. and Carey, S. (2003) Effects of plural syntax on number word learning: A Cross-linguistic study, Paper presented at the 28th Boston University Conference on Language Development.

Li, M. and Vitanyi, P. (1997) *An Introduction to Kolmogorov Complexity and its Applications*, Springer-Verlag, New York.

Liggins, D. (2006) Is there a good epistemological argument against platonism? *Analysis*, V. 66, No. 2, pp. 135–141.

Linchevski, L. and Herscovics, N. (1996) Crossing the cognitive gap between arithmetic and algebra: Operating on the unknown in the context of equations. *Educ Stud Math.*, V. 30, pp. 39–65.

Lindemann, F. (1882) *Über die Ludolph́sche Zahl, Sitzungsberichte der Akademie der Wiss.* Berlin, pp. 679–682.

Ling, C.-H. (1995) Representation of associative functions, *Publ. Math.*, V. 12, pp. 189–212.

Liouville, J. (1844) Sur les nombres transcendants, *C. R. Acad. Sci.*, V. 18, pp. 883–885.

Lipshitz, L. (1979) Diophantine correct models of arithmetic, *Proc. Amer. Math. Soc.*, V. 73, No. 1, pp. 107–108.

Lipton, J. S. and Spelke, E. (2003) Origins of number sense: Large number discrimination in human infants, *Psychol. Sci.*, V. 15, No. 5, pp. 396–401.

Lipton, J. S. and Spelke, E. (2004) Discrimination of large and small numerosities by human infants, *Infancy*, V. 5, No. 3, pp. 271–290.

Littlewood, J. E. (1953) *Miscellany*, Methuen, London.

Litvinov, G. L. (2007) The Maslov dequantization, idempotent and tropical mathematics: A brief introduction, *J. Math. Sci. (New York)*, V. 140, No. 3, pp. 426–444.

Litvinov, G. L., Maslov, V. P. and Shpiz, G. B. (2001) Idempotent functional analysis: an algebraic approach, *Math. Notes*, V. 69, No. 5–6, pp. 696–729.

Liu, Q. (2006) *Algebraic Geometry and Arithmetic Curves*, Oxford Graduate Texts in Mathematics 6, Oxford University Press.

Livanova, A. (1969) *Three Destinies — Comprehension of the World*, Znaniye, Moscow (in Russian).

Livesey, D. A. (1974) The importance of numerical algorithms for solving economic optimization problems, *Int. J. Syst. Sci.*, V. 5, No. 5, pp. 435–451.

Livio, M. (2009) *Is God a Mathematician?* Simon & Schuster, New York.

Livshits, L. Z., Ostrovskii, I. V. and Chistyakov, G. P. (1976) Arithmetic of probability laws, *J. Math. Sci.*, V. 6, pp. 99–122.

Locke, J. (1689/1975) *An Essay Concerning Human Understanding*, Clarendon Press, Oxford.

Lockhart, P. (2017) *Arithmetic*, Belknap Press.

Loeb, D. (1992) Sets with a negative numbers of elements, *Adv. Math.*, V. 91, pp. 64–74.

Löffler, E. (1919) *Ziffern und ziffernsysteme*, Leipzig,

Lonsdorf, E. V. (2007) The role of behavioral research in the conservation of chimpanzees and gorillas, *J. Appl. Anim. Welf. Sci.*, V. 10, No. 1, pp. 71–78.

Lorentz, H. A. (1895) *Versuch einer Theorie der electrischen und optischen Erscheinungen in bewegten Körpern*, Brill, Leiden.

Lorenzen, P. (1951) Algebraische und logistische Untersuchungen über freie Verbände, *J. Symbolic Logic*, V. 16, pp. 81–106.

Lorenzini, D. (1996) *An Invitation to Arithmetic Geometry*, Graduate Studies in Mathematics, V. 9, American Mathematical Society, Providence, RI.

Losev, A. F. (1994) *Myth, Number, Essence*, Mysl', Moscow (in Russian).

Lotito, P. Quadrat, J.-P. and Mancinelli, E. (2005) Traffic assignment & Gibbs–Maslov semirings. in *Idempotent Mathematics and Mathematical Physics*, Contemporary Mathematics, V. 377, American Mathematical Society, Providence, RI, pp. 209–220.

Löwenheim, L. (1915) Über Möglichkeiten im Relativkalkül, *Math. Ann.*, V. 76, No. 4, pp. 447–470.

Lozano-Robiedo, A. (2019) *Number Theory and Geometry: An Introduction to Arithmetic Geometry*, MAA, Washington, DC.

Lu, M. (2004) *Arithmetic and Logic in Computer systems*, Wiley, New York.

Luce, L. (1991) Literalism and the applicability of arithmetic, *British J. Philos. Sci.*, V. 42, No. 4, pp. 469–489.

Luce, R. D. (1964) The mathematics used in mathematical psychology, *Amer. Math. Monthly*, V. 71, pp. 364–378.

Luce, R. D. (1987) Measurement structures with Archimedean ordered translation groups, *Order*, V. 4, pp. 165–189.

Luce, R. D. (2002) A psychophysical theory of intensity proportions, joint presentations, and matches, *Psychol. Rev.*, V. 109, No. 3, pp. 520–532.

Luce, R. D. and Edwards, W. (1958) The derivation of subjective scales from just noticeable differences, *Psychol. Rev.*, V. 65, pp. 222–237.

Luce, R. D., Bush, R. R. and Galanter, E. (1963) *Handbook of Mathematical Psychology*, V. 1. Wiley, New York.

Luckenbill, D. (1927) *Ancient Records of Assyria and Babylonia*, V. 2, University of Chicago Press, Chicago.

Luttrull, E. G. (2013) *Arts & Numbers: A Financial Guide for Artists, Writers, Performers, and Other Members of the Creative Class*, Agate B2, Evanston, Illinois.

Lyons, I. M. and Beilock, S. L. (2013) Ordinality and the nature of symbolic numbers, *J. Neurosci.*, V. 33, pp. 17052–17061.

Ma, M., Friedman, M. and Kandel, A. (1999) A new fuzzy arithmetic, *Fuzzy Sets Syst.*, V. 108, pp. 83–90.

MacDuffee, C. C. (2004) Arithmetic, Britannica, https://www.britannica.com/science/arithmetic.

MacFarlane, A. (2010) *Physical Arithmetic*, Forgotten Books.

MacKenzie, D. (1993) Negotiating arithmetic, constructing proof: the sociology of mathematics and information technology, *Social Studies*, V. 23, No. 1, pp. 37–65.

Maclagan, D. and Sturmfels, B. (2015) *Introduction to Tropical Geometry*, Graduate Studies in Mathematics, V. 161, American Mathematical Society.

Mac Lane, S. (1939) The universality of formal power series fields, *Bull. Amer. Math. Soc.*, V. 45, pp. 880–890.

Mac Lane, S. (1996) Structure in mathematics, *Philos. Math.*, V. 4, No. 3, pp. 174–183.

Mac Lane, S. (1998) *Categories for the Working Mathematician*, Springer, New York.

Mac Lane, S. and Birkhoff, G. (1999) *Algebra*, AMS Chelsea Publishing Series, V. 330 American Mathematical Society

MacNeal, E. *Mathsemantics, Making Numbers Talk Sense*, Penguin Books, UK.

Macphee, K. (1999) The art of numbers, *Plus magazine!* https://plus.maths.org/content/art-numbers.

Madrid, M. J., Maz-Machado, A., López, C. and León-Mantero, C. (2020) Old arithmetic books: mathematics in Spain in the first half of the sixteenth century, *Int. Electron. J. Math. Education*, V. 15, No. 1, em0553.

Magueijo, J. (2003) New varying speed of light theories, *Rep. Prog. Phys.*, V. 66, p.2025.

Mahapatra, G. S. and Roy, T. K. (2013) Intuitionistic fuzzy number and its arithmetic operation with application on system failure, *Uncertain Sys.*, V. 7, No. 2, pp. 92–107.

Mahoney, M. S. (1994) *The Mathematical Career of Pierre de Fermat*, Princeton University Press, Princeton.

Mair, V. H. (Ed.) (1990) *Tao Te Ching: The Classic Book of Integrity and the Way*, Bantam Books, New York.

Malaisé, F. (1842) *Theoretisch-Praktischer Unterricht im Rechnen für die niederen Classen der Regimentsschulen der Königl. Bayer. Infantrie und Cavalerie [Theoretical and Practical Instruction in Arithmetic for the Lower Classes of the Royal Bavarian Infantry and Cavalry School]*, Munich, Germany.

Mallik, A. K. (2017) *The Story of Numbers*, World Scientific, Singapore.

Mancosu, P. (2009) Measuring the size of infinite collections of natural numbers: was cantor's theory of infinite number inevitable? *Rev. Symbolic Logic*, V. 2, No. 4, pp. 612–646.

Mancosu, P. (2015) In good company? On hume's principle and the assignment of numbers to infinite concepts, *Rev. Symbolic Logic*, V. 8, No. 2, pp. 370–410.

Mandler, G. and Shebo, B. J. (1982) Subitizing: An analysis of its component processes, *J. Exp. Psychol.: General*, V. 111, pp. 1–22.

Mane, R. (1952) Evolution of mutuality: one plus one equals three; formula characterizing mutuality, *La Revue Du Praticien*, V. 2, No. 5, pp. 302–304.

Manin, Y. I. (1991) *Course in Mathematical Logic*, Springer-Verlag, New York.

Manin, Y. (2007) *Mathematical Knowledge: Internal, Social and Cultural Aspects*, preprint in Mathematics History and Overview, arXiv:math/0703427 [math.HO].

Mannoury, G. (1909) *Methodologisches und Philosophisches zur Elementar Mathematik*, Visser, Haarlem.

Maor, E. (2018) *Music by the Numbers: From Pythagoras to Schoenberg*, Princeton University Press, Princeton.

Marcolli, M. (2005) *Lectures on Arithmetic Non-commutative Geometry*, Univercity Lecture Series 36, American Mathematical Society.

Marcolli, M. and Thorngren, R. (2011) *Thermodynamic Semirings*, preprint in Quantum Algebra, math.QA, arXiv:1108.2874.

Marichal, J.-L. (2000) On an axiomatization of the quasi-arithmetic mean values without the symmetry axiom, *Aequationes mathematicae*, V. 59, pp. 74–83.

Marichal, J.-L. (2009) *Aggregation Functions for Decision-Making*, Preprint math.ST, arXiv: 0901.4232.

Marie, K. L. (2007) One plus one equals three: joint-use libraries in urban areas — the ultimate form of library cooperation, *Library Leadership Manag.* (LL&M), V. 21, No. 1.

Marewski, J. N. and Bornmann, L. (2018) *Opium in Science and Society: Numbers*, preprint in Computer Science, [cs.DL], arXiv:1804.11210.

Markov, A. A. and Nagornii, N. M. (1988) *The Theory of Algorithms*, Springer, New York.

Marks, L. E. (1974) *Sensory Processes: The New Psychophysics*, Academic Press, London.

Marks, M. L. and Mirvis, P. H. (2010) Joining forces: making one plus one equal three in mergers, in *Acquisitions, and Alliances*, Jossey-Bass.

Marre, A. (Ed.) (1880) Le Triparty en la science des nombres par Maistre Nicolas Chuquet, parisien, d'après le manuscript fonds français, no. 1346 de la Bibliothèque nationale de Paris, *Bull. bibliografia Storia Sci. Mate. Fisiche*, V. 13 (1880), pp. 593–659, 693–814.

Marris, R. (1958) *Economic Arithmetic*, Macmillan, London.

Marshall, C. C. (1910) *Inductive Commercial Arithmetic: A Practical Treatise on Business Computation*, Goodyear-Marshall publishing company, Cedar Rapids, IA.

Marshall, O. R. (2018) The psychology and philosophy of natural numbers, *Philos Mathe.*, V. 26, No. 1, pp. 40–58.

Martin, G. (1985) *Arithmetic and Combinatorics*: *Kant and his Contemporaries*, Southern Illinois University Press, Carbondale and Edwardsville.

Martin, G. (1956) *Arithmetik und Kombinatorik bei Kant*, de Gruyter, Berlin.

Martin, J. C. (1991) *Introduction to Languages and the Theory of Computation*, McGraw-Hill, New York.

Martin, E. and Osherson, D. N. (2001) Induction by enumeration, *Inform. Comput.*, V. 171, No. 1, pp. 50–68.

Martinez, A. A. (2006) *Negative Math: How Mathematical Rules Can Be Positively Bent*, Princeton University Press, Princeton.

Marzocchi, G. M., Lucangeli, D., De Meo, T., Fini, F. R. and Comoldi, C. (2002) The disturbing effect of irrelevant information on arithmetic problem solving in inattentive children, *Developmental Neuropsychol.*, V. 21, No. 1, pp. S73–92.

Masi, M. (2007) On the extended Kolmogorov–Nagumo information-entropy theory, the $q \rightarrow 1/q$ duality and its possible implications for a non-extensive two-dimensional Ising model, *Physica A*, V. 377(1), pp. 67–78.

Maslov, V. P. (1987) *Asymptotic Methods for Solving Pseudo-differential Equations* (in Russian).

Matejko, A. A. and Ansari, D. (2015) Drawing connections between white matter and numerical and mathematical cognition: A literature review, *Neurosci. Biobehav. Rev.*, V. 48, pp. 35–52.

Matkowski, J. and Świątkowski, T. (1993) Subadditive functions and partial converses of Minkowski's and Mulholland's inequalities, *Fundam. Math.*, V. 143, No. 1, pp. 75–85.

Matson, J. (2009) The Origin of Zero, *Scientific American*, August 21, http://www.scientificamerican.com/article.cfm?id=history-of-zero.

Matsushita, S. (1951) On the foundations of orders in groups, *J. Inst. Polytechn. Osaka City Univ.*, A, V. 2, pp. 19–22.

Matsuzawa, T. (2009) Symbolic representation of number in chimpanzees, *Curr. Opin. Neurobiol.*, V. 19, pp. 92–98.

Mattessich, R. (1998) From accounting to negative numbers: A signal contribution of medieval India to mathematics, *Accounting Historians J.*, V. 25, No. 2, pp. 129–145.

Maugin, G. A. (2017) *Non-Classical Continuum Mechanics*, Advanced Structured Materials, V. 51, Springer Nature, Singapore.

Mazur, B. (1986) Arithmetic on curves, *Bull. Amer. Math. Soc*, V. 14, No. 2, pp. 207–259.

Mazur, B. (2008) When is one thing equal to some other thing? in B. Gold and R. Simons, (Eds.), *Proof and Other Dilemmas*: *Mathematics and Philosophy*, Mathematical Association of America, Washington, DC, pp. 221–243.

Mazur, B. (2008a) Mathematical Platonism and its opposites, *European Math. Society Newslett.*, No. 68, pp. 19–21.

Mazur, B. and Swinnerton-Dyer, P. (1974) Arithmetic of Weil curves, *Invent. Math.*, V. 25, pp. 1–61.

McCall, S. (2014) *The Consistency of Arithmetic and Other Essays*, Oxford University Press, New York.

McCartin, B. J. (2006) e: The Master of All, *Math.Intelligencer*, V. 28, No. 2, pp. 10–21.

McClenon, R. B. (1919) Leonardo of Pisa and his Liber Quadratorum, *Amer. Math. Monthly*, V. 26, No. 1, pp. 1–8.

McCloskey, M., Harley, W. and Sokol, S. M. (1991) Models of arithmetic fact retrieval: an evaluation in light of findings from normal and brain-damaged subjects, *J. Exp. Psychol.* Learn. Mem. Cogn., V. 17, No. 3, pp. 377–397.

McComb, K., Packer, C. and Pusey, A. (1994) Roaring and numerical assessment in contests between groups of female lions, Panthera leo, *Animal Behav.*, V. 47, pp. 379–387.

McCrink, K. and Wynn, K. (2004) Large-number addition and subtractrion by 9-month old infants, *Psychol. Sci.*, V. 15, No. 11, pp. 776–781.

McCulloch, W. S. (1961) What is a number that a man may know it, and a man, that he may know a number? (the Ninth Alfred Korzybski Memorial Lecture), *General Semantics Bulletin*, No. 26/27, Institute of General Semantics, pp. 7–18.

McEneaney, W. M. (1999) Exactly linearizing algebras for risk-sensitive filtering, in *Proc. Conf. Decision & Control, Phoenix.* Arizona, USA, pp. 137–142.

McCarty, C. (2014) Arithmetic, convention, reality, in *Castañeda and his Guises*, De Gruyter, Berlin, pp. 83–96.

McKenzie, R. (2000) Arithmetic of finite ordered sets: Cancellation of exponents, I, *Order*, V. 16, pp. 313–333.

McKenzie, R. (2000a) Arithmetic of finite ordered sets: Cancellation of exponents, II. *Order*, V. 17, pp. 309–332.

McLarty, C. (1993) Numbers can be just what they have to, *Nous*, V. 27, pp. 487–498.

McLeish, J. (1994) *The Story of Numbers: How Mathematics Has Shaped Civilization*, Fawcett Columbine, New York.

Meck, W. H. and Church, R. M. (1983) A mode control model of counting and timing processes, *J. Exp. Psychol. Anim. Behav. Process.*, V. 9, pp. 320–334.

Meiert, J. O. (2015) $1 + 1 = 3$: Explaining busyness and background noise on websites, in *On Web Development: Articles 2005–2015*, Kindle Edition.

Meinong, A. (1882) Zur Relationstheorie [On the theory of relations], *Sitzungsberichte der Phil.-Hist. Classe d. K. Akad. Wissenschaften zu Wien*, V. 101, pp. 573–752.

Melquiond, S. B. G. (2017) *Computer Arithmetic and Formal Proofs*, ISTE Press Elsevier.

Melzak, Z. A. (1961) An informal arithmetical approach to computability and computation, *Canad. Math. Bull.*, V. 4, pp. 279–293.

Mendell, H. (2017) Aristotle and Mathematics, *The Stanford Encyclopedia of Philosophy* (Spring 2017 Edition), E. N. Zalta (ed.), https://plato.stanford.edu/archives/spr2017/entries/aristotle-mathematics/.

Meng, D., Zhang, X. and Qin, K. (2011) Soft rough fuzzy sets and soft fuzzy rough sets, *Computers & Mathematics with Applications*, V. 62, No. 12, pp. 4635–4645.

Menger, K. (1942) Statistical metrics, *Proc. Natl. Acad. Sci. USA.*, V. 8, pp. 535–537.

Menger, K. (1951) Ensembles flous et fonctions aléatoires, *C. R. Acad. Sci.*, V. 37, pp. 2001–2003.

Menger, K. (1951a) Probabilistic theories of relations, *Proc. Natl. Acad. Sci. USA*, V. 37, pp. 178–180.

Menger, K. (1951b) Probabilistic geometry, *Proc. Natl. Acad. Sci. USA*, V. 37, pp. 226–229.

Menger, K. (1979) Geometry and positivism. A probabilistic microgeometry, in *Selected Papers in Logic and Foundations, Didactics, Economics*, Vienna Circle Collection, 10, D. Reidel Publ. Comp., Dordrecht, Holland, pp. 225–234.

Menninger, K. (1992) *Number Words and Number Symbols: A Cultural History of Numbers*, Dover Publications, New York.

Menon, V. (2010) Developmental cognitive neuroscience of arithmetic: implications for learning and education. *ZDM: Int. J. Math. Education*, V. 42, No. 6, pp. 515–525.

Menon, V. (2015) Arithmetic in the child and adult brain, in *Oxford Library of Psychology. The Oxford Handbook of Numerical Cognition*, Oxford University Press, New York, pp. 502–530.

Méray, C. (1869) Remarques sur la nature des quantitésdéfinies par lacondition de servir de limites à des variables données, *Revuedes Sociétés savantes, Sci. Math. phys. nat.*, V. 4, No. 2, pp. 280–289.

Méray Ch. (1872) *Nouveau précis d'analyse infinitesimal*, F. Savy, Paris.

Merritt, D. J. and Brannon, E. M. (2013). Nothing to it: Precursors to a zero concept in preschoolers, *Behav. Processes*, V. 93, pp. 91–97.

Merritt, D. J., Rugani, R. and Brannon, E. M. (2009) Empty sets as part of the numerical continuum: Conceptual precursors to the zero concept in rhesus monkeys, *J. Exp. Psychol. Gen.*, V. 138, pp. 258–269.

Mesiar, R. and Rybařik, J. (1993) Pseudo-aritmetical operations, *Tatra Mount. Math. Publ.*, V. 2, pp. 185–192.

Mesiar, R. and Rybařik, J. (1998) Entropy of fuzzy partitions: A general model, *Fuzzy Sets Syst.*, V. 99, No. 1, pp. 73–79.

Mesiar, R. and Pap, E. (1999) Idempotent integral as limit of g-integrals, *Fuzzy Sets Syst.*, V. 102, pp. 385–392.

Mesiar, R. and Pap, E. (2010) The Choquet integral as Lebesgue integral and related inequalities, *Kybernetika*, V. 46, pp. 1098–1107.

Mesiar, R. and Rybařik, J. (1993) Pseudo-arithmetical operations, *Tatra Mountains Math. Publ.*, V. 2, pp. 185–192.

Meyer, R. K. (1976) Relevant arithmetic, *Bulletin of the Section of Logic of the Polish Academy of Sciences*, V. 5, pp. 133–137.

Meyer, R. K. and Mortensen, C. (1984) Inconsistent models for relevant arithmetics, *J. Symbolic Logic*, V. 49, pp. 917–929.

Michalak, M. (2011) Rough Numbers and Rough Regression, in *Proc. 13th Int. Conf. Rough Sets, Fuzzy Sets, Data Mining Granular Computing, RSFDGrC 2011*, Moscow, Russia, pp. 68–71.

Michaux, C. and Villemaire, R. (1996) Presburger arithmetic and recognizability of sets of natural numbers by automata: new proofs of Cobham's and Semenov's theorems, *Ann. Pure Appl. Logic*. V. 77, No. 3, pp. 251–277.

Mikhalkin, G. (2004) Amoebas of algebraic varieties and tropical geometry, in *Different Faces of Geometry*, International Mathematical Series, V. 3, Kluwer Academic/Plenum Publishers, New York, pp. 257–300.

Mikhalkin, G. (2006) Tropical geometry and its applications, in *Proc. Int. Cong Mathematicians (ICM)*, Madrid, Spain, August 22–30, 2006. Volume II: Invited lectures, European Mathematical Society (EMS), Zürich, pp. 827–852.

Mikhalkin, G. and Rau, J. (2018) *Tropical geometry*, Internet resource, http://wwwmath.un-tuebingen.de/user/jora.

Mikusinski, J. (1983) Hypernumbers, Part I. Algebra, *Studia Math.*, V. 77, pp. 3–16.

Milea, S., Shelley, C. D. and Weissman, M. H. (2019) Arithmetic of arithmetic Coxeter groups, *Proc. Natl. Acad. Sci. USA*, V. 116, No. 2, pp. 442–449.

Miller, G. A. (1956) The magical number seven, plus or minus two: some limits on our capacity for processing information, *Psychol. Rev.*, V. 63, pp. 81–97.

Miller, G. R. (1966) *Speech Communication: A Behavioral Approach*, Bobbs-Merril, Indianapolis, IN.

Miller, J. P. (1982) *Numbers in Presence and Absence: A Study of Husserl's Philosophy of Mathematics*, Phaenomenologica, V. 90, Nijhoff, Den Haag.

Miller, M. (1999) *Time, Clocks and Causality*, Quackgrass Press, Calgary (electronic edition: http://www.quackgrass.com/home.html).

Milier, G. A. (1925) Arithmetization in the History of Mathematics, *Proc. Natl. Acad. Sci.* V. 11, No. 9, pp. 546–548.

Minsky, M. (1967) *Computation: Finite and Infinite Machines*, Prentice-Hall, Inc., Englewood Cliffs, NJ.

Mirimanoff, D. (1917) Les antinomies de Russell et de Burali-Forti et le problème fondamental de la théorie des ensembles, *Enseign. Math.* V. 19, pp. 37–52.

Mitchell, H. B. (2004) Ranking — Intuitionistic Fuzzy numbers, *Int. J. Uncert. Fuzz. Knowledge Based Systems*, V. 12, pp. 377–386.

Mix, K., Levine, S. C., and Huttenlocher, J. (2002) *Quantitative Development in Infancy and Early Childhood*, Oxford University Press, Oxford.

Miyamoto, S. (2001) Fuzzy Multisets and their Generalizations, in *Multiset Processing*, pp. 225–235.

Mizumoto, M. and Tanaka, K. (1976) Algebraic properties of fuzzy numbers, *Inform. Control*, V. 31, No. 3, pp. 312–341.

Mizumoto, M. and Tanaka, K. (1979) Some properties of fuzzy numbers, in *Advances in Fuzzy Set Theory and Applications*, North-Holland, Amsterdam, pp. 153–164.

Möbius, A. F. (1827) *Der Barycentrische Calcul*, Johan Ambrosius Barth, Leipzig.

Modell, W. and Place, D. J. (1957) Medical Arithmetic, in *The Use of Drugs*, Springer, Berlin.

Mollin, R. A. (2005) *Codes: The Guide to Secrecy from Ancient to Modern Times*, Discrete Mathematics and its Applications, Chapman & Hall/CRC, Boca Raton, FL.

Moltmann, F. (2013) Reference to numbers in natural language, *Philos. Studies*, V. 162, No. 3, pp. 499–536.

Moltmann, F. (2016) The number of planets, a number-referring term? in *Abstractionism: Essays in Philosophy of Mathematics*, Oxford University Press, Oxford, pp. 113–129.

Monk, J. D. (1976) *Mathematical Logic*, Graduate Texts in Mathematics, Springer-Verlag, Berlin/New York.

Monro, G. P. (1987) The Concept of Multiset, *Zeits. Math. Logik Grundlagen Math.*, V. 3, pp. 171–178.

Moore, R. E. (1966) *Interval Analysis*, Prentice-Hall, Englewood Cliff, NJ.

Moore, C. (1996) Recursion theory on the reals and continuous-time computation: real numbers and computers, *Theoret. Comput. Sci.*, 162, No. 1, pp. 23–44.

Moore, G. W. (2012) *Felix Klein Lectures: "Applications of the Six-dimensional (2,0) Theory to Physical Mathematics,"* 1–11 October, 2012 at the Hausdorff Institute for Mathematics, Bonn http://www.physics.rutgers.edu/_gmoore/FelixKleinLectureNotes.

Moore, G. W. (2014) *Physical Mathematics and the Future*, http://www.physics.rutgers.edu/~gmoore/PhysicalMathematicsAndFuture.pdf.

Mora, M., Córdova-Lepe, F. and Del-Vall, R. (2012) A non-Newtonian gradient for contour detection in images with multiplicative noise, *Pattern Recogn. Lett.*, V. 33, pp. 1245–1256.

Morgan, A. (1836) *The Connexion of Number and Magnitude: An Attempt to Explain the Fifth Book of Euclid*, Taylor Walton.

Morita, K. (2017) *Theory of Reversible Computing*, Springer, Japan, KK.

Morris, M. (2017) *Lessons in Math and Marriage: One plus One equals ONE — How to be Better Together*, Marriage Dynamics Institute, https://marriagedynamics.com/tips-growing-healthy-marriage-recognizing-rooting-selfishness/.

Morrow, M. (2012) Grothendieck's trace map for arithmetic surfaces via residues and higher adeles, *Algebra Number Theory J.*, V. 6–7, pp. 1503–1536

Mortensen, C. (1995) *Inconsistent Mathematics*, Kluwer Mathematics and Its Applications Series, Kluwer, Dordrecht.

Mortensen, C. (2000) Prospects for inconsistency, in *Frontiers of Paraconsistent Logic*, Research Studies Press, London, pp. 203–208.

Mosteller, F. (1977) *Data Analysis and Regression: A Second Course in Statistics*, Addison-Wesley., Reading, MA.

Mou, Y., and vanMarle, K. (2014). Two core systems of numerical representation in infants, *Developmental Rev.*, V. 34, No. 1, pp. 1–25.

Moulton, B. R. and Smith, J. W. (1992) Price indices, in *The New Palgrave Dictionary of Money and Finance*, V. 3, pp. 179–181.

Moyer, R. and Landauer, T. (1967) Time required for judgements of numerical inequality, *Nature*, V. 215, pp. 1519–1520.

Muchnik, A. A. (2003) The definable criterion for definability in Presburger arithmetic and its applications, *Theoret. Comput. Sci.*, V. 290, No. 3, pp. 1433–1444.

Mühlhölzer, F. (2010) Mathematical intuition and natural numbers: a critical discussion, *Erkenntniss*, V. 73, pp. 265–292.

Muir, J. (1996) *Of Men and Numbers: The Story of the Great Mathematicians*, Dover Books on Mathematics, Dover Publications.

Müller N. T. and Uhrhan C. (2012) Some Steps into Verification of Exact Real Arithmetic, in *NASA Formal Methods (NFM 2012)*, Lecture Notes in Computer Science, V. 7226. Springer, Berlin.

Multiplication Algorithms, Wikipedia, https://en.wikipedia.org/wiki/Multiplica tion_Algorithms.

Mumford, D. (2008) Why I am a Platonist, *EMS Newslett.*, No. 12, pp. 27–29.

Mundy, B. (1987) The metaphysics of quantity, *Philos. Studies*, V. 51, pp. 29–54.

Murawski, R. (1998) Undefinability of truth. The problem of the priority: Tarski vs. Gödel, *History Philos. Logic*, V. 19, pp. 153–160.

Murawski, R. (2001) On proofs of the consistency of arithmetic, *Studies Logic, Grammar Rhetoric*, V. 4, No. 17, pp. 41–50.

Murphy, J. (1999) Are your additives in synergy? (when 1+1=3), *Plastics, Additives Compounding*, V. 1, No. 7, pp. 20–24.

Murphy, M. and Miller, M. (2010) Making $1 + 1 = 3$, *FTI J.* http://www.ftijournal.com/article/making-1-1-3.

Nachtomy, O. (2011) A tale of two thinkers, one meeting, and three degrees of infinity: Leibniz and Spinoza (1675–8), *British J. History Philos.*, V. 19, No. 5, pp. 935–961.

Naets, J. (2010) How to define a number? A general epistemological account of Simon Stevin's art of defining, *Topoi*, V. 29, pp. 77–86.

Nagumo, M. (1930) Übereine Klasse der Mittelwerte, *Jap. J. Math.*, V. 7, pp. 71–79.

Nagumo, M. (1977) Quantities and real numbers, *Osaka J. Math.*, V. 14, pp. 1–10.

Napier, J. (1614) *Mirifici Logarithmorum Canonis Descriptio* [The Description of the Wonderful Rule of Logarithms], Andrew Hart, Edinburgh, Scotland, (in Latin).

Napier, J. (1990) *Rabdology*, (Richardson, W. F., Transl.) MIT Press, Cambridge, MA.

Narens, L. (1980) A qualitative treatment of Weber's law, *J. Math. Psychol.*, V. 13, pp. 88–91.

Narens, L. (1981) On the scales of measurement, *J. Math. Psychol.*, V. 24, No. 3, pp. 249–275.

Narens, L. (1981a) A general theory of ratio scalability with remarks about the measurement-theoretic concept of meaningfulness, *Theory Decision*, V. 13, pp. 1–70.

Narens, L. (1985) *Abstract Measurement Theory*, MIT Press, Cambridge, MA.

Narens, L. (1988) Meaningfulness and the Erlanger program of Felix Klein, *Math. Inform. Sci. Humaines,* V. 101, pp. 61–72.

Narens, L. and Luce, R. D. (1986) Measurement: the theory of numerical assignments, *Psychol. Bull.* V. 99, pp. 166–180.

Narens, L. and Mausfeld, R. (1992) On the relationship of the psychological and the physical in psychophysics, *Psychol.* Rev., V. 99, No. 3, pp. 467–479.

National Research Council, (2001) Number: What is there to know? in *Adding It Up: Helping Children Learn Mathematics,* The National Academies Press, Washington, DC.

Naudts J. (2011) *Generalised Thermostatistics,* Springer, London.

Naudts, J. (2002) Deformed exponentials and logarithms in generalized thermostatistics, *Physica* A, V. 316, pp. 323–334.

Naudts, J. and Czachor, M. (2001) Generalized thermostatistics and Kolmogorov–Nagumo averages, preprint, arXiv: cond-mat/0110077.

Nayagam, V. L. G., Venkateshwari, G., Sivaraman, G. (2008) Ranking of intuitionistic fuzzy numbers, *Proc. Int. Con. Fuzzy Syst.,* 2008, *Fuzz-IEEE.*

Nayak, S. and Chakraverty, S. (2017) Pseudo Fuzzy Set, *Ann. Fuzzy Math. Inform.,* v. 14, No. 3, pp. 237–248.

Nechaev, V. I. (1975) *Number Systems,* Nauka, Moscow (in Russian).

Nelson, E. (1986) *Predicative Arithmetic,* Mathematical Notes, V. 32, Princeton University Press, Princeton, NJ.

Neugebauer (1969) *The Exact Sciences in Antiquity,* Courier Dover Publications, New York.

Nevanlinna, R. (1939) Le Theoreme de Picard-Borel et la theorie des fonctions meromorphes, Gauthier-Villars, Paris.

Newman, J. R. (Ed.) (1956) *The World of Mathematics,* Simon and Schuster, New York.

Newton, I. (1728/1967) Universal Arithmetic: Or, a Treatise of Arithmetical Composition and Resolution, in *The mathematical Works of Isaac Newton,* D. T. Whiteside, (Ed.), Johnson Reprint Corp., New York, V. 2, pp. 3–134

Newton, I. (1736) *The Method of Fluxions and Infinite Series: With Its Application to the Geometry of Curve-Lines,* Translated from the Author's Latin Original Not Yet Made Publick, Henry Woodfall, London.

Nguen, N. T. (2008) Inconsistency of knowledge and collective intelligence, *Cybernetics Syst.,* V. 39, No. 6, pp. 542–562.

Nicolini, P. and E. Spallucci, E. (2011) Un-spectral dimension and quantum spacetime phases, *Phys. Lett. B,* V. 695, p. 290.

Nicomachus of Gerasa (1926) *Introduction to Arithmetic* (D'Ooge, M. L. Trans.), Macmillan Co., London.

Nieder, A. (2005) Counting on neurons: the neurobiology of numerical competence, *Nat. Rev. Neurosci,* V. 6, V. 177–190.

Nieder, A. (2016) The neuronal code for number, *Nature,* V. 17, pp. 366–382.

Nieder, A. (2016a) Representing something out of nothing: The dawning of zero. *Trends in Cognitive Sci.,* V. 20, No. 11, pp. 830–842.

Nieder, A. and Dehaene, S. (2009) Representation of number in the brain, *Ann. Rev. Neurosci.*, V. 32, pp. 185–208.

Nieder, A. and Miller, E. K. (2004) A parieto-frontal network for visual numerical information in the monkey, *Proc. Natl. Acad. Sci. USA*, V. 101, pp. 7457–7462.

Nieder, A., Freedman D. J. and Miller E. K. (2002) Representation of the quantity of visual items in the primate prefrontal cortex, *Science*, V. 297, pp. 1708–1711.

Nierderée, R. (1992) What do numbers measure? A new approach to fundamental measurement, *Math. Social Sci.* V. 24, pp. 237–276.

Nieuwentijt, B. (1695) *Analysis infinitorum, seu curvilineorum proprietates ex polygonorum natura deductae*, Johanes Walters, Amsterdam.

Nieuwmeijer, C. (2013) 1 + 1 = 3: *The Positive Effects of the Synergy Between Musician and Classroom Teacher on Young Children's Free Musical Play*, Dissertation, Roehampton University, London, May, 2013.

Niqui, M. (2005) *Formalising Exact Arithmetic in Type Theory, Conference on Computability in Europe (CiE 2005): New Computational Paradigms*, pp. 368–377.

Nirenberg, R. L. and Nirenberg, D. (2011) Badiou's number: A critique of mathematics as ontology, *Critical Inquiry*, V. 37, no. 4, pp. 583–614.

Nivanen, L., Le Mehaute, A., and Wang, Q. A. (2003) Generalized algebra within a nonextensive statistics, *Rep. Math. Phys*, V. 52, pp. 437–444.

Noma, E. and Baird, J. C. (1975) Psychophysical study of numbers: II. Theoretical models of number generation, *Psychol. Res.*, V. 38, pp. 81–95.

Noonan, H. (1978) Count nouns and mass nouns, *Analysis*, V. 38, No. 4, pp. 167–172.

Norrish, M. and Huffman, B. (2013) Ordinals in HOL: Transfinite arithmetic up to (and Beyond) $\omega 1$, in *Proc. Int. Conf. Interactive Theorem Proving (ITP 2013)*, pp. 133–146.

North, J. (2009) The "structure" of physics: A case study, *J. Philos.*, V. 106, pp. 57–88.

Norwich, K. H. (1993) *Information, Sensation, and Perception*, Academic Press, San Diego.

Number, (2016) *New World Encyclopedia*.

Nunn, W. H. (1952) Nationwide Numbering Plan, *Bell System Technical J.*, V. 31, No. 5, pp. 851–863.

Núñez, R. E. (2005) Creating mathematical infinities: Metaphor, blending, and the beauty of transfinite cardinals. *J. Pragmatics*, V. 37, No. 10, pp. 1717–1741.

Núñez, R. (2008). Proto-numerosities and concepts of number: biologically plausible and culturally mediated top-down mathematical schemas. *Behavi. Brain Sci.*, V. 31, No. 6, pp. 665–666.

Núñez, R. (2009) Numbers and arithmetic: neither Hardwired nor out there, *Biol. Theory*, V. 4, No. 1, pp. 68–83.

Núñez, R. (2011) No innate number line in the human brain, *J. Cross-Cultural Psychol.*, V. 45, No. 4, pp. 651–668.

Núñez, R. (2017) Number — Biological enculturation beyond natural selection [A response to A. Nieder], *Trends Cognitive Sci.*, V. 21, No. 6, pp. 403–404.

Núñez, R. (2017a) Is there really an evolved capacity for number? *Trends Cognitive Sci.*, V. 21, No. 6, pp. 409–424.

Núñez, R., Cooperrider, K., Wassmann, J. (2012). Number concepts without number lines in an indigenous group of Papua New Guinea, *PLoS ONE*, doi:10.1371/journal.pone.0035662

Núñez, R., Doan, D, Nikoulina, A. (2011). Squeezing, striking, and vocalizing: is number representation fundamentally spatial? *Cognition*, V. 120, pp. 225–235.

Núñez, R. and Lakoff, G. (1998) What did Weierstrass really define? The cognitive structure of natural and ε-δ continuity, *Math. Cognition*, V. 4, No. 2, pp. 85–101.

Núñez, R. and Lakoff, G. (2005) The cognitive foundations of mathematics: The role of conceptual metaphor, in *Handbook of Mathematical Cognition*, Campbell, J., (Ed), Psychology Press, New York, pp. 109–124.

Oakley, B. A. (2014) *A Mind For Numbers: How to Excel at Math and Science* (Even If You Flunked Algebra), Jeremy P. Tarcher/Penguin, New York.

Oberle, W. F. (1977) Numbers and mysticism, *Math. Teacher,* V. 70, No. 7, pp. 599–602.

O'Connor, J. J. and Robertson, E. F. (2000) Jaina mathematics, in *MacTutor History of Mathematics*, http://www-history.mcs.st-andrews.ac.uk/HistTopics/Jaina_mathematics.html.

O'Connor, J. J. and Robertson, E. F. (2003) Nicolas Chuquet (1445–1488) — Biography, *MacTutor History of Mathematics*, http://www-groups.dcs.st-and.ac.uk/history/HistTopics/Nine_chapters.html.

O'Connor, J. J. and Robertson, E. F. (2003) Nine chapters on the mathematical art, in *MacTutor History of Mathematics*, http://www-groups.dcs.st-and.ac.uk/history/HistTopics/Nine_chapters.html.

O'Connor, J. J. and Robertson, E. F. (2005) The function concept, *MacTutor History of Mathematics*, Archive, University of St Andrews.

Ohm, M. (1822) Versuch eines vollstandig konsequenten Systems der Mathematik,

Okuyama, S., Kuki, T., and Mushiake, H. (2015) Representation of the numerosity 'zero' in the parietal cortex of the monkey, *Scientific Rep.*, V. 5, pp. 1–9.

O'Meara, D. J. (1990) *Pythagoras Revived. Mathematics and Philosophy in Late Antiquity*, Oxford University Press, Oxford.

Ono, T. (1955) Arithmetic of orthogonal groups, *J. Math. Soc. Japan*, V. 7, No. 1, pp. 79–91.

Oppen, D. C. (1978) A $2^{2^{2^{pn}}}$ upper bound on the complexity of Presburger arithmetic, *J. Comput. Syst. Sci.*, V. 16, No. 3, pp. 323–332.

Oppenheim, A. V. and Schafer, R. W. (1975) *Digital Signal Processing*, Prentice-Hall, Englewood Cliffs, NJ.

Oppenheim, A. V., Schafer, R. W. and T. G. Stockham, (1968) Nonlinear filtering of multiplied and convolved signals, *Proc. IEEE*, V. 56, pp. 1264–1291.

Ore, O. (1935) On the foundation of abstract algebra, I, *Ann. of Math.*, V. 36, pp. 406–437.

Ore, O. (1948) *Number Theory and Its History*, McGraw-Hill, New York.

Orsi, R. (2007) When 2+2=5, *The American Scholar*, March 1, 2007, https:// theamericanscholar.org/when-2-2-5/#.W7GLXWhKjIU.

Orwell, G. (1949) *Nineteen Eighty-Four*, Secker & Warburg, London.

Osborn, A. D. (1934) The Philosophy of Edmund Husserl in Its Development from His Mathematical Interests to His First Concept of Phenomenology in Logical Investigations, International Press, New York.

Osborn, D. (2005) The discovery of numbers, in *Vedic Science*, https:// vedicsciences.net/articles/history-of-numbers.html.

Ostashevsky, E. (2000) *Quintessence From Nothingness: Zero, Platonism, and the Renaissance*, Dissertation, Stanford University.

Ostrin, G. E. and Wainer, S. S. (2005) Complexity in predicative arithmetic, in *Conf. Computability in Europe (CiE 2005): New Comput. Paradigms*, pp. 378–384.

Ostrovskii, I. V. (1987) The Arithmetic of probability distributions, *Theory Probab. Appl.*, V. 31, No. 1, pp. 1–24.

Otte, M. (1990) Arithmetic and geometry: some remarks on the concept of complementarity, *Stud. Philos. Ed.*, V. 10, pp. 37–62.

Overmann, K. A. (2013) Material scaffolds in numbers and time, *Cambridge Archaeological J.*, V. 23, No. 1, pp. 19–39.

Overmann, K. A. (2017) The material difference in human cognition, *J. Cognition Culture*, V. 17, No. 3–4, pp. 354–373.

Overmann, K. A. (2019) *The Material Origin of Numbers: Insights From the Archaeology of the Ancient Near East*, Gorgias Press, Piscataway, NJ.

Ozyapιcι, A. and Bilgehan, B. (2016) Finite product representation via multiplicative calculus and its applications to exponential signal processing, *Numer. Algorithms*, V. 71, No. 2, pp. 475–489.

Pacioli, L. (1494) *Summa de Arithmetica, Geometria, Proportioni et Proportionalita*, Paganino de' Paganini, Venice.

Pacioli, L. (1509) *Divina Proportione*, A. PaganiusPaganinus, Venetiis.

Palaiseau, J. F. G. (1816) Métrologie universelle, ancienne et moderne: ou rapport des poids et mesures des empires, royaumes, duchés et principautés des quatre parties du monde *[Universal, ancient and modern metrology: or report of weights and measurements of empires, kingdoms, duchies and principalities of all parts of the world]*, Bordeaux.

Palladio, A. (1965) The *Four Books of Architecture*, Dover Publications, New York.

Palmer, S. H. (1977) *Economic Arithmetic: A Guide to the Statistical Sources of English Commerce, Industry, and Finance, 1700–1850*, Routledge.

Palmer, C. and Stavrinou, P. N. (2004) Equations of motion in a non-integer-dimensional space, *J. Phys. A: Math. Gen.*, V. 37, pp. 6987–7003.

Pandit, S. N. N. (1961) A new matrix calculus, *SIAM J. Appl. Math.*, V. 9, pp. 632–639.

Pantsar, M. (2014). An empirically feasible approach to the epistemology of arithmetic. *Synthese*, V. 191, No. 17, pp. 4201–4229.

Pantsar, M. (2015). In search of $\aleph_0$: How infinity can be created, *Synthese*, V. 192, No. 8, pp. 2489–2511.

Pap, E. (1993) g-calculus, Univ. u Novom Sadu, *Zb. Rad. Prirod.-Mat. Fak. Ser. Mat.*, V. 23, No. 1, pp. 145–150.

Pap, E. (1995) *Null-Additive Set Functions*, Kluwer Academic Publishers, Dordrecht.

Pap, E. (2002) Pseudo-additive measures and their applications, in *Handbook of Measure Theory*, V. 2, Elsevier, pp. 1403–1465.

Pap, E. (2008) Generalized real analysis and its applications, *Int. J. Approx. Reasoning*, V. 47, pp. 368–386.

Pap, E. (Ed) (2021) *Artificial Intelligence: Theory and Applications*, Studies in Computational Intelligence, V. 973, Springer, New York.

Pap, E. and Štrboja, M. (2010) Generalization of the Jensen inequality for pseudo-integral, *Inf. Sci.*, V. 180, pp. 543–548.

Pap, E. and Štrboja, M. (2013) Generalizations of integral inequalities for integrals based on nonadditive measures, in *Intelligent Systems: Models and Application*, Springer, pp. 3–22.

Pap, E., Štrboja, M. and Rudas, I. (2014) Pseudo-*Lp* space and convergence, *Fuzzy Sets Syst.*, V. 238, pp. 113–128.

Pap, E., Takaci, D. and Takaci, A. (2002) The g-operational calculus, *International J. Uncertainty, Fuzziness and Knowledge-Based Syst.*, V. 10, Supplement, pp. 75–88.

Paré, R. and Román, L. (1989) Monoidal categories with natural numbers object, *Studia Logica*, V. 48, No. 3, pp. 361–376.

Parhami, B. (2002) Number representation and computer arithmetic, in *Encyclopedia of Information Systems*, Academic Press.

Parhami, B. (2010) *Computer Arithmetic: Algorithms and Hardware Designs*, Oxford University Press, New York.

Parikh, R. (1969) A conservation result, in *Application of Model Theory*, Holt, Rinehart and Winston, New York, pp. 107–108.

Parikh, R. (1971) Existence and feasibility in arithmetic, *J. Symbolic Logic*, V. 36, No. 3, pp. 494–508.

Paris, J. and Wilkie, A. (1985) Counting problems in bounded arithmetic. in *Methods in Mathematical Logic*, Springer-Verlag. pp. 317–340.

Paris, J. and Wilkie, A. (1987) On the scheme of induction for bounded arithmetic formulas. *Ann. Pure Appl. Logic*, V. 35, pp. 261–302.

Parshin, A. (1976) On the arithmetic of two dimensional schemes. I, Distributions and residues, *Math. USSR Izv.*, V. 10, pp. 695–729.

Parshin, A. (1978) Abelian coverings of arithmetic schemes, *Dokl. Akad. Nauk SSSR*, V. 243, No. 4, pp. 855–858.

Parker, T. H. and Baldridge, S. J. (1999) *Elementary Mathematics for Teachers*, Sefton-Ash Publ., Okemos, MI.

Parsons, C. (1964) Infinity and Kant's conception of the 'possibility of experience', *Philos. Rev.*, V. 73, No. 2, pp. 182–197.

Parsons, C. (1969) Kant's philosophy of arithmetic, in *Philosophy, Science and Method: Essays in Honor of Ernest Nagel*, S. Morgenbesser, P. Suppes, and M. White (Eds.), St. Martin's Press, New York.

Parsons, C. (1984) Arithmetic and the categories, *Topoi*, V. 3, No. 2, pp. 109–121.

Parsons, C. (1990) The structuralist view of mathematical objects, *Synthese*, V. 84, pp. 303–346.

Parsons, C. (1990a) The uniqueness of the Natural Numbers, *Iyyun*, V. 13, pp. 13–44.

Parsons, T. (2000) *Indeterminate Identity: Metaphysics and Semantics*, Clarendon Press, Oxford.

Parvathi, R. and Malathi, C. (2012) Arithmetic operations on symmetric trapezoidal intuitionistic fuzzy numbers, *Int. J. Soft Comp. Eng.*, V. 2, pp. 268–273.

Pascoe, M. (2017) One plus one makes more than two: Our overlooked immigration benefits, *The Sydney Morning Herald*, 16 November 2017.

Paseau, A. (2009) Reducing arithmetic to set theory, in *New Waves in Philosophy of Mathematics*, Palgrave Macmillan, Basingstoke, pp. 35–55.

Pawlak, Z. (1982) Rough sets, *Int. J. Comput. and Inform. Sci.*, V. 11, pp. 341–356.

Pawlak, Z. (1991) *Rough Sets — Theoretical Aspects of Reasoning about Data*, Kluwer Academic Publishers, Boston.

Peacock, G. (1842) *A Treatise on Algebra: Arithmetical Algebra*, J. & J. J. Deighton, London.

Peano, G. (1888) *Calcolo Geometrico secondo l'Ausdehnungslehre di H. Grassmann preceduto dalle Operazioni della Logica Deduttiva*, Fratelli Bocca Editori, Bocca, Turin.

Peano, G. (1889) *Arithmetices Principia, Nova Method Exposita*, Ediderunt Fratres Bocca, Torino.

Peano, G. (1890) Sur une courbe, qui remplit toute une aire plane, *Math. Ann.*, V. 36, No. 1, pp. 157–160.

Peano, G. (1908) *Formulario Mathematico*, Bocca, Torino.

Peano, G. (1973) *Selected works of Giuseppe Peano*, Kennedy, H. C. (Ed.), Allen & Unwin, London.

Pegg, D. (2014) *These are 25 Famous Numbers and Why They Are Important*, https://list25.com/25-famous-numbers-and-why-they-are-important/.

Peirce, C. S. (1881) On the logic of number, *Amer. J. Math.*, V. 4, pp. 85–95.

Peirce C. S. (1931–1935) *Collected Papers*, V. 1-6, Cambridge University Press, Cambridge, UK.

Penn, D. C., Holyoak, K. J. and Povinelli, D. J. (2008) Darwin's mistake: Explaining the discontinuity between human and nonhuman minds, *Behav. Brain Sci.*, V. 31, pp. 109–178.

Penrose, R. (1972) On the nature of quantum geometry, in *Magic Without Magic*, J. Klauder (Ed), Freeman, San Francisco, pp. 333–354.

Pepperberg, I. M. (2006) Grey parrot numerical competence: a review, *Anim. Cogn.*, V. 9, V. 377–391.

Pereira Dimuro, G. (2011) On interval fuzzy numbers, in *WEIT'11 Proceedings of the 2011 Workshop-School on Theoretical Computer Science*, pp. 3–8.

Perez Velazquez, J. L. (2005) Brain, behavior and mathematics: are we using the right approaches? *Physica D*, V. 212, pp. 161–182.

Pesenti, M., Zago L, Crivello F, Mellet E, Samson D, Duroux B, Seron, X., Mazoyer, B. and Tzourio-Mazoyer, N. (2001) Mental calculation in a prodigy is sustained by right prefrontal and medial temporal areas, *Nat. Neurosci.*, V. 4, No. 1, pp. 103–107.

Peters, L. and De Smedt, B. (2018) Arithmetic in the developing brain: A review of brain imaging studies, *Developmental Cognitive Neurosci.*, V. 30, pp. 265–279.

Peters, E., Hibbard, J., Slovic, P. and N. Dieckmann, N. (2007) Numeracy skill and the communication, comprehension, and use of risk-benefit information, *Health Affairs*, V. 26, No. 3, pp. 741–748.

Peterson, J. L. (1981) *Petri Net Theory and the Modeling of Systems*, Prentice-Hall, Englewood Cliffs, NJ.

Phillips, J. (2016) When one plus one equals three, *Wellness Universe*, April 10, 2016, http://blog.thewellnessuniverse.com/when-one-plus-one-equals-three.

Phillips, M. (2008) Desperately seeking synergy: an often promised, rarely delivered outcome, *Coastal Business J.*, V. 7, No. 1, pp. 21–26.

Philo (1978) *Questions and Answers on Genesis* (Quaest. in Gen.), Petit.

Piaget, J. (1941) *La genèse du nombre chez l'enfant,* Delachaux et Niestlé, Neuchâtel (English trans.: *The Child's Conception of Number,* Routledge and Kegan, London, 1952).

Piaget, J. (1964) Mother structures and the notion of number, in *Cognitive Studies and Curriculum Development*, School of Education, Cornell University, pp. 33–39.

Piazza, M. and Dehaene, S. (2004) From number neurons to mental arithmetic: The cognitive neuroscience of number sense, in *The Cognitive Neurosciences*, M. S. Gazzaniga; (Ed.), MIT Press, Cambridge, MA, pp. 865–875.

Piazza, M., Izard, V., Pinel, P., Le Bihan, D. and Dehaene, S. (2004) Tuning curves for approximate numerosity in the human intraparietal sulcus, *Neuron*, V. 44, No. 3, pp. 547–555.

Pica, P., Lemer, C., Izard, V. and Dehaene, S. (2004) Exact and approximate arithmetic in an Amazonian indigene group, *Science*, V. 306, pp. 499–503.

Piegat, A. and Landowski, M. (2017) Is Fuzzy Number the Right Result of Arithmetic Operations on Fuzzy Numbers? in *Proc. Conf. European Society for Fuzzy Logic and Technology*, Advances in Intelligent Systems and Computing, V. 643, pp. 181–194.

Pierce, D. (2011) *Numbers*, preprint in Logic (math.LO), arXiv:1104.5311.

Pierce, R. C. (1977) A brief history of logarithm, *Two-Year College Math. J.*, V. 8, No. 1, pp. 22–26.

Pierce, R. S. (1982) *Associative Algebras*, Springer-Verlag, New York.

Pierpont, J. (1899) On the arithmetization of mathematics, *Bull. Amer. Math. Soc.*, pp. 394–406.

Pilipchuk, V. N. (2010) *Nonlinear Dynamics: Between Linear and Impact Limits.* Lecture Notes in Applied and Computational Mechanics, V. 52, Springer, Berlin.

Pilipchuk, V. N. (2011) Non-smooth spatio-temporal coordinates in nonlinear dynamics, preprint arXiv:1101.4597.

Pillay, A. (1981) Models of Peano arithmetic: A survey of basic results, in *Model Theory and Arithmetic*, Lecture Notes in Mathematics, V. 890, pp. 263–269.

Pin, J.-E. (1998) Tropical semirings, in *Idempotency. Publications of the Newton Institute*, V. 11, Cambridge University Press, Cambridge, pp. 50–69.

Pinel, P., Dehaene, S., Riviere, D. and Le Bihan, D. (2001) Modulation of parietal activation by semantic distance in a number comparison task, *NeuroImage*, V. 14, No. 5, pp. 1013–1026.

Pinel, P., Piazza, M., Le Bihan, D. and Dehaene, S. (2004) Distributed and overlapping cerebral representations of number, size, and luminance during comparative judgments, *Neuron*, V. 41, No. 6, pp. 983–993.

Pines, S. (1968). Thabit B. Qurra's conception of number and theory of the mathematical infinite, in *Actes du Onzième Congrès International d'Histoire des Sciences, Sect. III: Histoire des Sciences Exactes (Astronomie, Mathématiques, Physique)* (Wroclaw, 1963), pp. 160–166.

Plaisier, M. A., Bergmann Tiest, W. M. and Kappers, A. M. L. (2009) One, two, three, many — Subitizing in active touch, *Acta Psychologica*, V. 131, No. 2, pp. 163–170.

Plaisier, M. A., Bergmann Tiest, W. M. and Kappers, A. M. L. (2010) Range dependent processing of visual numerosity: similarities across vision and haptics, *Exp. Brain Res.*, V. 204, No. 4, pp. 525–537.

Plaisier, M. A. and Smeets, J. B. J. (2011) Haptic subitizing across the fingers, *Attention, Perception, Psychophys.*, V. 73, No. 5, pp. 1579–1585.

Plato, (1961) *The Collected Dialogues of Plato*, Princeton University Press, Princeton.

Plofker, K. (2009) *Mathematics in India*, Princeton University Press, Princeton.

Plotinus, (2018) *The Enneads*, Cambridge University Press, Cambridge.

Poincaré, H. (1901) Sur les propriétés arithmétiques des courbes algébriques, *J. Math., 5éme série*, t. 7, fasc. III, pp. 161–233.

Poincaré, H. (1902) *La Science et l'hypothèse*, Flammarion, Paris.

Poincaré, H. (1905) *La valeur de la science*, Flammarion, Paris.

Poincaré, H. (1908) *The future of mathematics*, Address delivered April 10, 1908, at the general session of the Fourth International Congress of Mathematicians, Rome.

Poincaré, H. (1913) *The Foundations of Science*. Science Press.

Pollack, C. and Ashby, N. C. (2018) Where arithmetic and phonology meet: The meta-analytic convergence of arithmetic and phonological processing in the brain, *Developmental Cognitive Neurosci.*, V. 30, pp. 251–264.

Poonen, B. (2009) *Introduction to arithmetic geometry*, MIT Press, Cambridge, MA.

Poonen, B. and Tschinkel, Y. (Eds.) (2004) *Arithmetic of Higher-Dimensional Algebraic Varieties*, Birkhaüser, Boston.

Popper, K. R. (1959) *The Logic of Scientific Discovery*, Basic Books, New York.

Popper, K. R. (1974) Replies to my critics, in *The Philosophy of Karl Popper*, P. A. Schilpp, (Ed.), Open Court, La Salle, IL, pp. 949–1180.

Popper, K. R. (1979) *Objective Knowledge: An Evolutionary Approach*, Oxford University Press, New York.

Porter, T. M. (1995) *Trust in Numbers: The Pursuit of Objectivity in Science and Public Life*, Princeton University Press, Princeton, NJ.

Potter, M. (2002) *Reason's Nearest Kin: Philosophies of Arithmetic from Kant to Carnap*, Oxford University Press, Oxford.

Prado, J., Noveck, I. A. and van der Henst, J. B. (2010) Overlapping and distinct neural representations of numbers and verbal transitive series, *Cereb. Cortex*, V. 20, No. 3, pp. 720–729.

Prakash, A., Suresh, M. and Vengataasalam, S. (2016) A new approach for ranking of intuitionistic fuzzy numbers using a centroid concept, *Math. Sci.*, V. 10, pp. 177–184.

Presburger, M. (1929) Über die Vollstaendigkeit eines gewissen Systems der Arithmetik ganzer Zahlen, in welchem die Addition als einzige Operation hervortritt, *C.R. du I Congr. des Math. des pays Slaves*, Warszawa, pp. 92–101.

Prestet, J. (1675) *Elemens des Mathematiques*, André Pralard, Paris.

Price, G. B. (1991) *An Introduction to Multicomplex Spaces and Functions*, Marcel Dekker.

Price, G. R. and Ansari, D. (2011) Symbol processing in the left angular gyrus: Evidence from passive perception of digits, *Neuroimage*, V. 57, No. 3, pp. 1205–1211.

Price, G. R. Mazzocco, M. M. M. and Ansari, D. (2013) Why mental arithmetic counts: brain activation during single digit arithmetic predicts high school math scores, *J. Neurosci.*, V. 33, No. 1, pp. 156–163.

Priest, G. (1994) Is arithmetic consistent? *Mind*, V. 103, No. 420, pp. 337–349.

Priest, G. (1994) What could the least inconsistent number be? *Logique et Analyse*, V. 37, pp. 3–12.

Priest, G. (1996) On inconsistent arithmetics: reply to Denyer, *Mind*, V. 105, pp. 649–659.

Priest, G. (1997) Inconsistent models for arithmetic: I, Finite models, *J. Philos. Logic*, V. 26, pp. 223–235.

Priest, G. (1998) Number, in *Routledge Encyclopedia of Philosophy,* V. 7, Taylor Francis, London/NewYork, pp. 47–54.

Priest, G. (2000) Inconsistent models for arithmetic: II, The general case, *J. Symbolic Logic*, V. 65, pp. 1519–1529.

Priest, G. (2003) On alternative geometries, arithmetics, and logics; a tribute to Lukasiewicz, *Studia Logica*, V. 74, No. 3, pp. 441–468.

Priest, G. (2003) Inconsistent arithmetics: issues technical and philosophical, in *Trends in Logic*, Studia Logica Library, V. 21, Springer, Dordrecht, pp. 273–299.

Proverbio, A. M., De Benedetto, F., Ferrari, M. V. and Ferrarini, G. (2018) When listening to rain sounds boosts arithmetic ability, *PLOS One*, https://doi.org/10.1371/journal.pone.0192296.

Pudlák, P. A note on bounded arithmetic, *Fundamenta Mathematica*, V. 136, pp. 85–89.

Putnam, H. (1960) Minds and machines, in *Dimensions of Mind*, pp. 148–179.

Putnam H. (1975) What is mathematical truth? *Historia Mathematica*, V. 2, No. 4, pp. 529–533.

Pycior, H. M. (2011) *Symbols, Impossible Numbers, and Geometric Entanglements: British Algebra through the Commentaries on Newton's Universal Arithmetick*, Cambridge University Press, Cambridge.

Rabi, L. (2016) Ortega y gasset on Georg Cantor's theory of transfinite numbers, *Kairos. J. Philos. & Sci.*, V. 15, No. 1, pp. 46–70.

Rabin, M. (1966) Diophantine equations and non-standard models of arithmetic, *Studies Logic Found. Math.*, V. 44, pp. 151–158.

Rabinovitch, N. (1970) Rabbai Hasdai Crescas (1340–1410) on numerical infinities. *Isis*, V. 61, pp. 224–230.

Rado, R. (1975) The cardinal module and some theorems on families of sets, *Ann. Mat. pura Appl.*, V. 102, No. 4, pp. 135–154.

Radu, M. (2003) A debate about the axiomatization of arithmetic: Otto Hölder against Robert Graßmann, *Historia Math.*, V. 30, No. 3, pp. 341–377.

Radzikowska, A. M. and Kerre, E. E. (2004) On L-fuzzy rough sets, in *Proc. 7th Inte. Conf. on Artificial Intelligence and Soft Computing (ICAISC 2004)*, Zakopane, Poland.

Rall, L. B. (1986) The arithmetic of differentiation, *Math. Magazine*, V. 59, No. 5, pp. 275–282.

Ramirez-Cardenas, A., Moskaleva, M. and Nieder, A. (2016) Neuronal representation of numerosity zero in the primate parieto-frontal number network, *Current Bio.*, V. 26, No. 10, pp. 1285–1294

Ramsey, D. (2011) *Entre Leadership: 20 Years of Practical Business Wisdom from the Trenches*, Howard Books.

Rappaport, J.(2002) The social life of numbers: A quechua ontology of numbers and philosophy of arithmetic (review), *Ethnohistory*, V. 49, No. 2, Spring, pp. 430–433.

Rasch, G. (1934) Zur Theorie und anwendung des produktintegrals, *J. Reine Angew. Math.*, V. 191, pp. 65–119.

Rashed, R. (1994) *The development of Arabic Mathematics:Between Arithmetic and Algebra*, Springer, New York.

Rashed, R. (2011) *D'Al-Khwārizmī à Descartes: étude sur l'histoire des mathématiques classiques*, Hermann, Paris.

Rashevsky, P. K. (1973) On the Dogma of the natural numbers. *Russian Math. Surv.*, V. 28, No. 4, pp. 243–246.

Rasiowa, H. and Sikorski, R. (1963) *The Mathematics of Metamathematics*, Panswowe Wydawnictwo Naukowe, Warszawa.

Razborov, A. A. (1993) An equivalence between second order bounded domain bounded arithmetic and first order bounded arithmetic, in *Arithmetic, Proof Theory and Computational Complexity*, Oxford University Press, Oxford, pp. 247–277.

Rechter, O. (2006) The view from 1763: Kant on the arithmetical method before intuition, in *Intuition and the Axiomatic Method*, Springer, Dordrecht.

Reck, E. (2017) Dedekind's contributions to the foundations of mathematics, *The Stanford Encyclopedia of Philosophy*, https://plato.stanford.edu/archives/win2017/entries/dedekind-foundations/.

Reck, E. H. and Price, M. P. (2000) Structures and structuralism in contemporary philosophy of mathematics, *Synthese*, V. 125, pp. 341–416.

Recorde, R. (1543) *The Grounde of Artes*, Tho Harper, London.

Recorde, R. (1557) *Arithmetike: Containyng the Extraction of Rootes: The Coßike Practise, with the Rule of Equation and the Woorkes of Surde Nombers*, Jhon Kyngstone, London.

Reddy, C. R., and Loveland, D. W. (1978) Presburger arithmetic with Bounded quantifier alternation, in *ACM Symposium on Theory of Computing*, pp. 320–325.

Reid, C. (2006) *From Zero to Infinity*: *What Makes Numbers Interesting*, A K Peters, Ltd., Natick. MA.

Renner, L. (2013) $1 + 1 = 3$ *The New Math of Business Strategy: How to Unlock Exponential Growth through Competitive Collaboration*, Telemark Publishing.

Requena, I., Delgado, M. and Verdegay, J. (1994) Automatic ranking of fuzzy numbers with the criterion of decision-maker learnt by an artificial neural network. *Fuzzy Sets Syst.* V. 64, 1–19.

Resnick, M. (1997) *Mathematics as a Science of Structures*, Oxford University Press, Oxford.

Resnik, M. D. (1999) *Mathematics as a Science of Patterns*, Clarendon Press, Oxford.

Resnikov, A. (1997) Three-manifolds class field theory, Sel. Math., New Ser, V. 3, pp. 361–399.

Reves, G. E. (1951) Outline of the history of arithmetic, *School Sci. Math.*, V. 51, No. 8, pp. 611–617.

Ribenboim, P. (1985) Review of the book Weil, A. *Number theory: an approach through history from Hammurapi to Legendre* (Boston, 1984), *Bull. Amer. Math. Soc.*, V. 13, No. 2, pp. 173–182.

Ribenboim, P. (2000) *My Numbers, My Friends*, Springer, New York.

Richards, M. (1967) *The BCPL Reference Manual*, Massachusetts Institute of Technology, Boston.

Richeson, A. W. (1947). The first arithmetic printed in English, *Isis*, V. 37, No. 1/2, pp. 47–56.

Riedell, W. E., Pikul, J. L. and Carpenter-Boggs, L. (2002) *One Plus One Equals Three: The Synergistic Effects Of Crop Rotation On Soil Fertility And Plant Nutrition*, UNL's Institutional Repository, https://digitalcommons.unl.edu/usdaarsfacpub/1061/.

Ries, A. (1518) *Rechenung nach auf den linihen*, Matheas Maler, Erfurt.

Ries, A. (1522) Rechenung nach auf den linihen und feder, Matheas Maler, Erfurt.

Ries, A. (2014) *In the Marketing World, One Plus One Equals Three-Fourths* http://adage.com/article/al-ries/marketing-world-equals-fourths/295251/.

Riggs, K. J., Ferrand, L., Lancelin, D., Fryziel, L., Dumur, G. and Simpson, A. (2006). Subitizing in tactile perception, *Psychol. Sci.*, V. 17, No. 4, pp. 271–272.

Riley, M. S., Greeno, J. G. and Heller, J. I. (1983) Development of children's problem-solving ability in arithmetic, in *The Development of Mathematical Thinking*, Academic, Orlando, FL, pp. 153–196.

Rips, L. J. (2013) How many is a zillion? Sources of number distortion, *J. Exp. Psychol.: Learning, Memory, Cognition*, V. 39, pp. 1257–1264.

Rips, L. J. (2015) Beliefs about the nature of numbers, in *Mathematics, Substance and Surmise: Views on the Meaning and Ontology of Mathematics*, Springer, Berlin, pp. 321–345.

Rips, L. J., Asmuth, J. and Bloomfield, A. (2006) Giving the boot to the bootstrap: how not to learn the natural numbers, *Cognition*, V. 101, pp. B51–B60.

Rips, L. J., Asmuth, J. and Bloomfield, A. (2008) Do children learn the integers by induction? *Cognition*, V. 106, pp. 940–951.

Rips, L. J., Bloomfield, A. and Asmuth, J. (2008) From numerical concepts to concepts of number, *Behav. Brain Sci.*, V. 31, pp. 623–687.

Ritchie, J.(2014) $1 + 1 = 3$ — *How Partner Marketing Defies the Laws of Math*, https://blog.marketo.com/2014/08/113-how-partner-marketing-defies-the-laws-of-math.html.

Roberts, B. W. (2011) Group structural realism, *British J. Philos. Sci.*, V. 62. No. 1, pp. 47–69.

Roberts W. A. (1995) Simultaneous numerical and temporal processing in the pigeon, *Curr. Dir. Psychol. Sci.*, V. 4, pp. 47–51.

Roberts W. A. and Boisvert M. J. (1998) Using the peak procedure to measure timing and counting processes in pigeons, *J. Exp. Psychol. Anim. Behav. Process.*, V. 24, pp. 416–430.

Roberts, F. and Luce, R. (1968) Axiomatic thermodynamics and extensive measurements, *Synthese*, V. 18, pp. 311–326.

Robins, G. and Shute, C. (1987) *The Rhind Mathematical Papyrus: an Ancient Egyptian Text*, British Museum Publications Limited, London.

Robinson, R. M. (1950) An essentially undecidable axiom system, in *Proc. Int. Congr. Mathematics*, pp. 729–730.

Robinson, A. (1961) Non-standard analysis, *Indagationes Math. v.* 23, pp. 432–440.

Robinson, A. (1966) *Non-standard Analysis*, Studies of Logic and Foundations of Mathematics, North-Holland, New York.

Robinson, A. (1967) Nonstandard arithmetic, *Bull. Amer. Math. Soc.*, V. 73, No. 6, pp. 818–843.

Robinson, A. (1973) Numbers — What are they and what are they good for? *Yale Scientific*, May 1973, pp. 14–16.

Robinson, R. M. (1937) The theory of classes: A modification of von Neumann's system, *J. Symbolic Logic*, V. 2, No. 1, pp. 29–36.

Robinson, K. M., Arbuthnott, K. D., Rose, D., McCarron, M. C., Globa, C. A. and Phonexay, S. D. (2006) Stability and change in children's division strategies, *J. Exp. Child. Psychol.*, V. 93, No. 3, pp. 224–238.

Rogers, H. (1958) Gödel numberings of partial recursive functions, *J. Symb. Logic*, V. 23, pp. 331–341.

Rogers, H. (1987) *Theory of Recursive Functions and Effective Computability*, MIT Press, Cambridge, MA.

Rogers, L. (2008) The history of negative numbers, in *Enriching Mathematics*, http://Enrich.maths.org/5961.

Rogers, A. (2007) *Supermanifolds*: *Theory and Applications*, World Scientific, Singapore.

Roitman, J. D., Brannon E. M. and Platt M. L. (2007) Monotonic coding of numerosity in macaque lateral intraparietal area, *PLoS. Biol.*, V. 5, e208; 10.1371/journal.pbio.0050208.

Román, L. (1989) Cartesian categories with natural numbers object, *J. Pure Appl. Algebra*, V. 58, No. 3, pp. 267–278.

Romig, H. G. (1924) Early History of Division by zero, *Amer. Math. Monthly*, V. 31, No. 8, pp. 387–389.

Rose, H. E. (1961) On the consistency and undecidability of recursive arithmetic, *Zeit. Math. Logik Grundlagen Math.*, V. 7, pp. 124–135.

Rose, G. J. (2018) The numerical abilities of anurans and their neural correlates: insights from neuroethological studies of acoustic communication, *Philos. Trans. R. Soc. Lond. B, Biol. Sci.*, V. 373, 20160512.

Rosen, G. (2006) What are numbers? *Philosophy Talk*, March 14, 2006, http://www.philosophytalk.org/pastShows/Number.html.

Rosen, G. (2011) The reality of mathematical objects, in *Meaning in Mathematics*, J. Polkinghorne (Ed.), Oxford University Press, Oxford, pp. 113–132.

Rosinger, E. E. (2008) *On the Safe Use of Inconsistent Mathematics*, preprint in mathematics, math. GM, arXiv.org, 0811.2405v2.

Rosner, R. S. (1965) The power law and subjective scales of number, *Perceptual Motor Skills*, V. 21, p. 42.

Ross, (1996) K. A. *Elementary Analysis*: *The Theory of Calculus*, Springer-Verlag, New York.

Roth, P. (1608) *Arithmetica Philosophica, Oder schöne wolgegründete Uberauß Kunstliche Rechnung der Coß oder Algebrae: In drey unterschiedliche Theil getheilt*, Lantzenberger, Nürnberg.

Rotman, B. (1996) Counting information: A note on physicalized numbers, *Minds Machines*, V. 6, No. 2, pp. 229–238.

Rotman, B. (1997) The truth about counting, *Sciences*, No. 11, pp. 34–39.

Rotman, B. (2003) Will the digital computer transform classical mathematics? *Philos. Trans. R. Soc. Lond. A*, V. 361, pp. 1675–1690.

Rotman, J. (1996) *A First Course of Abstract Algebra*, Prentice-Hall, Upper Saddle River, NJ.

Roubach, M. (2008) *Being and Number in Heidegger's Thought*, Continuum.

Roux, A. (2007) Un plus un égale trois: La perte de l'objet primaire comme condition de l'apparition du symbole. *Rev. Française Psychanal.*, V. 71, No. 1, pp. 153–168.

Rovelli, K. (2017) Michelangelo's Stone: An Argument Against Platonism in Mathematics, *European J. Philos. Sci.*, V. 7, No. 2, pp. 285–297.

Rowen, L. H. (1988) *Ring Theory*, Academic Press, Boston, MA.

Rozeboom, W. W. (1966) Scaling theory and the nature of measurement, *Synthese*, V. 16, No. 2, pp. 170–233.

Rozenfel'd, B. A., Rozhanskaya, M. M. and Skolovskaya, Z. K. (1973) *Abu'l-Rayhan al-Biruni (973–1048)*, Nauka, Moscow, (in Russian).

Rucker, R. (1987) *Mind Tools: The Five Levels of Mathematical Reality*, Houghton Mifflin Co., Boston.

Rudman, P. S. (2007) *How Mathematics Happened: The First* 50,000 *Years*, Prometheus Books.

Rugani, R. (2017) Towards numerical cognition's origin: insights from day-old domestic chicks, *Philos. Trans. R. Soc. Lond. B Biol. Sci.*, V. 373, 20160509.

Rugani, R., Vallortigara, G., Priftis, K. and Regolin, L. (2015) Number-space mapping in the newborn chick resembles humans' mental number line. *Science*, V. 347, pp. 534–536 (in *Numbers and Nerves: Information, Emotion, and Meaning in a World of Data*, Oregon State University Press, Corvallis, OR).

Russell, B. (1897) On the Relations of Number and Quantity, *Mind*, V. 6, pp. 326–341.

Russell, B. (1903) *Principles of Mathematics*, George Allen and Unwin, London.

Russell, B. (1919) *Introduction to Mathematical Philosophy*, George Allen and Unwin, London.

Russell, B. (1923) Vagueness, *Aust. J. Psychol. Philos.*, V. 1, pp. 84–92.

Ryle, G., Lewy, C. and Popper, K. R. (1946) Symposium: Why are the calculuses of logic and arithmetic applicable to reality? *Proc. Aristotelian Soc.*, V. 20, Logic and Reality, pp. 20–60.

Ryll-Nardzawski, C. (1952) The role of the axiom of induction in the elementary Arithmetic, *Fund. Math.*, V. 39, pp. 239–263.

Saibian, S. (2021) *One to Infinity: A Guide to the Finite, A Web Book on Large Numbers*, https://sites.google.com/site/largenumbers.

Salmon, N. (2008) Numbers versus nominalists, *Analysis*, V. 68, No. 3, pp. 177–182.

Sambuc, R. (1975) Fonctions φ-floues: Application al'aide au diagnostic enpathologie thyroidienne, Ph. D. Thesis, Univ. Marseille, France.

Santcliment, F. (1482) *Summa de l'artd'Aritmètica,* Pere Posa, Barcelona.

Sapir, E. (1963) *Selected Writings in Language Culture and Personality*, University of California Press, Berkeley.

Sarala, N. and Jothi, S. (2018) Pseudo arithmetic operations on fuzzy measure, *Int. J. Math. Trends Technology (IJMTT)*. V. 57, No. 2, pp. 78–84.

Sarnecka, B. and Carey, S. (2008) How counting represents number: What children must learn and when they learn it, *Cognition*, V. 108, No. 3, pp. 662–674.

Sarnecka, B. W., Kamenskaya, V. G., Ogura, T., Yamana, Y. and Yudovina, J. B. (2007) From grammatical number to exact numbers: Early meanings of "one," "two," and "three," in English, Russian, and Japanese. *Cognitive Psychol.*, V. 55, No. 2, pp. 136–168.

Sawamura H., Shima K. and Tanji J. (2002) Numerical representation for action in the parietal cortex of the monkey, *Nature*, V. 415, pp. 918–922.

Sawamura, H., Shima K. and Tanji J. (2010) Deficits in action selection based on numerical information after inactivation of the posterior parietal cortex in monkeys, *J. Neurophysiol.*, V. 104, pp. 902–910.

Saxena, B. and Pal, S. K. (2010) Some new concepts in fuzzy arithmetic, *J. Discrete Math. Sci. Cryptography*, V. 13, No. 3, pp. 257–270.

Sayward, C. (2002) A conversation about numbers and knowledge, *Amer. Philos. Quarterly*, V. 39, No. 3, pp. 275–287.

Sazanov, V. Y. (1980) A logical approach to the problem "P=NP?", in *Mathematics Foundations of Computer Science*, Lecture Notes in Computer Science, V. 88, Springer-Verlag, pp. 562–575.

Sazanov, V. Y. (1981) On existence of complete predicate calculus in matemathematics without exponentiation, in *Mathematics Foundations of Computer Science*, Lecture Notes in Computer Science, V. 118, Springer-Verlag, pp. 383–390.

Sazonov, V. Yu. (1995) On feasible numbers, in *Logic and Computational Complexity, Lecture Notes in Computer Science*, V. 960, Springer, New York, pp. 30–51.

Schappacher, N. (1998) Wer war Diophant?, *Math. Semesterberichte* V. 45, No. 2, pp. 141–156.

Schepler, H. C. (1950) The chronology of π, *Mathe. Magazine*, No. 1, pp. 169–170; No. 2, pp. 216–228; No. 3, pp. 279–283.

Schervish, M. J. (1995). *Theory of Statistics*, Springer, New York.

Schimmel, A. (1993) *The Mystery of Numbers*, Oxford University Press.

Schirn, M. (1995) Axiom V and Hume's principle in Frege's foundational project, *Dialogos*, V. 66, pp. 7–20.

Schlote, K.-H. (1996) Hermann Guenther Grassmann and the theory of hypercomplex number systems, *Boston Studies Philos. Science*, V. 187, pp. 165–174.

Schmandt-Besserat, D. and Hays, M. (1999) *The History of Counting*, Harper Collins.

Schmieden, C. and Laugwitz, D. (1958) Eine Erweiterung der Infinitesimalrechnung, *Math. Zeitschr.*, V. 69, pp. 1–39.

Schneider, B., Parker, S., Ostrosky, D., Stein, D. and Kanow, G. (1974) A scale for the psychological magnitude of numbers, *Perception Psychophys.*, v.16, pp. 43–46.

Scholl, B. J. and Pylyshyn, Z. W. (1999). Tracking multiple items through occlusion: Clues to visual objecthood, *Cognitive Psychol.*, V. 38, pp. 259–290.

Schröder, E. (1873) *Lehrbuch der Arithmetik und Algebra für Lehrer und Studirende,* Teubner, Leipzig.

Schroeder, M. R. (1985) Number theory and the real world, *Math. Intelligencer*, V. 7, No. 4, pp. 18–26.

Schroeder, M. R. (1988) The unreasonable effectiveness of number theory in science and communication (1987 Rayleigh Lecture), *IEEE ASSP Magazine*, V. 5, No. 1, pp. 5–12.

Schütte, K. (1951) Beweistheoretische Erfassung der unendlichen Induktion in der Zahlentheorie, *Math. Ann.*, V. 122, pp. 369–389.

Schütte, K. (1977) *Proof Theory*, Springer-Verlag, Berlin.

Schwartz, L. (1950/1951) *Théorie des Distributions*, v. I–II, Hermann, Paris.

Schweizer, B. and Sklar, A. (1960) Statistical metric spaces, *Pacific J. Math.*, V. 10, pp. 313–334.

Scott, D. (1969) *Boolean* models and nonstandard analysis, in *Applications of Model Theory to Algebra, Analysis, and Probability*, Holt, Reinehart and Winston, New York, pp. 87–92.

Scott, W. F. (2007) *Logic in Arithmetic*, Outskirts Press.

Scriba, C. J. (1968) *The Concept of Number*, Bibliographisches Institut, Mannheim-Zurich.

Sebenius, J. K. (1983) Negotiation arithmetic: adding and subtracting issues and parties, *Int. Organizations*, V. 37, No. 2, pp. 281–316.

Seghers, M. J., Longacre, J. J. and Destefano, G. A. (1964) The golden proportion and beauty, *Plastic Reconstructive Surg.*, V. 34, No. 4, pp. 382–386.

Segre, C. (1892) Le rappresentazioni real idelle forme complesse e glientii per algebrici, *Math. Ann.*, V. 40, pp. 413–467.

Segre, M. (1994) Peano's Axioms in their Historical Context, *Arch. History Exact Sci.*, V. 48, No. 3/4, pp. 201–342.

shCherbak V. (2003) Arithmetic inside the universal genetic code, *Biosystems*, V. 70, No. 3, pp. 187–209.

shCherbak V. (2008) The arithmetical origin of the genetic code, in *The Codes of Life, Biosemiotics*, V. 1, Springer, Dordrecht, pp. 153–185.

Seising R. (2007) Pioneers of vagueness, haziness, and fuzziness in the 20th century, in *Forging New Frontiers: Fuzzy Pioneers* I, Studies in Fuzziness and Soft Computing, V. 217. Springer, Berlin, pp. 55–81.

Seising R. (2008) On the absence of strict boundaries — vagueness, haziness, and fuzziness in philosophy, science, and medicine, *Appl. Soft Compu.*, V. 8, No. 3, pp. 1232–1242.

Selin, H. (Ed.) (1997) *Encyclopaedia of the History of Science, Technology, and Medicine in Non-western Cultures*, Springer, New York.

Sella, F., Sader, E., Lolliot, S. and Cohen Kadosh, R. (2016) Basic and advanced numerical performances relate to mathematical expertise but are fully mediated by visuospatial skills, *J. Exp. Psychol. Learn. Mem. Cogn.*, V. 42, No. 9, pp. 1458–1472.

Sellars W (1962) Naming and saying, *Philos. Sci.*, V. 29, pp. 7–26.

Semadeni, Z. (1982) *Schauder Bases in Banach Spaces of Continuous Functions*, Lecture Notes in Mathematics, V. 918, Springer, Berlin.

Sereda, K. (2017) *Leibniz on the Concept, Ontology, and Epistemology of Number*, Dissertation, University of California, San Diego.

Seresht, N. G. and Fayek, A. R. (2018) Fuzzy arithmetic operations: theory and applications, in *Construction Engineering and Management*, Emerald Publishing Limited, pp. 111–147.

Sergeyev, Y. (2009) Numerical computations and mathematical modelling with infinite and infinitesimal numbers, *J. Appl. Math. Comput.*, V. 29, pp. 177–195.

Serre, J.-P. (1973) *A Course in Arithmetic*, Graduate Texts in Mathematics, No. 7, Springer-Verlag, New York.

Sethe, K. *Von Zahlen und Zahlwortenbeidenalten Ägypten*, Strassburg.

Shabel, L. (2016) Kant's philosophy of Mathematics, *The Stanford Encyclopedia of Philosophy* (Spring 2016 Edition), E. N. Zalta (Ed.), https://plato.stanford.edu/archives/spr2016/entries/kant-mathematics/.

Shaffner, G. (1999) *The Arithmetic of Life*, Ballantine Books.

Shannon, C. (1948) A mathematical theory of communication, *Bell Syst. Techn. J.*, V. 27, pp. 379–423.

Shannon, C. E. (1993) *Collected Papers*, N. J. A. Sloane and A. D. Wyner (Eds) IEEE Press, New York.

Shapiro, S. (1983) Mathematics and reality, *Philos. Sci.*, V. 50, pp. 523–548.

Shapiro, S. (1991) *Foundations Without Foundationalism*, Oxford University Press, New York.

Shapiro, S. (1996) Space, number, and structure: A tale of two debates, *Philos. Math.*, V. 4, No. 3, pp. 148–173.

Shapiro, S. (1997) *Philosophy of Mathematics: Structure and Ontology*, Oxford University Press, New York.

Sharma, G. (2011) Color imaging arithmetic: physics ∪ math > physics + math, in *CCIW 2011*, Lecture Notes in Computer Science, V. 6626, Springer. pp. 31–46.

Shelah, S. (1994) *Cardinal Arithmetic*, Oxford Logic Guides, Clarendon Press.

Shields, P. (1997) Peirce's axiomatization of arithmetic, in (Eds.) Houser, N., Roberts, D. D. and Van Evra, J. *Studies in the Logic of Charles Sanders Peirce*, Indiana University Press, Bloomington, pp. 43–52.

Shields, P. (2012) *Peirce on the Logic of Number*, Docent Press.

Shiozawa, Y. (2015) International trade theory and exotic algebras, *Evolutionary Institutional Econo. Rev.*, V. 12, pp. 177–212.

Shirley, J. W. (1951) Binary numeration before Leibniz, *Amer. J. Phy.*, V. 19, No. 8, pp. 452–454.

Shoenfield, J. R. (2001) *Mathematical Logic*, Association for Symbolic Logic, K Peters, Ltd., Natick, MA.

Shulman, M. (2016) *Homotopy Type Theory: A Synthetic Approach to Higher Equalities*, preprint in Mathematics, arXiv:1601.05035v3.

Shum, J., Hermes, D., Foster, B. L., Dastjerdi, M., Rangarajan, V., Winawer, J., Miller, K. J. and Parvizi, J. (2013) A brain area for visual numerals, *J. Neurosci.*, V. 33, No. 16, pp. 6709–6715.

Shureshjani, R. A. and Darehmiraki, M. (2013) A new parametric method for ranking fuzzy numbers, *Indagationes Math.*, V. 24, pp. 518–529.

Sicha, J. F. (1970) Counting and the natural numbers, *Philos. Sci.*, V. 37, No. 3, pp. 405–416.

Sierpiński, W. (1958) *Cardinal and Ordinal Numbers*, Państwowe Wydawnictwo Naukowe, Warsaw.

Šikić, Z. (1996) What are numbers? *Int. Studies Philos. Sci.*, V. 10, No. 2, pp. 159–171.

Sikorski, R. (1948) On an ordered algebraic field, *Soc. Sci. Lett. Varsovie, Cl. III, Sci. Math. Phys.*, V. 41, pp. 69–96.

Silver, R. A. (2010) Neuronal arithmetic, *Nat. Rev. Neuroscience*, V. 11, No. 7, pp. 474–489.

Silverman, J. H. (1986) *The Arithmetic of Elliptic Curves*, Graduate Texts in Mathematics, V. 106, Springer-Verlag, New York.

Silverman, J. H. (1994) *Advanced Topics in the Arithmetic of Elliptic Curves*, Graduate Texts in Mathematics, V. 151, Springer-Verlag, New York.

Silverman, J. H. (2007) *The Arithmetic of Dynamical Systems*, Graduate Texts in Mathematics 241, Springer-Verlag.

Simon, J. (1977) On feasible numbers, in *Proc. 9th Annual ACM Sympo. Theory of Computing (STOC '77)*, pp. 195–207.

Simon, I. (1988) *Recognizable Sets with Multiplicities in the Tropical Semiring, Mathematical Foundations of Computer Science*. Springer, Berlin, pp. 107–120.

Simon, T. J. (1997) Reconceptualizing the origins of number knowledge: A "nonnumerical" account. *Cognitive Develop.*, V. 12, pp. 349–372.

Simpson, S. G. (2009) *Subsystems of Second Order Arithmetic*, Cambridge University Press, Cambridge.

Simpson, S. G. (2009) *Subsystems of Second Order Arithmetic, Perspectives in Logic*, Cambridge University Press.

Singh, D. (1994) A note on the development of multiset theory, *Modern Logic*, V. 4, pp. 405–406.

Singh, S. (2012) *Fermat's Last Theorem*, Harper Press.

Singh, D., Ibrahim, A. M., Yohanna, T. and Singh, J. N. (2007) An overview of the applications of multisets, *Novi Sad Journ. Math.*, V. 37, No. 2, pp. 73–92.

Sinkov, M. V. and Gubareni, N. M. (1979) *Nonposional Representations in Multidimensional Number Systems*, Naukova Dumka, Kiev (in Russian)

Siorvanes, L. (1986) *Proclus on the Elements and the Celestial Bodies* (physical thought in late Neoplatonism), A Thesis submitted for the Degree of Doctor of Philosophy to the Dept. of History and Philosophy of Science, Science Faculty, University College London.

Sipser, M. (1997) *Introduction to the Theory of Computation*, PWS Publishing Co., Boston.

Sitomer, M. and Sitomer, H. (1976) *How Did Numbers Begin?* Crowell, New York.

Skelley, A. (2004) A Third-Order Bounded Arithmetic Theory for PSPACE, in Proceedings of the18th International Workshop *"Computer Science Logic"* (CSL 2004), 13th Annual Conference of the EACSL, Karpacz, Poland.

Skolem, T. (1922/1923) Einige Bemerkungen zur axiomatischen Begründung der Mengenlehre, Matematikerkongressen I Helsingfors den 4–7 Juli 1922, *Den femte skandinaviska matematikerkongressen*, Redogörelse, Akademiska Bokhandeln, Helsinki, 1923, pp. 217–232.

Skolem, T. (1923/1976) Begründung der elementaren Arithmetik durch die rekurrierende Denkweise ohne Anwendung scheinbarer Veränderlichen mit unendlichem Ausdehnungsbereich, *Skrifter utgit av Videnskapsselskapet i Kristiania*. I, *Matematisk-naturvidenskabelig klasse*, V. 6, pp. 1–38 (in English: The foundations of elementary arithmetic established by means of the recursive mode of thought, without the use of apparent variables ranging over infinite domains, in *From Frege to Gödel* (Jean van Heijenoort, Ed.), Harvard University Press, pp. 306–333).

Skolem, T. (1934) Über die Nicht-charakterisierbarkeit der Zahlen reihe mittels endlich oder abzählbar unendlich vieler Aussagen mit ausschließlich Zahlenvariablen, *Fundamenta Mathematicae*, V. 23, No. 1, pp. 150–161.

Skolem, T. (1955) Peano's axioms and models of arithmetic, in *Mathematical interpretation of formal systems*, North-Holland, Amsterdam, pp. 1–14.

Slaveva-Griffin, S. (2009) *Plotinus on Number*, Oxford University Press, Oxford.

Slavík, A. (2007) *Product integration, its history and applications*, History of Mathematics, V. 29, Matfyzpress, Prague.

Slembek, S. (2008) On the arithmetization of algebraic geometry, *Historia Math.*, V. 35, pp. 285–300.

Smeltzer, D. (1959) *Man and Number*, Emerson Books, New York.

Smith, D. E. (1908) *Rara Arithmetica: A Catalogue of the Arithmetics Written Before the Year MDCI*, with Description of Those in the Library of George Arthur Plimpton, of New York, Ginn, Boston.

Smith, D. E. (1924) The first printed arithmetic (Treviso, 1478), *Isis*, V. 6, No. 3, pp. 311–331.

Smith, D. E. (1939) *Addenda to Raraarithmetica Which Described in 1908 Such European Arithmetics Printed Before1601 as Were Then in the Library of the Late George Arthur Plimpton*, Ginn, Boston.

Smith, D. E. (1923) *Mathematics*, Marshall Jones Co, Boston, MA.

Smith, D. E. and Ginsberg, J. (1937) *Numbers and Numerals*, Columbia University, New York.

Smith, D. E. and Karpinski, L. C. (1911) *The Hindu–Arabic Numerals*, Ginn, Boston.

Smorynski, C. (1984) Lectures on nonstandard models of arithmetic: Commemorating Giuseppe Peano, in *Logic Colloquium'82, Proc. Colloquium Held in Florence*, 23–28 August, 1982, Elsevier Science Pub. Co., Amsterdam, pp. 1–70.

Smullyan, R. M. (1962) *Theory of Formal Systems*, Princeton University Press, Princeton.

Snow, J. E. Views on the real numbers and the continuum, *Rev. Modern Logic* 9 (2003), No. 1–2, pp. 95–113.

Soare, R. I. (1987) *Recursively Enumerable Sets and Degrees*, Springer-Verlag, Berlin.

Sobczyk, G. (1995) The hyperbolic number plane. *College Math. J.*, V. 26, No. 4, pp. 268–280.

Soderstrand, M. A., Jenkins, W. K., Jullien, G. A. and Taylor, F. J. (Eds.) (1986) *Residue Number System Arithmetic*, IEEE Press.

Sol, M. (2021) *Synchronicity, Symbolism and the Meaning of Numbers, Lonerwolf*, lonerwolf.com/meaning-of-numbers.

Solovay, R. M. (1963) Independence results in the theory of cardinals. I, II, *Notices Amer. Math. Soc.*, V. 10, p. 595.

Sondheimer, E. and Rogerson, A. *Numbers and Infinity: A Historical Account of Mathematical* Concepts, Dover Books on Mathematics, Dover Publications, New York, 1981.

Sonnen, M. (2013) Merger math: Can $1+1 = 3$? *Investment News*, Dec 11, 2013. https://www.investmentnews.com/article/20131211/FREE/131219982/merger-math-can-1-1-3.

Sophian, C. and Wood, A. (1996) Numbers, thoughts, and things: The Ontology of numbers for children and adults, *Cognitive Development*, V. 11, No. 3, pp. 343–356.

Spector, L. (2021) The math page, skill in arithmetic, https://themathpage.com/ Arith/fractions.htm.

Spelke, E. and Barth, H. (2003) The construction of large number representations in adults, *Cognition*, V. 86, No. 3, pp. 201–221.

Spelke, E. S. and Tsivkin, S. (2001) Language and number: A bilingual training study, *Cognition*, V. 78, No. 1, pp. 45–88.

Spengler, O. (1922) *Der Untergang des Abendlandes. Umrisse Einer Morphologie der Weltgeschichte*, Oskar Beck, München.

Speyer, D. and Sturmfels, B. (2009) Tropical mathematics, *Math. Magazine*, V. 82, No. 3, pp. 163–173.

Spiegel, M. R. (1971) *Calculus of Finite Differences and Difference Equations*, McGraw-Hill Education, New York.

Spinoza, B. (1988) *Collected Works*, V. 1, Princeton University Press, Princeton.

Stähle, K. (1931) *Die Zahlenmystik bei Philon von Alexandreia*, Teubner, Tuebingen.

Stanley, D. (1999) A multiplicative calculus, *Primus*, V. 9, No. 4, pp. 310–326.

Starkey, P. and Cooper, R. (1980) Perception of numbers by human infants, *Science*, V. 210, pp. 1033–1035.

Steen, S. W. P. (1972) *Mathematical Logic with Special Reference to the Natural Numbers*, Cambridge University Press, Cambridge.

Steenrod, N. (1951) *The Topology of Fibre Bundles*, Princeton University Press.

Stefanini, L., Sorini, L. and Guerra, M. L. (2006) Parametric representation of fuzzy numbers and application to fuzzy calculus, *Fuzzy Sets Sys.*, V. 157, No. 18, pp. 2423–2455.

Stefanini, L., Sorini, L., Guerra, M. L., Pedrycz, W., Skowron, A. and Kreinovich, V. (2008) Fuzzy numbers and fuzzy arithmetic, in *Handbook of Granular Computing*, pp. 249–284.

Stein, H. (1990) Eudoxos and Dedekind: on the ancient Greek theory of ratios and its relation to modern mathematics, *Synthese*, V. 84, pp. 163–211.

Stein, S. K. (1999) *Strength in Numbers: Discovering the Joy and Power of Mathematics in Everyday Life*, Wiley.

Steindl, F. (1778) *Institutiones Arithmeticae in usum Gymnasiorum et Scholarum Gramaticarum per Regnum Hungariae et Provinciase odemad nexas [Basic Arithmetic for the Use in Gymnasia and Grammar Schools in the Kingdom of Hungary and Associated Provinces]*, Typis RegiaeUniversitatis, Budapest (in Latin).

Steinitz, E. (1910) Algebraische theorie der Körper, *J. Reine Angew. Math.*, V. 137, pp. 167–309.

Stevens, S. S. (1935) The operational definition of psychological concepts, *Psychol. Rev.*, V. 42, pp. 517–527.

Stevens, S. S. (1946) On the theory of scales of measurement, *Science*, V. 103, No. 2684, pp. 677–680.

Stevens, S. S. (1951) Mathematics, measurement and psychophysics, in *Handbook of Experimental Psychology*, Wiley, New York, pp. 1–49.

Stevens, S. S. (1975) *Psychophyscis: Introduction to its Perceptual, Neuronal, and Social Prospects*, Wiley, New York.

Stevin, S. (1585) *De Thiende*, Chriftoffel Plantijn, Leyden, (in Flemish).

Stevin, S. (1585/1634). L'Arithmetique, in *Les Oeuvres Mathematiques de Simon Stevin*, A. Girard (Ed.), Leyde, pp. 1–101.

Stevin, S. (1958) *The Principal Works of Simon Stevin*, C. V. Swets & Zeitlinger, Amsterdam.

Stewart, I. (1986) Meaning from numbers, *Nature*, V. 324, pp. 519–520.

Stewart, I. (2008) *Nature's Numbers: The Unreal Reality of Mathematics*, Basic Books.

Stewart, I. (2017) *The Beauty of Numbers in Nature: Mathematical Patterns and Principles from the Natural World*, MIT Press.

Stewart, I. (2018) Number symbolism, *Encyclopedia Britannica*, Gale Cengage.

Stewart, I. and Tall, D. (2015) *Algebraic Number Theory and Fermat's Last Theorem*, CRC Press, New York.

Stifel, M. (1544) *Arithmetica Integra*, Iohan Petreium, Nuremberg.

Stillinger, F. H. (1977) Axiomatic basis for spaces with noninteger dimension, *J. Math. Phys.*, V. 18, pp. 1224–1234.

Stillwell, J. (2018) *Reverse Mathematics*, Princeton University Press.

Stevens, S. S. (1946) On the theory of scales of measurement, *Science*, V. 103, No. 2684, pp. 677–680.

Stoianov, I. and Zorzi, M. (2012) Emergence of a "visual number sense" in hierarchical generative models, *Nat. Neurosci.*, V. 15, pp. 194–196.

Stokes, J. (2007) *Inside the Machine: An Illustrated Introduction to Microprocessors and Computer Architecture*, No Starch Press.

Stolz, O. (1881) B. Bolzano's Bedeutung in der Geschichte der Infinitesimalrechnung, *Math. Ann.*, V. 18, pp. 255–279.

Stolz, O. (1882) Zur Geometrie der Alten, insbesondere über ein Axiom des Archimedes, *Berichtedes Naturwissenschaftlich-Medizinischen Vereines in Innsbruck*, V. 12, pp. 74–89.

Stolz, O. (1883) Zur Geometrie der Alten, insbesondere über ein Axiom des Archimedes, *Math. Ann.* V. 22, pp. 504–519.

Stolz, O. (1885) *Vorlesungen über Allgemeine Arithmetik, Erster Theil: Allgemeines und Arithhmetik der Reelen Zahlen*, Teubner, Leipzig.

Stolz, O. (1886) *Vorlesungen über Allgemeine Arithmetik, Zweiter Theil: Arithhmetik der Complexen Zahlen*, Teubner, Leipzig.

Stolz, O. (1888) Ueber zwei Arten von unendlich kleinen und von unendlich grossen Grössen, *Math. Ann.* 31, pp. 601–604.

Stolz, O. (1891) Ueber das Axiom des Archimedes, *Math. Ann.* 39, pp. 107–112.

Stolz, O. and Gmeiner, J. A. (1902) *Theoretische Arithmetik*, Teubner, Leipzig.

Stolz, O. and Gmeiner, J. A. (1915) *Theoretische Arithmetik, Abteilung II: Die Lehren Von Den Reelen Und Von Den Komplexen Zahlen*, Zweite Auflage, Teubner, Leipzig.

Stonemill, P. (2016) *Synergy: The Premium for Successor* 1 +1 = 3, https://stonemillpartners.com/wp-content/uploads/2015/12/Article-%E2%80%93-Synergy.pdf.

Strauss, D. F. M. (2006) The concept of number: multiplicity and succession between cardinality and ordinality, *South African J. Philos.*, V. 25, No. 1, pp. 7–47.

Štrboja, M., Pap, E. and Mihailović, B. (2018) Discrete bipolar pseudo-integrals, *Inform. Sci.*, V. 468, pp. 72–88.

Štrboja, M., Pap, E. and Mihailović, B. (2019) Transformation of the pseudo-integral and related convergence theorems, *Fuzzy Sets Syst.*, V. 355, pp. 67–82.

Struik, D. J. (1987) *A Concise History of Mathematics*, Dover Publications, New York.

Suber, P. (2011) *Geometry and Arithmetic are Synthetic*, https://legacy.earlham.edu/~peters/writing/synth.htm#arithmetic.

Sugeno, M. and Murofushi, T. (1987) Pseudo-additive measures and integrals, *J. Math. Anal. Appl.*, V. 122, pp. 197–222.

Sulkowski, G. M. and Hauser, M. D. (2001) Can rhesus monkeys spontaneously subtract? *Cognition*, V. 79, pp. 239–262.

Sunaga, T. (1958) Theory of interval algebra and its application to numerical analysis, *RAAG Memoirs*, No. 2, pp. 29–46.

Sunehag, P. (2007) *On a Connection Between Entropy, Extensive Measurement and Memoryless Characterization*, Preprint in Physics [physics.data-an], arXiv:0710.4179.

Sunehag, P. (2007) *On a Connection between Entropy, Extensive Measurement and Memoryless Characterization*, Preprint in Data Analysis, Statistics and Probability (arXiv 0710.4179).

Suppes, P. (1957) *Introduction to Logic*, Litton Educational Publishing.

Suroweicki, J. (2004) *The Wisdom of Crowds: Why the Many Are Smarter Than the Few and How Collective Wisdom Shapes Business, Economies, Societies and Nations*, Little & Brown, Boston.

Sutherland, D. (2006) Kant on arithmetic, algebra, and the theory of proportions, *J. the History Philos.*, V. 44, No. 4, pp. 533–558.

Sutherland, D. (2017) Kant's conception of number, *Philos. Rev.*, V. 126, No. 2, pp. 147–190.

Sutton Lennes, N. J. and C. W. (1961) *Economic Mathematics: Business Arithmetic for the Consumer*, Allyn & Bacon Inc., Boston.

Svoboda, D. and Sousedík, P. (2014) Mathematical one and many: aquinas on number, in *The Thomist: A Speculative Quarterly Review*, The Catholic University of America Press, V. 78, No. 3, pp. 401–418.

Swamy, P. N. (1999) *Infinities in Physics and Transfinite numbers in Mathematics*, preprint in Mathematical Physics, math-ph/9909033, 16 pp. http://arXiv.org.

Swartzlander, E. E., Jr., (1990) *Computer Arithmetic*, v. I & II, IEEE Computer Society Press.

Swetz, F. J. (1987) *Capitalism & Arithmetic: The new Math of the Fifteenth Century, Including the Full Text of the Treviso Arithmetic of 1478*, Open Court, La Salle.

Swift, J. D. (1956) Diophantus of Alexandria, *Amer. Math. Monthly*, V. 63, pp. 163–170.

Synergy, (2011) Cambridge Business English Dictionary, Cambridge University Press, London.

Szelepcsényi, R. (1988) The method of forced enumeration for nondeterministic automata, *Acta Inform.*, V. 26, pp. 279–284.

Szpiro, G. G. (2006) *The Secret Life of Numbers: 50 Easy Pieces on How Mathematicians Work and Think*, Joseph Henry Press, Washington, DC.

Szuba, T. (2001) *Computational Collective Intelligence*, Wiley, New York.

Tait, W. W. (1996) Frege versus cantor and Dedekind: on the Concept of number, in *Frege: Importance and Legacy*, M. Schirn (Ed.), de Gruyter, Berlin, pp. 70–113.

Takeuti, G. (1979) Boolean valued analysis, in *Proc. Applications of Sheaves*, Springer, Berlin, pp. 714–731.

Tall, D. O. (1980) The notion of infinite measuring number and its relevance in the intuition of infinity, *Educational Studies Math.* V. 11, pp. 271–284.

Tall, D. O. (2004) The three worlds of mathematics, *For the Learning Math.*, V. 23, No. 3, pp. 29–33.

Tall, D. O. (2004) Thinking through three worlds of mathematics, in *Proc. 28th Conference of the International Group for the Psychology of Mathematics Education*, Bergen, Norway, V. 4, pp. 281–288.

Tall, D. (2008) The transition to formal thinking in mathematics, *Math. Education Res. J.*, V. 20 No. 2, pp. 5–24.

Tandon, A., Song, Y., Mitta, S. B.,Yoo, S., Park, S., Lee, S., Raza, M. T., Ha, T. H. and Park, S. H. (2020) Demonstration of arithmetic calculations by DNA tile-based algorithmic self-assembly, *ACS Nano*, V. 14, No. 5, pp. 5260–5267.

Tang, Y., Zhang, W., Chen, K., Feng, S., Ji, Y., Shen, J., Reiman, E. M. and Liu, Y. (2006) Arithmetic processing in the brain shaped by cultures, *Proc. Natl. Acad. Soc. USA*, V. 103, No. 28, pp. 10775–10780.

Tannery, P. (Ed.) (1893/1895) *Diophantus Alexandrinus*, Opera Omnia, I & II, Leipzig.

Tao, T. C. and Vu, V. H. (2006) *Additive Combinatorics*, Cambridge Studies in Advanced Mathematics, V. 105, Cambridge University Press, Cambridge.

Tappenden, J. (1995) Geometry and generality in Frege's philosophy of arithmetic, *Synthese*, V. 102, pp. 319–361.

Tappenden, J. (2006). The Riemannian background to Frege's philosophy, in *The Architecture of Modern Mathematics*, Oxford University Press, pp. 107–150.

Tarski, A. (1936) Der Wahrheitsbegriff in den formalisierten Sprachen, *Studia Philosophica*, V. 1, pp. 261–405.

Tarski, A. and Vaught, R. L. (1957) Arithmetical extensions of relational systems, *Compos. Math.*, V. 13, pp. 81–102.

Taetaglia, N. (1556–1560) *General trattato di numeri, et mesuri*, 6 parti, Curcio Trolano de'Navo, Venice.

Taves, E. H. (1941) Two mechanisms for the perception of visual numerousness, *Arch. Psychol.*, V. 37, pp. 1–47.

Taylor, T. (1816) *Theoretic Arithmetic*, A. J.Valpt, Walwarth, UK.

Taylor, R. E. (1942) *No Royal Road, Luca Pacioli and his Times*, University of North Carolina Press, Chapel Hill.

Taylor, R. and Wiles A. (1995) Ring theoretic properties of certain Hecke algebras, *Ann. Math.*, V. 141, N. 3, pp. 553–572.

Tegmark, M. (2014) *Our Mathematical Universe: My Quest for the Ultimate Nature of Reality*, Alfred A. Knopf, New York.

Tennant, N. (1997) On the necessary existence of numbers, *Noûs*, V. 31, No. 3, pp. 307–336.

Taylor, T. (1816) *The Six Books of Proclus, the Platonic Successor, on the Theology of Plato by Proclus*, Printed by A. J. Valpy, Tooke's Court, London.

Thesleff, H. (Ed.) (1965) *The Pythagorean Texts of the Hellenistic Period.Åbo: ÅboAkademi*, in ActaAcademiaeAboensis. Ser. A. Humaniora. HumanistikaVetenskaper. Social-Vetenskaper.Teologi, V. 30, No. 1.

Thomae, J. (1870) *Abriss einer Theorie der complexen Functionen und der Thetafunctionen einer Veränderlichen*, Verlag von Louis Nebert, Halle.

Thomae, J. (1872) Sur les limites de la convergence et de la divergence des séries infinies à termes positifs, *Ann. Mat. Pura Appl.*, V. 5, No. 2, pp. 121–129.

Thomae, J. (1880) *Elementare Theorie der analytischen Functionen einer complexen Veränderlichen*, Verlag von Louis Nebert, Halle.

Thomas, K. V. and Nair, L. S. Rough intuitionistic fuzzy sets in a lattice, *Int. Math. Forum*, V. 6, 2011, No. 27, pp. 1327–1335.

Thomson, J. (2001) Comments on Professor Benacerraf's Paper, in *Zeno's Paradoxes*, Hackett, Indianapolis, IN.

Thompson, R. F., Mayers, K. S., Robertson, R. T. and Patterso, C. J. (1970) Number coding in association cortex of the cat, *Science*, V. 168, No. 3928, pp. 271–273.

Thorne, J. A. (2012) *The Arithmetic of Simple Singularities*, Doctoral dissertation, Harvard University.

Tobin, R. (1975) The Canon of Polykleitos, *American J. Archaeol.*, V. 79, No. 4, pp. 307–321.

Tod, M. N. (1979) *Ancient Greek Numerical Systems*, Ares Publishers, Chicago.

Tolksdorf, R. (1995) *A Machine for Uncoupled Coordination and Its Concurrent Behavior*, Lecture Notes in Computer Science, V. 924, Springer, New York, pp. 176–193.

Tolksdorf, R. (1996) *Coordinating Services, in Open Distributed Systems with LAURA, Coordination Languages and Models*, Lecture Notes in Computer Science, V. 1061, Springer, New York, pp. 386–402.

Tolpygo, A. (1997) Finite infinity, in *Methodological and Theoretical Problems of Mathematics and Information Sciences*, Kiev, Ukrainian Academy of Information Sciences, pp. 35–44 (in Russian)

Toscano, F. (2020) *The Secret Formula*, Princeton University Press, Princeton.

Trabacca, A., Moro, G., Gennaro, L. and Russo, L. (2012) When one plus one equals three: the ICF perspective of health and disability in the third millennium, *European J. Phys. Rehabilitation Med.*, V. 48, No. 4, pp. 709–710.

Träff, U. (2013) The contribution of general cognitive abilities and number abilities to different aspects of mathematics in children, *J. Exp. Child. Psychol.*, V. 116, No. 2, pp. 139–156.

Trafton, P. (1992). Using number sense to develop mental computation and computational estimation, in *Challenging Children to Think When They Compute*, Centre for Mathematics and Science Education, Queensland University of Technology, Brisbane, pp. 78–92.

Trick L. M. and Pylyshyn Z. W. (1994) Why are small and large numbers enumerated differently? A limited-capacity preattentive stage in vision, *Psychol. Rev.*, V. 101, No. 1, pp. 80–102.

Troelstra, A. S. (Ed) (1973) *Metamathematical Investigation of Intuitionistic Arithmetic and Analysis*, Springer, New York.

Trott, D. (2015) *One Plus One Equals Three: A Masterclass in Creative Thinking*, Macmillan Publishing Company, New York.

Trudeau, R. J. (1987) *The Non-Euclidean Revolution*, Birkhauser, Boston.

Tsallis, C. (1988) Possible generalizations of Boltzmann-Gibbs statistics, *J. Stat. Phys.*, V. 52, pp. 479–487.

Tschentscher N. and Hauk, O. (2014) How are things adding up? Neural differences between arithmetic operations are due to general problem solving strategies, *NeuroImage*, V. 92, pp. 369–380.

Tubbs, R. (2009) *What is a Number?: Mathematical Concepts and Their Origins*, Johns Hopkins University Press, Baltimore, MD.

Tufte, E. R. (1990) *Envisioning Information*, Graphics Press, Cheshire, CT.

Tunstall, S. L. Connecting Numbers with emotion: review of "Numbers and Nerves: Information, Emotion, and Meaning in a World of Data by Scott Slovic and Paul Slovic (2015)", *Numeracy*, V. 10, No. 1 (2017): Article 9, DOI: http://dx.doi.org/10.5038/1936-4660.10.1.9.

Turing, A. (1936) On computable numbers with an application to the Entscheidungs-problem, *Proc. London Math. Soc.,* Ser.2, V. 42, pp. 230–265.

Turner P. (1993) *Modern Macroeconomic Analysis*, McGraw-Hill.

Ueno, Y (2003) Kronecker's idea of arithmetization of mathematics, *Acad. Rep. Fac. Eng. Tokyo Polytechnic Univ.*, V. 26, No. 1, pp. 71–76.

Uller, C., Carey, S., Huntley-Fenner, G. and Klatt, L. (1999) What representations might underlie infant numerical knowledge. *Cognitive Development*, V. 14, pp. 1–36.

Ulrych, S. (2005) Relativistic quantum physics with hyperbolic numbers, *Phys. Lett. B* V. 625, No. 3–4, pp. 313–323.

Ulrych, S. (2010) Considerations on the hyperbolic complex Klein-Gordon equation, *J. Math. Phys.*, V. 51, No. 6, 063510.

Ulivi, E. (2015) Masters, questions and challenges in the abacus schools, *Arch. History Exact Sci.*, V. 69, No. 6, pp. 651–670.

Umarov, S. Tsallis, C. and Steinberg, S. (2006) *A generalization of the Central Limit Theorem Consistent with Nonextensive Statistical Mechanics*, preprint cond-mat/0603593.

Unger, F. (1888) *Die Methodik der praktischen Arithmetik in historischer Entwickelung vom Ausgange des Mittelaltersbis auf die Gegenwart* [*The*

Methodology of Practical Arithmetic in Its Historical Development from the End of the Middle Ages until the Present Day], B.G. Teubner, Leipzig.

Urbaniak, R. (2012) Numbers and propositions versus nominalists: yellow cards for Salmon & Soames, *Erkenntnis*, V. 77, No. 3, pp. 381–397.

Urton, G. (1997) *The Social Life of Numbers: A Quechua Ontology of Numbers and Philosophy of Arithmetic*, University of Texas Press.

Uzer, A. (2010) Multiplicative type complex calculus as an alternative to the classical calculus, *Comput. Math. Appls.*, V. 60, No. 10, pp. 2725–2737.

Väänänen, J. (2021) Second-order and Higher-order Logic, *The Stanford Encyclopedia of Philosophy*, https://plato.stanford.edu/archives/fall2021/entries/logic-higher-order/.

Väänänen, J. and Wang, T. (2015) Internal categoricity in arithmetic and set theory, *Notre Dame Journal of Formal Logic*, v. 56, No.1, pp. 121–134.

Vallortigara, G. Foundations of number and space representations in non-human species, in *Evolutionary Origins and Early Development of Number Processing*, Academic Press, London, pp. 35–66.

Van Bendegem, J. P. (1994/1996) Strict finitism as a viable alternative in the foundations of mathematics, *Logique Anal.*, V. 37, No. 145, pp. 23–40.

van Dantzig, D. (1956) Is $10^{10^{10}}$ a finite number? *Dialectica*, No. 9.

Van de Voorde, N. (2017) One plus one equals three? The electoral effect of multiple office-holding in national and local elections, *ECPR General Conference 2017*, https://biblio.ugent.be/publication/8534419.

Van der Corput, J. G. (1938) Sur l'hypothèse de Goldbach, *Proc. Akad. Wet. Amsterdam*, V. 41, pp. 76–80.

Van der Waerden, B. L. (1971) *Algebra*, Springer-Verlag, Berlin.

Van der Waerden, B. L. (1985) *A History of Algebra from al-Khwārizmī to Emmy Noether*, Springer-Verlag, New York.

vanMarle, K. (2015) Foundations of the formal number concept: how preverbal mechanisms contribute to the development of cardinal knowledge, in *Evolutionary Origins and Early Development of Number Processing*, Academic Press, London pp. 175–199.

Varadarajan, V. S. (2002) Some remarks on arithmetic physics, *J. Statis. Planning Inference*, V. 103, No. 1–2, pp. 3–13.

Varadarajan, V. S. (2004) Arithmetic quantum physics: why, what, and whither, *Tr. Mat. Inst. Steklova*, V. 245, pp. 273–280.

Varley, R. A., Klessinger, N. J., Romanowski, C. A. and Siegal, M. (2005) Agrammatic but numerate, *Proc. Natl. Acad. Sci. USA*, V. 102, No. 9, pp. 3519–3524.

Vasas, V. and Chittka, L. (2018) Insect-inspired sequential inspection strategy enables an artificial network of four neurons to estimate numerosity, *Science*, V. 11, pp. 85–92.

Verguts, T. and Fias, W. (2004) Representation of number in animals and humans: A neural model, *J. Cogn. Neurosci.*, V. 16, pp. 1493–1504.

Verguts, T. and Fias, W. (2008) Symbolic and nonsymbolic pathways of number processing, *Philos. Psychol.*, V. 21, pp. 539–554.

Vermij, R. H. (1989) Bernard Nieuwentijt and the Leibnizian Calculus, *Studia Leibnitiana*, Bd. 21, H. 1, pp. 69–86.

Vernaeve, H. (2010) Ideals in the ring of Colombeau generalized numbers, *Comm. Alg.*, V. 38, No. 6, pp. 2199–2228.

Veronese, G. (1889) Il continuo rettilineo e l'assioma V di Archimede, *Memorie della Reale Accademia dei Lincei, Atti della Classe di Scienze Naturali, Fisiche e Matematiche*, V. 6, No. 4, pp. 603–624.

Vitruvius, (1914) *The Ten Books on Architecture*, Harvard University Press, Cambridge, MA.

Vladimirov, V. S., Volovich, I. V. and Zelenov, E. I. (1994) *p-Adic Analysis and Mathematical Physics*, World Scientific, Singapore.

Voigt, J. (1994) Negotiation of mathematical meaning and learning mathematics, educational studies in mathematics, learning mathematics, *Constructivist Interactionist Theories Math. Development*, V. 26, No. 2/3, pp. 275–298.

Volodarsky, A. I. (1977) *Aryabhata*, Nauka, Moscow (in Russian).

Volovich, I. V. (1987) *Number Theory as the Ultimate Physical Theory*, preprint CERN-TH 87, pp. 4781–4786.

Volterra, V. (1887) Sui fondamenti della teoria delle equazioni differenziali lineari, *Memorie Soc. Italiana Sci.*, V. 3, VI.

Volterra, V. (1887a) Sulle equazioni differenziali lineari, *Rend. Acad. Lincei*, V. 3, pp. 393–396.

von Helmholtz, H. Zahlen und Messen, in *Philosophische Aufsatze*, Fues's Verlag, Leipzig, pp. 17–52 (Translated by C. L. Bryan, "*Counting and Measuring*", Van Nostrand, 1930).

von Glasersfeld, E. (1981) An attentional model for the conceptual construction of units and number, *J. Res. Math. Education*, V. 12, No. 2, pp. 83–94.

von Glasersfeld, E. (1995) *Radical Constructivism: A Way of Knowing and Learning*, Studies in Mathematics Education Series, V. 6, Falmer Press, Taylor & Francis Inc., Bristol, PA.

von Koch, H. (1904) Sur une courbe continue sans tangente, obtenue par une construction géométrique élémentaire, *Ark. Mat.*, V. 1, pp. 681–704.

von Neumann, J. (1923) Zur Einführung der transfiniten Zahlen, *Acta Litterarum AC Scientiarum Ragiae Universitatis Hungaricae Francisco-Josephinae, Sectio Scientiarum Mathematicarum*, V. 1, pp. 199–208.

von Neumann, J. (1925) Eine Axiomatisierung der Mengenlehre, *J. Reine Angew. Math.*, V. 154, pp. 219–240.

von Neumann, J. (1927) Zur Hilbertschen Beweistheorie, *Math. Zeitschrift*, V. 26, pp. 1–46.

von Neumann, J. (1928) Die Axiomatisierung der Mengenlehre, *Math. Zeitschrift*, V. 27, pp. 669–752.

von Neumann, J. (1929) Über eine Widerspruchsfreiheitsfrage in der axiomatischen Mengenlehre, *J. Reine Angew. Math.*, V. 160, pp. 227–241.

von Neumann, J. (1955) *Mathematical Foundations of Quantum Mechanics*, Princeton University Press, Princeton, NJ.

von Neumann, J. (1961) *Collected Works*, V. 1, *Logic, Theory of Sets and Quantum Mechanics*, Pergamon Press, New York/.

von Plato J. (2017) Husserl and Grassmann, in *Essays on Husserl's Logic and Philosophy of Mathematics*, Centrone S. (Ed.), Synthese Library (Studies in Epistemology, Logic, Methodology, and Philosophy of Science), V. 384. Springer, Dordrecht.

Vopěnka, P. (1964) The independence of the continuum hypothesis, *Comment. Math. Univer. Carolinae*, V. 5, Supplement I, pp. 1–48.

Vopěnka, P. (1979) *Mathematics in the Alternative Set Theory*, Teubner-Verlag, Leipzig.

Vorobjev, N. N. (1963) The extremal matrix algebra, *Soviet Math. Dokl.*, V. 4, pp. 1220–1223.

Vrandečić, D., Krötzsch, M., Rudolph, S. and Lösch, U. (2010) Leveraging non-lexical knowledge for the linked open data web, in *The Fifth RAFT'2010, The Yearly Bilingual Publication on Nonchalant Research*, V. 5, No. 1, pp. 18–27.

Wade, J. E. (1872) *The Merchants Mechanics' Commercial Arithmetic; or, Instantaneous Method of Counting*, Russell Brothers, New York.

Waddell, J. (1899) *The Arithmetic of Chemistry*, Macmillan, New York.

Wagner, U. (1482) *Rechnung in Mancherleyweys*, Heinrich Petzensteiner, Bamberg.

Walkingame, F. (1849) *The Tutor's Companion; or, Complete Practical Arithmetic*, Webb, Millington & Co, Leeds.

Wallis, J. (1656) *Arithmetica infinitorum sive Nova Methodus Inquirendi in Curvilineorum Quadraturam, aliaque difficiliora Matheseos Problemata*, Oxonii. Typis Leon Lichfield Academiae Typographi Impensis Tho. Robinson.

Wallis, J. (1685) *A Treatise of Algebra*, John Playford, London.

Walsh, S. (2010) Arithmetical Knowledge and Arithmetical Definability: Four Studies, Dissertation, University of Notre Dame, Notre Dame, Indiana.

Walsh, S. (2012) Comparing Peano arithmetic, Basic Law V, and Hume's Principle, *Ann. Pure Appl. Logic*, V. 163, No. 11, pp. 1679-1709.

Walsh, S. (2014) Logicism, interpretability, and knowledge of arithmetic, *Rev. Symbolic Logic*, V. 7, No. 1, pp. 84–119.

Walsh, V. (2003) A theory of magnitude: Common cortical metrics of time, space and quantity, *Trends Cognitive Sci.*, V. 7, pp. 483–488.

Walsh, V. (2015) A theory of magnitude: the parts that sum to number, in *The Oxford Handbook of Numerical Cognition*, Cohen Kadosh R. and Dowker A., (Eds.), Oxford University Press, Oxford, pp. 552–565.

Wang, H. (1956) Arithmetic translations of axiom systems, *J. Symbolic Logic*, V. 21, No. 4, pp. 402–403.

Wang, H. (1957) The axiomatization of arithmetic, *J. Symbolic Logic*, V. 22, No. 2, pp. 145–158.

Wang, H. (1974) *From Mathematics to Philosophy*, Routledge, Kegan & Paul, London.

Wang, X. (2008) Fuzzy number and intuitionistic fuzzy arithmetic aggregation operators, *Int. J. Fuzzy Syst.*, V. 10, No. 2, pp. 104–111.

Wang, X. and Kerre, E. E. (2001) Reasonable properties for the ordering of fuzzy quantities (I), *Fuzzy Sets Syst.*, V. 118, pp. 375–385.

Wang, Z. and Klir, G. J. (2009) *Generalized Measure Theory*, Springer, Boston.

Wang, L., Uhrig, L., Jarraya, B. and Dehaene, S. (2015) Representation of numerical and sequential patterns in macaque and human brains, *Curr. Biol.*, V. 25, No. 15, pp. 1966–1974.

Wapnick, K. (2013) *When* 2 + 2 = 5: *Reflecting Love in a Loveless World*, Foundation for A Course in Miracles, Henderson, NV.

Ward, M. (1927) General Arithmetic, *Proc. Natl. Acad. Sci. USA*, V. 13, No. 11, pp. 748–749.

Ward, M. (1928) Postulates for an abstract arithmetic, *Proc. Natl. Acad. Sci. USA*, V. 14, pp. 907–911.

Ward, M. (1928a) A simplification of certain problems in arithmetical division, *Amer. Math. Monthly*, V. 35, No. 1, pp. 9–14.

Ward, M. (1928) *The Foundations of General Arithmetic.*, (Ph.D.), Dissertation, California Institute of Technology, doi:10.7907/XDSZ-KF26.

Wartofsky, M. (1976) The relation between philosophy of science and history of science, in *Essays in memory of Imre Lakatos*, Boston Studies in the Philosophy of Science XXXIX, D. Reidel Publishing, Dordrecht, pp. 717–737.

Wartofsky, M. (1977) Nature, number and individuals: Motive and method in Spinoza's philosophy, an *Interdisciplinary J. Philos.*, V. 20, No. 1–4, pp. 457–479.

Waszkiewicz, W. (1971) The notions of isomorphism and identity for many-valued relational structures, *Studia Logica*, V. 27, pp. 93–98.

Watzlawick, P, Beavin, J. H., and Jackson, D. D. (1967) *Pragmatics of Human Communication: A Study of Interactional Patterns, Pathologies, and Paradoxes*, W W Norton, New York.

Webb, J. C. (1997), Hilbert's formalism and arithmetization of mathematics, A symposium on David Hilbert, *Synthese*, V. 110, No. 1, pp. 1–14.

Weber, S. (1984) $\perp$-Decomposable measures and integrals for Archimedean *t*-conorm, *J. Math. Anal. Appl.*, V. 101, pp. 114–138.

Wedderburn, J. (1908) On hypercomplex numbers, *Proc. London Math. Soc.*, V. 6, pp. 77–118.

Weihrauch, K. (2000) *Introduction to Computable Analysis*, Springer, Berlin.

Weil, A. (1929) L'arithmétique sur les courbes algébriques, *Acta Math.*, V. 52, pp. 281–315.

Weil, A. (1984) *Number Theory: An Approach Through History from Hammurapi to Legendre*, Birkhauser, Boston.

Weil, A. (1995) *Basic number theory*, Springer, New York.

Weiss, A. (2005) The power of collective intelligence, netWorker — Beyond file-sharing, *Collective Intelligence*, V. 9, No. 3, pp. 16–24.

Weierstrass, K. (1880) Zur functionenlehre, *Monatsberichte der K. Akad der Wiss.*, s. 201–230.

Weierstrass, K. (1884) Zur Theorie der aus *n* Haupteinheitengebildeten komplexen Grössen, *Nachrichten Königl. Ges. Wiss. zu Göttingen*, pp. 395–410.

Weierstrass, K. (1884) Zur Theorie der aus *n* Haupteinheitengebildeten komplexen Grössen, *Math. Werke*, V. 2, pp. 311–332.

Weierstrass, K. (1885) *Zu Lindemanns̓ Abhandlung "Üeber die Ludolphsche Zahle"*, K. Weierstrass, Berichte der Berliner Akademie, s. 1067–1085.

Weissmann, S. M., Hollingsworth, S. R. and Baird, J. C. (1975) Psychophysical study of numbers (III): Methodological applications, *Psychol. Res.*, V. 38, pp. 97–115.

Wellman H. M. and Miller K. F. (1986) Thinking about nothing: developmental concepts of zero, *Br. J. Dev. Psychol.*, V. 4, pp. 31–42.

Wells, D. G. (1998) *The Penguin Dictionary of Curious and Interesting Numbers*, Penguin Books, London, UK.

Wells, H. L. (2018) *A Text-book of Chemical Arithmetic*, Forgotten Books.

Were, E. (2019) Mathematics in Ancient Africa: Lebombo Bone & Others, *Buyu*, https://buyuafrika.wordpress.com/2019/03/19/mathematics-in-ancient-africa-lebombo-bone-others/.

Wessel, C. (1799) Om Directionens analytiske Betegning, et Forsog, anvendt fornemmelig til plane og sphæriske Polygoners Oplosning [On the analytic representation of direction, an effort applied in particular to the determination of plane and spherical polygons], *Nye Samling af det Kongelige Danske Videnskabernes Selskabs Skrifter*, Royal Danish Academy of Sciences and Letters, Copenhagen, V. 5, pp. 469–518 (in Danish).

Weyl, H. (1918) *Raum-Zeit-Materie*, Springer, Berlin.

Weyl, H. (1949) *Philosophy of Mathematics and Natural Science*, Princeton University Press, Princeton, NJ.

Weyl, H. (1985) Axiomatic versus constructive procedures in mathematics, *Math. Intelligencer*, V. 7, No. 4, pp. 10–17, 38.

Whalen, J., Gallistel, C. R. and Gelman, R. (1999) Nonverbal counting in humans: the psychophysics of number representation, *Psychol. Sci.*, V. 10, pp. 130–37.

White, M. (1999) Incommensurables and Incomparables: On the Conceptual Status and the Philosophical Use of Hyperreal Numbers, *Notre Dame J. Formal Logic*, V. 40, pp. 420–446.

Whitehead, A. N. (1934) Indication, classes, number, validation, *Mind, New Ser.*, V. 43, pp. 281–297, 543.

Whitehead, A. N. and Russell, B. (1910–1913) *Principia Mathematica*, 3 vols., Cambridge Unversity Press.

Whitney, H. (1933) Characteristic functions and the algebra of logic, *Ann. of Math.*, V. 34, pp. 405–414.

Whorf, B. L. (1965) *Language, Thought and Reality*, MIT Press, Cambridge, MA.

Widmann, J. (1489) *Behende und hüpsche Rechnunguffallen Kauffmanschafften*, Leipzig.

Widmann, J. (ca. 1492) *Algorithmusl inealis*, Martin Landsberg, Leipzig.

Wigner, E. (1960) The unreasonable effectiveness of mathematics, *Commun. Pure Appl. Math.*, V. 13, No. 1, pp. 1–14.

Wildberger, N. J. (2003) *A new Look at Multisets*, preprint, University of New South Wales, Sydney, Australia.

Wiles, A. (1995) Modular elliptic curves and Fermat's last theorem, *Ann. of Math.*, V. 142, No. 3, pp. 443–551.

Wilkie, A. (1975) On models of arithmetic — answers to two problems raised by
    H. Gaifman, *J. Symbolic Logic*, V. 40, No. 1, pp. 41–47.

Williams, T. D. (2012). The earliest English printed arithmetic books, *The
    Library*, V. 13, No. 2, pp. 164–184.

Wilson, A. M. (1850/1996) *The Infinite in the Finite*, Oxford University Press,
    Oxford.

Wittgenstein, L. (1921) *Tractatus Logico-Philosophicus*, Routledge and Kegan
    Paul, London.

Wittgenstein, L. (1953) *Philosophical investigations*, Blackwell, Oxford.

Wittgenstein, L. (1983) *Remarks on the Foundations of Mathematics*, MIT Press,
    Cambridge, MA.

Woepcke, F. (1863) *Memoire sur la propagation des chiffres Indiens*, Impr.
    Impériale, Paris.

Woleński, J. (2005/2004) Godel, Tarski and Truth, *Revue int. Philos.*, No. 234,
    pp. 459–490.

Wolff P. and Holmes, K. J. (2011) Linguistic relativity, *WIREs Cogni Sci*, V. 2,
    pp. 253–265.

Wolpert, D. H. and Tumer, K. (2000) An introduction to collective intelligence,
    in *Handbook of Agent Technology*, AAAI/MIT Press.

Wood, J. and Spelke, E. (2005) Infants' enumeration of actions: Numerical
    discrimination and its signature limits, *Developmental Sci.*, V. 8, No. 2,
    pp. 173–181.

Wardley, P. and White, P. (2003) The arithmeticke project: A collaborative
    research study of the diffusion of Hindu–Arabic numerals, *Family Community
    History*, V. 6, No. 1, pp. 5–17.

Worrall, J. (1989) Structural Realism: The best of both worlds? *Dialectica*, V. 43,
    pp. 99–124.

Wright, C. (1983) *Frege's Conception of Numbers as Objects*, Aberdeen University
    Press, Aberdeen.

Wynn, K. (1990) Children's understanding of counting, *Cognition*, V. 36,
    pp. 155–193.

Wynn, K. (1992) Children's acquisition of number words and the counting system,
    Cogn. Psychol., V. 24, pp. 220–251.

Wynn K. (1992a) Addition and subtraction by human infants, *Nature*, V. 358,
    pp. 749–750.

Wynn K. (1995) Origins of numerical knowledge, *Math Cogn.*, V. 1, pp. 35–60.

Wynn, K. (1998) Psychological foundations of number: Numerical competence in
    human infants, *Trends Cognitive Sci.*, V. 2, No. 8, pp. 296–303.

Xiao, Z. (2014) Application of Z-numbers in multi-criteria decision making, in
    *Proc. 2014 International Conference on Informative and Cybernetics for
    Computational Social Systems* (ICCSS), Qingdao, pp. 91–95.

Xu, F. and Spelke, E. S. (2000) Large number discrimination in 6-month-old
    infants, *Cognition*, V. 74, pp. B1–B11.

Xu, F., Spelke, E. S. and Goddard, S. (2005) Number sense in human infants,
    Dev. Sci., V. 8, pp. 88–101.

Yablo, S. (2005) The myth of the seven, in *Fictionalism in Metaphysics*, Oxford University Press, New York, pp. 88–115.

Yadin, A. (2016) *Computer Systems Architecture*, Chapman & Hall/CRC Textbooks in Computing, CRC Press, New York.

Yager, R. R. (1980) On the lack of inverses in fuzzy arithmetic, *Fuzzy Sets Sys.*, V. 4, pp. 73–82.

Yager, R. R. (1986) On the theory of bags, *International J. General Sys.*, V. 13, pp. 23–37.

Yaglom, I. M. (1979) *A Simple Non-Euclidean Geometry and Its Physical Basis*. Heidelberg Science Library. Springer-Verlag, New York.

Yalcin, N., Celik, E. and Gokdogan, A. (2016) Multiplicative Laplace transform and its applications, *Optik*, V. 127, No. 20, pp. 9984–9995.

Yamano, T. (2002) Some properties of q-logarithm and q-exponential functions in Tsallis statistics, *Physica A*, V. 305, No. 3, pp. 486–496.

Young, R. C. (1931). The algebra of many-valued quantities, *Math. Ann.*, V. 104, No. 1, pp. 260–290.

Yanovskaya, S. A. (1963) Are the problems known as Zeno's aporia solved in contemporary science? in *Problems of Logic*, Nauka, Moscow, pp. 116–136 (in Russian).

Yeldham, F. (1936) *The Teaching of Arithmetic Through Four Hundred Years*, Harrap, London.

Yellot, J. I., Jr. (1977) The relationship between Luce's Choice Axiom, Thurstone's theory of comparative judgement, and the Double exponential distribution, *J. Math. Psych.*, V. 15, pp. 109–144.

Yesenin-Volpin, A. S. (1960) On the Grounding of Set Theory, in *Application of Logic in Science and Technology*, Nauka, Moscow, pp. 22–118 (in Russian).

Yesenin-Volpin, A. S. (1970) The ultra-intuitionistic criticism and the antitraditional program for foundations of mathematics, in Intuitionism and Proof Theory, North-Holland, Amsterdam, pp. 3–45.

Yong, L. L. (1996) The development of Hindu–arabic and traditional Chinese arithmetic, *Chinese Sci.*, V. 13, pp. 35–54.

Young, R. C. (1931) The algebra of many-valued quantities, *Math. Ann.*, V. 104, No. 1, pp. 260–290.

Zadeh, L. (1965) Fuzzy sets, *Inform. Control*, V. 8, No. 3, pp. 338–353.

Zadeh, L. (1965) Fuzzy sets and systems, in *System Theory*, Polytechnic Press, Brooklyn, NY, pp. 29–39.

Zadeh, L. (1974) Fuzzy logic and its application to approximate reasoning, in *Information Processing*, Proc. IFIP Congr., V. 74, Part. 3, pp. 591–594.

Zadeh, L. A. (1975) The concept of a linguistic variable and its application to approximate reasoning, pt. 1, *Inf. Sci.*, V. 8, pp. 199–249.

Zadeh, L. A. (1976) The concept of a linguistic variable and its application to approximate reasoning, pt. 2, *Inf. Sci.*, V. 9, pp. 43–80.

Zadeh, L. A. (1984) Fuzzy probabilities, *Inform. Process. Manag.*, V. 20, No. 3, pp. 363–372.

Zadeh, L. A. (2011) A note on Z-numbers, *Inform. Sci.*, V. 181, pp. 2923–2932.

Zakon E. (1955) Fractions of ordinal numbers, *Israel Inst. Tecnol. Sci. Publi.*, V. 6, pp. 94–103.

Zalta, E. N. (1999) Natural Numbers and Natural Cardinals: A Partial Reconstruction of Frege's Grundgesetze in Object Theory, *J. Philos. Logic*, V. 28, No. 6, pp. 619–660.

Zalta, E. N. (2014) Frege's theorem and foundations for arithmetic, in *Stanford Encyclopedia of Philosophy*.

Zamyatin, E. (1924) *We*, Dutton, New York (Zilboorg, G. trans.).

Žarnić, B. (1999) Mathematical platonism: From objects to patterns, *Synthesis Philos*. No. 27–28, pp. 53–64.

Zaslavsky, C. (1973) *Africa Counts: Number and Pattern in African Culture*, Prindle, Weber & Schmidt, Boston.

Zebian, S. (2005) Linkages between number concepts, spatial thinking, and directionality of writing: The SNARC effect and the REVERSE SNARC effect in English and Arabic monoliterates, biliterates, and illiteratee Arabic speakers, *J. Cogn. Culture*, V. 5, No. 1–2, pp. 165–190.

Zeldovich, Ya. B., Ruzmaikin, A. A. and Sokoloff, D. D. (1990) *The Almighty Chance*, World Scientific Lecture Notes, V. 20, World Scientific, Singapore.

Zellini, P. (2005). *A Brief History of Infinity*, Penguin Global, London.

Zermelo, E. (1908) Untersuchungen über die Grundlagen der Mengenlehre, *Math. Annalen*, V. 65, pp. 261–281.

Zermelo, E. (1930/1996) On bounded numbers and domains of Sets, in *From Kant to Hilbert: A Source Book in Mathematics*, V. 2, Oxford University Press, Oxford, pp. 1208–1233.

Zhang, W.-R. (1998) Yin-Yang bipolar fuzzy sets, in *Proc. 1998 IEEE International Conf. Fuzzy Systems*, IEEE World Congress on Computational Intelligence, IEEE. pp. 835–840.

Zhang, L., Naitzat, G. and Lim, L.-H. (2018) Tropical Geometry of Deep Neural Networks, in *Proc. 35th Inte. Conf. Machine Learning*, Stockholm, Sweden, PMLR, V. 80, pp. 5824–5832.

Zheng, X. (1999) Binary enumerability of real numbers, computing and combinatorics, in *Proc. COCOON'99*, Tokyo, Japan, pp. 300–309.

Zheng, X. (2000) Weakly computable real numbers, *J. Complexity*, V. 6, pp. 676–679.

Zheng, X. and Weihrauch, K. (1999) The arithmetical hierarchy of real numbers, in *Int. Symp. Mathematical Foundations of Computer Science (MFCS 1999)*, Szklarska Poreba, Poland, pp. 23–33.

Zhmud', L. J. (1989) "All is Number"? "basic doctrine" of Pythagoreanism reconsidered, *Phronesis*, V. 34, No. 3, pp. 270–292.

Zhou, J., Yang, F. and Wang, K. (2016) Fuzzy arithmetic on LR fuzzy numbers with applications to fuzzy programming, *J. Intelligent Fuzzy Syst.*, V. 30, No. 1, pp. 71–87.

Zimmermann, U. (1981) Linear and combinatorial optimization in ordered algebraic structures, *Ann. Discrete Math.*, V. 10, pp. 1–380.

Zimmermann, K.-J. (1991) *Fuzzy Set Theory and Its Applications*, Kluwer Academic Publishers, Dordrecht.

Zolotareff, E. I. (1880) Sur la thèorie des nombres complexes, *J. Math. Pures Appl.*, V. 6, pp. 51–84, 129–166.

Zuckerman, M. (1973) Natural sums of ordinals, *Fundam. Math.*, V. 77, pp. 289–294.

Zur, O. and Gelman, R. (2004) Doing arithmetic in preschool by predicting and checking, *Early Childhood Quarterly Rev.*, V. 19, pp. 121–137.

Index

Hilbert cube, 12
Hilbert cyrve, 12
Hilbert matrix, 12
Hilbert ring, 12
Hilbert space, 12
Hilbert spectrum, 12
Hindu-Arabic, 1
Hindu-Arabic numerals, 43
Hippasus, 19
Hippo, 28
Hisab al-jabr w'al-muqabala, 45
historian, 27, 148
historical evidence, 10
history, vi, xviii, 23
history of mankind, 170
history of numbers, 122
homogeneity, 133
homomorphic filtering, 197
House of Wisdom, 44
Hui, 14
human culture, xviii
human history, 50
Humboldt University, 144
hybrid sets, 115
hydrostatics, 26
Hypatia, 30
hypercomplex numbers, 84–85, 87, 105, 109
hypernumbers, 26, 105, 167
hyperreal numbers, 105, 109
hypotenuse, 186
Hypsicles, 30

I

Iamblichus, 16
Iberian Peninsula, 48
idea, 142
ideal structures, 22
ideas/forms, 22
idempotent analysis, 124, 135, 191, 196, 199
idempotent calculus, 198
idempotent functional analysis, 197, 199

idempotent linear algebra, 199
idempotent semirings, 199
idempotent spectral theory, 199
identity element, 190
identity function, 167
identity mapping, 192
ideology, 18
image analysis, 195
immaterial numbers, 28
imperial archive, 29
Imperial University of Tokyo, 188
impetus, 199
inaccessible number, 175
inconsistent mathematics, 178
inconsistent system, 97, 179
increasing function, 192
India, 6
Indian mythology, 18
Indo-European root, 4
induction, 126
inductive Turing machines, 126
industry, xvii
inequality, 74, 150
inference, 19
inference mechanism, 127
infinite, 11
infinite addition, 26
infinite numbers, 11
infinite sets, 102
infinite whole hypernumbers, 12
infinitely big numbers, 103
infinitesimals, 24, 90, 108, 113
infinity, 107, 175
infinity puzzle, 110
infinity quantization, 108
information, v, 10, 18, 131
information acquisition, processing and management, 50
information processing technology, 46
information technology, xviii
information theory, 12, 124, 189
inlets, 12
innovations, 20
Innsbruck, 188
innumerable numbers, 11

pseudo fuzzy numbers, 118
pseudo fuzzy sets, 118
pseudo-addition, 195
pseudo-analysis, 195
pseudo-integral, 193
pseudo-multiplication, 195
pseudo-positional, 7
pseudoscience, 19
psychology, xviii, 124, 144
psychophysics, 191
Ptolemaic model, 182
Ptolemy, 14
public finance, 56
publication, 71, 164
pure arithmetics, 65
pyramids, 5
Pythagoras, 15
Pythagorean theorem, 6

Q

al-Qalaṣādī, 48
q-Hammersley–Clifford Theorem, 194
Qin, 118
quadratic and cubic equations, 7
quadratic equations, 56, 62
quadratic forms, 70
quadratic reciprocity, 72, 130
Quadrivium, 51
quantitative, v
quantitative literacy, v
quantity, 89
quantization, 110
quantum field theories, 42, 70
quantum mechanics, 12, 42
quantum physics, 111, 177
quantum theory, 146
quasi-arithmetic means, 189
quasiaddition, 190
quasilinear spaces, 199
quasimultiplication, 190, 193
quaternions, 55, 84, 114
queer arithmetic, 160

R

rabbit, 145
radius, 7
Radzikowska, Anna Maria, 118
raindrop, 144
random evolution, 152
Rashevsky, 159, 163, 198
ratio, 151
rational number, 1, 9, 34, 54
rationality, v, xvi
ratios, 10
ratios of numbers, 32
rats, 132
Ravenna, 43
real function, 111
real hypernumbers, 76, 110
real line, 195
real numbers, 11, 51, 54, 65, 107
real valued multisets, 115
real world, 160, 171
real-valued multinumbers, 115
realm, 18
reason, 77
reciprocity theorem, 72
rectangle, 153
recurrence, 102
recurrence relation, 56
recursive algorithms, 125, 128
recursive axioms, 128
Red rods, 40
reduction, 93
reduction–expansion method, 93
reformer, 41
regular analysis, 196
regular non-Diophantine arithmetics, 172
regularity, 119
reincarnations, 16
relations, 12
relatively composite numbers, 31
relatively prime numbers, 31
relativity theory, 146
renaissance, 55, 60
Rennes, 67
Republic of Venice, 62

Printed in the United States
by Baker & Taylor Publisher Services